THE GAM / DP THEORY
of Personality and Creativity

Volume IV

THE GAM / DP THEORY
of Personality and Creativity

Volume IV

WILLIAM A. THERIVEL

Kirk House Publishers
Minneapolis, Minnesota

THE GAM / DP THEORY
of Personality and Creativity
Volume IV

First Edition

The opinions expressed herein are solely those of the author.

This book is printed on acid-free paper and the binding materials have been chosen for strength and durability.

Library of Congress Cataloging-in-Publication Data

Therivel, William, 1928-

The GAM/DP theory of personality and creativity / William A. Therivel.

p.cm.

Includes bibliographical references and index.

ISBN 1-886513-53-8

1. Personality and creative ability. 2. Personality and creative ability–Cross cultural studies. 3. Nature and nurture. I. Title.

BF698.9.C74 T48 2001

155.2—dc21

20011038972

Kirk House Publishers, PO Box 390759, Minneapolis, MN 55439

Manufactured in the United States of America

To Brigitte and Silvia

CONTENTS

Preface

The current volume is a direct continuation of the studies at the base of the previous three volumes. As in volumes 1 to 3, in view of the interdisciplinary nature of the study, I have made use of all the help I could get from the expertise of friends and colleagues. While gratefully acknowledging their assistance, I want to emphasize that any shortcomings are entirely my own, and so are all unconventional opinions.

In addition to continued gratitude to my wife Brigitte, I want to extend my thanks to my daughter Silvia for all her help in bringing the first four volumes to completion, and it is to both of them that this book is dedicated.

As before, references and footnotes follow the system adopted by the American Psychological Associations which separates the references (at the end of the book) from the footnotes (at the bottom of each pertaining page). This separation allows for substantive footnotes that are completely devoted to expanding and clarifying the main text. If at all possible, I would ask the reader not to skip the footnotes.

Quotes are reported verbatim, not paraphrased, to allow the reader a direct contact with the sources. Whenever possible I have used existing translations of foreign texts, otherwise the translation is mine.

By now, readers have become used to my method of constantly referring to what I said in previous volumes. At first many of those "as said in chapter x of volume y" will be a disturbance and not a help. However, in the long run, these cross-references should transform a string of seemingly unrelated stories into a strong web, in which each piece helps to explain those nearby, and also many at a distance. As discussed in detail in chapter 28 of this volume, it is my conviction that the GAM/DP theory is true, scientific, and approaching the level of a natural law. And as other natural laws are understood and verified by the multiplicity of the phenomena they explain and predict, so it should be with the GAM/DP theory, thanks in part to those many "as said in chapter x, volume y".

Introduction

There are many ways to read this fourth volume. The first and most natural one is as a continuation of the previous three volumes, in which chapter after chapter presents the results of applying the GAM/DP Theory to interesting subjects, hopefully saying something new, important and valid about each. Already at the end of volume three, but even more so at the end of this fourth volume, most readers will have concluded that what has been presented so far is only the mapping of the coastal area of a huge continent open to exploration.

The second approach—in reality an expansion of the first—is to read each chapter as part of an integrated world, crisscrossed by roads of various importance, from six-lane highways to small winding paths: it is always possible to go from one place to the other. The trip is always rewarding, such as meeting Isaiah Berlin as *leadsman* in chapter 1 of this volume, and meeting him again in chapter 6 as *arch-fox*; or having understood the DP vs. UP of the evolution of the representation of Jesus (chapter 10), which we can relate to the kind of saints that were honored and loved along the centuries (chapter 12). Many connections are positive comparisons, while others are contrasting, such as chapter 18 on courtly love and chapter 19 on the *Ritter* reaction to courtly love.

The third approach consists of studying each chapter as a joint discovery between reader and author, like that, in 1846, of the planet Neptune by the mathematician Le Verrier and the astronomer Galle. Le Verrier did the Newtonian analysis of the irregular orbit of the planet Uranus, which had to be caused by the influence of another, so-far unknown, planet. Then, Galle searched the sky for a planet where Le Verrier said it should be and found it on the first attempt.

Clearly, the GAM/DP Theory cannot achieve this precision, yet my approach is the same as that of Le Verrier and Galle. For instance, I knew a long time ago that the practice of human dissections for scientific purposes could only have begun under strong DP conditions. Having written a short chapter on dissection in volume 1, I returned to this subject, did extensive research, and discovered the "planet Neptune of dissection", now described in the long chapter 14. Here, the "discovery" focused first on the four DP steps (forensic/post mortem, simple university, formal university, scientific university) which led to the wide adoption of dissection for anatomical studies in the West; then on the political and cultural changes, in Hellenistic Alexandria of Egypt, which favored dissections only for a short time.

After these general comments, let me introduce volume 4 the way I did the previous three, starting by saying that this fourth volume is somewhat different from the preceding three:

Three chapters have illustrations, in line with the Chinese proverb of "One picture is worth more than ten thousand words": #10 "DP Christian Iconography Versus UP Christian Iconography", #11 "Contrasting Imperial Iconographies (Dwarf or Semigod) Paved the Way for the DP of Canossa", and #15 "The DP Contribution to the Abolition of Slavery in the 18th and 19th Centuries". There is also a Part III—beside the usual two, on GAM and DP—dealing with matters beyond the strict range of the two main parts of my theory.

Part I—GAM consists of nine chapters.

The first, "GAM *Leadsmen* are not Parkian *Marginal Men*" is a detailed discussion of the 5th *challenged personality family*, the *leadsmen*, who are not, or only in part, *marginal men* as described by Robert Park in his seminal article of 1928.

The next four chapters are devoted to four great writers and one great painter, each from a different country: #2 "Nathaniel Hawthorne: GAM *Radiologist-Universalist-Miner*", #3 "Aleksandr Pushkin, GAM *Miner* and *Trapper*, therefore Berlinian *Arch-Fox*", #4"Alberto Moravia, Down-to-Earth *Radiologist*", and #5 Deliberately More *Hedgehog* (Solzhenitsyn), or More *Fox* (Goethe and Picasso). The chapter on Hawthorne is also an additional piece of evidence for the validity of the GAM part of the theory. There, I report how I have been able to successfully predict (as was the case with Nietzsche, in volume 2) that many important sections of his writings had to be *radiological* given the severe physical infirmity of which he suffered in his youth.

Two chapters, #8 and #9, are based on a recent book, by Professor Richard S. Tedlow, of Harvard Business School, entitled *Giants of Enterprise*. The first gives evidence for the GAM part of the theory: five of the seven giants studied by Tedlow are *challenged personali-*

ties, especially the greatest among them: Henry Ford, Andrew Carnegie, and George Eastman.

Part II—DP consist of seventeen chapters.

Chapters 10 and 12 deal with closely related matters: in #10 the subject is the evolution of the visual representation of Jesus in function of the ruling power, DP or UP; in #12, "UP Saints Versus DP Saints", the subject is the parallel evolution of which kind of saints was honored under DP or UP.

Two more chapters are directly related, but this time contrasting: #18 "The DP Origins of Courtly Love" and #19 "The *Ritter*'s Reaction to Courtly Love". Here, I recommend that the chapters are read in sequence, to confirm our appreciation for that wonderfully civilizing creation of *courtly love*, among other reasons.

While chapter 13 "Why Did Venice Have No Great Writers When it Had Great Painters, Architects, and Musicians?", by its very nature, has a somber tone (we want a beloved city to have all beauties); chapter 14 "The DP Origins of Western Medicine: On the History of the Scientific Use of Dissection" has a more positive tone (like its Part I, in volume 1) because it deals with the beginnings of "the greatest benefit of mankind", to use the title of Roy Porter's medical history of humanity.

This volume also contains an ethnopsychological *pièce de résistance*, # 16 "An Ethnopsychological Explanation of Argentina's Difficulties", a companion chapter to that on the Germans (volume 1), the French, the Spaniards, and the Chinese (volume 2), and the Japanese (volume 3).

In turn, chapter 20, "Anticipating More Changes: A Chinese I*nsular Virtue* from Necessity" should be read as an expansion of the just mentioned chapter on the Chinese in volume 2. The same applies to chapter 21 "Interpreting Three Recent Studies on the Contrast Between the Japanese and American Ethnopsychology" which should be read as an expansion of chapter 9 of volume 3 on the Japanese, and of chapters 15 and 16 of volume 1, on the Americans. Chapter 26 is a discussion of the famed imperturbability of the English and Japanese and its causes.

The four chapters of **Part III** extend the range of my studies:

Chapter 27 presents my explanation of the mechanism of the act of creation, starting with the role of induction; chapter 28 "Truth" is epistemological and explains both my trust in science and my hope that what the GAM/DP Theory says is true, valid and important; important also for fostering the *positive psychology* of empowering people as discussed in chapter 29.

The last chapter 30, "Synopsis of the GAM/DP Theory" should have been written a long time ago, and published at the beginning of volume 1, because it gives a much needed overall view of theory, methods, and results. But I—who am human like my readers—first had to do the research, and then the derived writing, before summarizing them.

This time, I will not repeat my mantra of "Please do not read this volume before having read volumes 1, 2, and 3": there is only so much that an author can do to maintain control of the process. I must accept a DP with my readers who do what they want, including reading the four volumes in backward order. One way or the other—and in this I keep the initiative—I wish the readers a most pleasant journey.

List of Pictures

PART I

GAM

1

GAM *Leadsmen* Are Not Parkian *Marginal Men*

Leadsman: In nautical parlance, a man who uses a lead line to take soundings; in GAM parlance a member of the 5th *challenged personality family*.

This chapter is divided into the following sections:

I. Introduction
II. V. S. Naipaul
  A. The making of a leadsman
  B. As leadsman
III. Edward Said
  A. The making of a leadsman
  B. As leadsman
IV. Eminent Leadsmen of Jewish Origin
  A. Cervantes
  B. Montaigne
  C. Marx
  D. Freud
  E. Veblen's 1919 comments
V. Napoleon
  A. The making of a leadsman
  B. As leadsman
VI. Shusaku Endo
  A. The making of a leadsman
  B. As leadsman
VII. In the End

I. Introduction

Marginal men, as described by Robert Park in his seminal article "Human migration and the marginal man" of 1928, are those whose personality is "a cultural hybrid, a man living and sharing intimately in the cultural life and traditions of two distinct peoples; never quite willing to break, even if he were permitted to do so, with his past and his traditions, and not quite accepted, because of racial prejudice, in the new society in which he now sought to find a place. He was a man on the margin of two cultures and two societies, which never completely interpenetrated and fused" (p. 892). The *marginal man* was characterized as suffering from "spiritual instability, intensified self-consciousness, restlessness, and *malaise*" (ib., p. 893). Then, as stressed by Leslie, Larson, and Gorman, 1973: "The lot of the *marginal man* is, thus, a lonely one. Unable to share fully in the way of any group, he may lack deep friendships and otherwise feel cut off from his fellows.... In extreme cases, these symptoms may be defined as those of incipient mental illness" (p. 88).

Related to the *marginal man*, the GAM *leadsman* is a person with a high GxAxMxDP, for whom the misfortune of youth M (see table 1 of volume 1) is "rootlessness or uprootedness (e.g., forced conversion or assimilation of parents to the religious or ethnic ways of the majority; major changes of abode; major religious or cultural differences between parents)." The interaction of this misfortune with a high GxAxDP fosters the development of personalities who are: "detached; critical thinkers; relativistic; who often hold cosmopolitan and pragmatic attitudes." (Therivel, 2001, p. 41)

I called *leadsmen* the members of this personality family because of their basic skepticism, of their constant need to check if things are as they are said to be. In this the *leadsman* is like the sailor who, at the prow of the ship, is responsible for sounding the depth of the water with a weighted line. He must ensure that the ship will avoid hidden rocks or treacherous sandbanks. Only after his "OK", will the ship go forward.[1]

This constant probing by the *leadsmen* makes them relativists, as was Einstein who discovered the relativity of our measurements in physics: space, speed and mass are interrelated, not fixed or independent values; as was Marx who asserted the relativity of our values, which depend on the means by which we make a living; and as was Freud who

---

[1] The right words should be *leadsperson* and *marginal person*, because what I said applies equally to men and women. For simplicity I have kept Park's term and the common name for the sailor who takes soundings.

claimed the relativity of our motivations, based on our unconscious drives and forgotten youth.[2]

Closer to our times, Karl Popper said that "we must regard all laws or theories as hypothetical or conjectural; that is as guesses" (1975, p. 5),[3] and Thomas Kuhn spoke of science(s) rooted in incommensurable and unprogressive paradigms.[4]

The *leadsman* is similar to the *marginal man* in some respects: "a cultural hybrid [or synthesis], a man living and sharing intimately in the cultural life and traditions of two distinct people ... a man on the margin of two cultures and two societies, which never completely interpenetrated and fused." However, for the *leadsman* it is not true that he is "never quite willing to break, even if he were permitted to do so, with his past and his traditions, and not quite accepted, because of racial prejudice, in the new society in which he now sought to find a place."

First, the *leadsman* rarely feels the need to break from his past and his traditions because, thanks to his high GxAxDP, he built his own life, his own past and traditions. Secondly, his talents and creativity make him accepted, if not immediately and not completely, still to such a high degree that he does not need more. Thirdly, the lot of the leadsman is not a lonely one, with the exception of the solitude he has chosen for himself, in order to be creative. Yes, from time to time, he may feel cut off from his fellows, but he also knows that that is the price he has to pay to be creative.

---

2 Relativism: "The doctrine that statements in a given area of discourse can be true for one group of people but false for another: e. g., that ethical statements can be true for the members of one society or culture but false for those of another" (Lowe, 1999, p. 743). "Relativists tend to see whole outlooks as determined by forces of which people are unaware" (Lukes, 1998, p. 9).

3 For him and other *falsificationists* "science has nothing to do with the quest for certainty or probability or reliability. We are not interested in establishing scientific theories as secure, or certain, or probable. Conscious of our fallibility we are only interested in criticizing them and testing them, hoping to find out where we are mistaken; of learning from our mistakes; and, if we are lucky, of proceeding to better theories" (Popper, 1963/1972, p. 229).

4 In the words of Maxwell and Delaney: "We share with many a concern regarding what appears to be Kuhn's relativism. The reading of his work by a number of critics is that Kuhn maintains that there is no fixed reality of nature for science to attempt to more accurately describe. For example, he writes 'We may, to be precise, have to relinquish the notion, explicit or implicit, that changes of paradigm carry scientists and those who learn from them closer and closer to the truth....The developmental process of evolution described in this essay has been a process of evolution *from* primitive beginnings—a process whose successive stages are characterized by an increasingly detailed and refined understanding of nature. But nothing that has been or will be said makes it a process of evolution *toward* anything' [Kuhn, 1970, p. 170]" (1990, p. 20).

The list of eminent *leadsmen* of the past is a long one, including Cervantes, Napoleon, Marx, and Freud, whom I will discuss further-on. However, there are two eminent contemporary *leadsmen* whom we should study first, because they have written extensively about both their youth and the world around them: V. S. Naipaul and Edward Said.

II. V(idiadhar) S(urajprasad) Naipaul

British-educated Trinidadian writer of Indian descent known for his pessimistic novels set in Third World nations. Born in 1932, V. S. Naipaul was knighted in 1989, and awarded the Nobel Prize for Literature in 2001.

A. The Making of a *Leadsman*

In Naipaul's own words:

> We were an immigrant Asian community on a small plantation island in the New World. To me India seemed very far away, mythical, but we were at that time, in all the branches of our extended family, only about forty or fifty years out of India. We were still full of the instincts of people of the Gangetic plain, though year by year the colonial life around us was drawing us in.... Mangled bits of old India (very old, the India of nineteenth-century villages, which would have been like the India of earlier centuries) were still with me, not only in the enclosed life of our extended family, but also in what came to us sometimes from our community outside....
>
> The island was small, 1,800 square miles, half a million people, but the population was very mixed and there were many separate worlds.[5] When my father got a job on the local paper we went to live in the city. It was only twelve miles away, but it was like going to another country. Our little rural Indian world, the disintegrating world of a remembered India, was left behind. I never returned to it; lost touch with the language; never saw another *Ramlila*.[6]

---

[5] Trinidad was a British colony until 1962; and its official language is still English. Its 1960 ethnic composition was Black 43.3%, East Indian 36.4%, Mixed 16.3%, Chinese 1.0%, White 1.9%, Other 1.1%.

[6] The pageant-play based on the *Ramayana*, the epic about the banishment of Rama, the Hindu divinity.

> In the city we were in a kind of limbo. There were few Indians there, and no one like us on the street. Though everything was very close, and houses were open to every kind of noise, and no one could really be private in his yard, we continued to live in our old enclosed way, mentally separate from the more colonial, more racially mixed life around us....
>
> It was like that for the twelve years I was to stay in the city before going to England. I never ceased to feel a stranger. I saw people of other groups only from the outside; school friendships were left behind at school or in the street. I had no proper understanding of where I was, and really never had the time to find out: all but nineteen months of those twelve years were spent in a blind, driven kind of colonial studying. (1999, pp. 13-14)

It would be hard to find a more rootless, and stronger *leadsman*, than Naipaul.[7]

B. As *Leadsman*

> • Content to be a colonial without a past, without ancestors.
> (Naipaul, quoted by Buruma, 1991, p. 3)

---

[7] Sad—and yet fascinating for our study on the impact of rootlessness—are the similar comments from the pen of the thirteen-years-younger brother of V.S.: Shiva Naipaul (born in 1945; died in 1985), and a novelist in his own right (*Fireflies*, *The Chip-chip Gatherers*. *Fireflies* takes up the story of the family at the point where V.S.'s *A House for Mr. Biswas* ends.)

"Indians began to arrive in Trinidad around the middle of the 19th century—brought over, like the Africans before them, to work on the sugar-cane plantations. Trinidad was about as far away from India as it was possible to get. It was only around the turn of the century that my maternal grandfather arrived on the island, yet my links with India are, to say the least, tenuous. Today, Indians make up one-third or more of the island's population.... It was out of that landscape that my father emerged, managed somehow to get an education—English was a language he'd had to acquire—and eventually was able to turn himself into a journalist and writer of short stories. Looking back, it seems a miraculous achievement....

Trinidad calls itself a nation. It has a flag, a flower, a seat in the UN. But we have no genuinely collective existence. We are merely a random collection of races thrown together by outside forces. What did the Indian of my mother's generation have in common with the black? They were separated by religion, by geography, by occupation, by sentiment—even to some extent by language. The calypso and the steel band,

- Nearly all my adult life had been spent in countries where I was a stranger. I couldn't as a writer go beyond that experience.

(Naipaul, 1999, p. 17)

Referring to himself, Nirad Chaudhuri, the cosmopolitan Bengali writer, now living in Britain, once wrote: "To be *déraciné,* is to be on the road forever." This could serve nicely as V. S. Naipaul's motto, too. Naipaul, the grandson of uprooted Indians, uprooted himself to come to England. He is a man continuously on the road, and continuously fretting about roots, his own and those of the people he meets. One sometimes has the impression of a man traveling through the dark and rainy night, stopping at houses on the way, pressing his nose against the windows. Peering at the people inside, cozily sitting around the family hearth, he is reminded of his own rootlessness. The assumption is that those others, seen through the window, are at home, rooted and whatever the opposite is of *déraciné*. This must account for Naipaul's nostalgia for what he has called "whole and single societies." He has often used such words as "damaged" or "wounded" for societies that are fragmented and apparently rootless. (Buruma, 1991, p. 5)

[Indeed] He portrayed Trinidad as a damaged society, feeding itself on deceit, fantasy, lame excuses; a colony that had been artificially created, maladministered, then abandoned. Above all, he found there no sense of history, no way of thinking about history. He found a blindness to the past: this is his grievance also, in the writing on the Caribbean and Central America included here.... Naipaul has a care for the marginal, an eye for the traces of those almost erased. The small town of Chaguanas, where he was born, was named after an extinct Indian tribe.

---

celebrated today as the twin hearts of Trinidadian culture, belonged entirely to the town, to the urban black. The Indian was not a prominent part of that milieu. When political awareness came, it was black in complexion.... The term 'Trinidadian', aside from its geographical reference, has for me very little meaning. I live in London now. The fervours of patriotism are unknown to me. I gaze in wonder at those who wave flags. Sometimes I think it would be nice to have a flag to wave—but that mood soon passes. Yet the ragged landscape of Trinidad lives on in me. Its smells, its textures, its squalor and its beauties run in my blood. I cannot disown it and it cannot disown me. Whatever I am, whatever I may become, it and I are yoked together. I accept the paradox" (1982, p. 14).

> Every day he and his fellow townfolks walked, unknowingly, over the land of this dispossessed people. It was only in 1967 that he came across a reference to them, in a letter from the King of Spain to the governor of Trinidad. He was shocked not just because these people had been obliterated, but because he was ignorant of it. The culture into which he was born did not favor this kind of self-knowledge; it was not analytical, it was not inquiring. There was a vacuum in how the people of the Caribbean imagined themselves and—as he sees it—the crudest form of US influence rushed in to fill the space....However, even those developing societies that are sharply self-critical do not like their failings and uncertainties held up to scrutiny by outsiders. From the first, Naipaul's sardonic and fastidious approach distinguished him from those who write about the underdeveloped world in eggshell pieties. He has a sharp eye for the intellectually fraudulent, and is a scourge of self-delusion; he gives the underdog as bad a name as his master. Oppression, he notices, doesn't make people saintly, it makes them potential killers; all victims are dangerous.... Since then he has gone about the world equipped with the *Devil's Dictionary,*[8] redefining and challenging every received idea he encountered in his travels through Africa and the Caribbean, South America, the non-Arab Muslim world. (Mantel, 2002, pp. 10-11)

III. Edward Said

A. The Making of a *Leadsman*

On the first page of his memoirs, Edward Said, professor of English and Comparative Literature at Columbia University, a man well established in American academic and intellectual life, and author of many books, including *Orientalism*, declares that he has always had an overriding feeling of being "out of place":

> There was always something wrong with how I was invented and meant to fit in with the world of my parents and four sisters...Yet the overriding sensation I had was of always being out of place. Thus it took me about fifty years to become accus-

---

[8] A 1906 volume of ironic definitions by the newspaperman, misanthrope, wit, satirist and author of sardonic short stories, Ambrose Bierce.

tomed to or, more exactly, to feel less uncomfortable with, "Edward," a foolishly English name yoked forcibly to the unmistakably Arabic family name Said. True my mother told me that I had been named Edward after the Prince of Wales, who cut so fine a figure in 1935, the year of my birth, and Said was the name of various uncles and cousins. But the rationale of my name broke down both when I discovered no grandparents called Said and when I tried to connect my fancy English name with its Arabic partner. For years, and depending on the exact circumstances, I would rush past "Edward" and emphasize "Said"; at other times I would do the reverse, or connect these two to each other so quickly that neither would be clear....I have retained this unsettled sense of many identities—mostly in conflict with each other—all of my life, together with an acute memory of the despairing feeling that I wish we could have been all-Arab, or all-European and American, or all-Orthodox Christian, or all-Muslim, or all-Egyptian, and so on. I found I had two alternatives with which to counter what in effect was the process of challenge, recognition, and exposure, questions and remarks like "What are you?"; "But Said is an Arab name"; "You're American?"; "You're American without an American name, and you've never been to America"; "You don't look American!"; "How come you were born in Jerusalem and you live *here*?"; "You're an Arab after all, but what kind are you? A Protestant?" (1999, pp. 3-6)

As commented by Amos Elon:

Said's problems with "alienation," "place," and "identity" started before his arrival in the United States as a boy of sixteen.... His mother, who read Shakespeare's *Hamlet* aloud with him when he was nine (she was Gertrude, he Hamlet), told him that she had named him after the Prince of Wales. In Cairo, Said grew up in the Shami, or Syrian, community and attended elite English schools. He was an Anglican Christian in a sea of Muslims, one minority within another, along with members of many other Levantine minorities in Egypt—Greeks, Copts, Jews, Italians, Maltese, Turks, Lebanese, and Armenians. He seems to have struggled with his several identities for years. He sometimes stressed the Edward of his name, at other times the Said. Arabic was his nursery's language, English his school's. The family used both languages among themselves. Most of his prodigious reading was in English. He says that he has never known which was his first and which his second language, that he never felt at home in either but dreams in both....

Wherever he refers to his father's politics he gives the impression of subdued anger. There is in Said's book a sadness about his father's total lack of interest in Palestine but anger also at his own early indifference. That it took Said himself more than twenty years to shed his equanimity over the loss of Palestine suggests to him how overprotected and ignorant he was. The 1967 war changed him. Until then his only interest was English literature. His father had become almost immediately reconciled with the loss of Palestine in 1948. He himself, he remembers, never really confronted his father about this. "I had no available vocabulary for the question," he writes. (1999, p. 14)

B. As *Leadsman*

Edward W. Said is a man of singular achievements. As Professor of English and Comparative Literature at Columbia, he is the outstanding representative of the post-structuralist Left in America. His most influential book, *Orientalism* (1978), has been credited with changing the drift of scholarship in several disciplines, opening opportunities for intellectuals from what used to be called the "East" to make careers out of transmitting, interpreting and debating representations of the non-Western "other". (Ruthven, 1999, p. 29)

[In Said's own words] Much of the personal investment in this study [*Orientalism*] derives from my awareness of being an "Oriental" as a child growing up in two British colonies. All of my education, in those colonies (Palestine and Egypt) and in the United States, has been Western, and yet that deep early awareness has persisted. In many ways my study of Orientalism has been an attempt to inventory the traces upon me, the Oriental subject, of the culture whose domination has been so powerful a factor in the life of all Orientals. This is why for me the Islamic Orient has had to be the center of attention....

> The nexus of knowledge and power creating "the Oriental" and in a sense obliterating him as a human being is therefore not for me an exclusively academic matter. Yet it is an *intellectual* matter of some very obvious importance. I have been able to put to use my humanistic and political concerns for the analysis and description of a very worldly matter, the rise, development, and consolidation of Orientalism. Too often literature and culture are presumed to be politically, even historically innocent; it has regularly seemed otherwise to me, and certainly my study of Orientalism has convinced me (and I hope will convince my literary colleagues) that society and literary culture can only be understood and studied together. (1978, pp. 25-27)[9]

IV. Eminent *Leadsmen* of Jewish Origin[10]

1. Cervantes

---

[9] However, "In the eyes of his critics, his work is flawed by contradictions. As Aijaz Ahmad has pointed out in his penetrating critique of *Orientalism,* the very terms that are deployed to undo what Said presents as a 'centuries-old tie between narratives of High Humanism and the colonial project' are those most familiar in the humanist lexicon, namely 'tolerance, accommodation, cultural pluralism and relativism, and those insistently repeated words, *sympathy, constituency, affiliation, filiation*. What is remarkable', Ahmad concludes, 'about this at times very resounding affirmation of humanist value is that humanism-as-ideality is invoked precisely at the time when humanism-as-history has been rejected so unequivocally'" (Ruthven, 1999, p. 29;); [Ahmad, 1992, p. 164].

Another interesting critique (interesting because it is indirect) was raised by David Arnold in his review of Charles Allen's *The Buddha and the Sahibs: The men who discovered India's lost religion*: "[Allen's book] is a fascinating story of East and West, and Allen tells it well. His reason in doing so is clear—he wants, as he puts it, to set the record straight. He asks where India and the world would be today without the labours of the men who initiated the recovery of India's 'lost past' and whose work led directly to Buddhism's resurgence. Deploring the 'heavy pejorative overtones' that have come, since Edward Said, to haunt the very word Orientalism, he sees instead evidence not for knowledge as an instrument of imperialism, but rather of Enlightenment understanding, proof of the ability of Westerners to access and appreciate an 'alien culture' and selflessly share its history with others...Orientalists and archaeologists did what they did, Allen suggests, not because imperialism egged them on, but from curiosity, heightened perhaps by personal ambitions, professional advancement and patriotic anxiety lest French or German Indologists found the answers first" (Arnold, 2003, p. 22).

[10] Cervantes and Montaigne were Christians, and so in his youth Marx (after age six). Freud remained a Jew.

Modern scholars (e.g., Américo Castro, 1966, 1972; William Byron, 1979; Pierre Guenoun, 1971; Jean Canavaggio, 1992) agree on a probable Jewish origin of Cervantes which, given the times, he kept hidden.

As leadsman (in his case, as a non fully assimilated *new-Christian*) he shows a high level of religious and ethnic relativity. In his stories "the inter-racial and inter-confessional loves, in short anti-statutory, are too many not to draw attention.... [His *Don Quixote]* carries in itself the traditions of the three Spains: the Jewish, the Christian, and the Muslim, the story of the knight errant of impure blood [i.e., New Christian], and of his loyal squire of statutory blood [Old Christian]" (Guenoun, 1971, pp. 66, 183); "Cervantes expressed the Christian point of view as well as the Jewish" (Castro, 1966, p. 87).

2. Montaigne

> His mother was, by blood, a Jewess. She was born Antoinette de Louppes [originally Lopez de Villanueva] of a Toulon and Bordeaux family of converted Jews—part of a considerable group of Spanish refugees who settled north of the Pyrenees after the establishment of the Inquisition in Spain and the expulsion of the race in 1492. She was a convert to Protestantism, and two of her children—a brother and a sister of Michel—were likewise followers of the new religion. (Lowenthal, 1935/1999, pp. xx-xxi).
>
> Montaigne's sharp focus on the injustice done to the Jews and his low estimate of the value of their forced conversion are what the *Essays* lead us to expect. His conclusion, curiously, suggests doubt about the sincerity of 'New Christians' in general, and thus by implication of the Louppes and others of Bordeaux. (Frame, 1965, p. 19)[11]

---

[11] In 1580-1581 Montaigne travelled through France, Germany, Switzerland, Austria to Italy. He left a 300-page record of his trip, his *Travel Journal*, not intended for publication and published only in 1774. In it one reads that while in Rome, "Among the sermons he heard in Lent was that of a converted rabbi who preached to a captive audience of sixty Jews who were obliged to hear him every Saturday. [The Christian] Montaigne admired the learning with which he turned their own texts and arguments against them; but it is interesting that he refers to him not as a *converted*, but as a *renegade* rabbi" (Frame, 1965, p. 18).

Montaigne's personal rule of "*Il faut tout voir, ne croire personne sur la parole*"[12] (Moreau, 1966, p. 39) could not be more *leadsman*, and so are his monumental *Essays*, "the exploration of skeptical and relativistic arguments" (Cave, 1994, p. 25).

3. Marx

> Thoroughly Jewish in their origins, Protestant by necessity yet living in a Catholic region, his family could never regard their social integration as complete. The sense of alienation was heightened in Marx's personal case by his subsequent inability to obtain a teaching post in a university system that had no room for dissident intellectuals. (McLellan, 1973, p. 1)

Marx was baptized at the age of six, after his father converted, solely in order to be able to continue his profession of counsellor-at-law at the High Court of Appeals in Trier. Nearly identical to Montaigne's rule, was Marx's favorite motto of "*De omnibus dubitandum*",[13] again a very strong *leadsman* motto.

4. Freud

Freud, himself, in his 1926 address to the Society of B'nai B'rith[14] said:

> That you were Jews could only be agreeable to me; for I was myself a Jew, and it had always seemed to me not only unworthy but positively senseless to deny the fact. What bound me to Jewry was (I am ashamed to admit) neither faith nor national pride, for I have always been an unbeliever and was brought up without any religion.... I cannot tell whether I have been a genuine B.B. in your sense. I am almost inclined to doubt it.... But of this I can assure you—that you meant much to me and did much for me during the years in which I belonged to you" (1959, pp. 273-74).

---

[12] "One must see everything, and never believe anybody on their word."

[13] "All things must be doubted".

[14] Also known as *Sons of the Covenant*, it is the oldest and largest Jewish service organization (cultural, intellectual and charitable) in the world.

Before this last point, Freud said something important for the present chapter:

> Because I was a Jew I found myself free from many prejudices which restricted others in the use of their intellect; and as a Jew I was prepared to join the opposition and to do without agreement with the "compact majority". (ib., p. 274)

Strictly speaking, Freud should have said: "Because of my Jewish roots but lack of Jewish faith, because I have been brought up without any religion,[15] I found myself free from many Christian and Jewish prejudices, and could disagree with both compact majority and compact minority"; but that would have been offensive to his B'nai B'rith listeners.

Then, as noted by Peter Gay: "[Freud] took pleasure in being the leader of the opposition, the unmasker of shams, the nemesis of self-deception and illusions.... In a much-quoted early letter, Freud had told his fiancée, "One would hardly know to look at me, but already in school I was always a bold man of the opposition, was always where one could avow an extreme and, as a rule, had to atone for it." (1988, p. 604)

5. Isaiah Berlin[16]

In 1983, Roger Hausheer concluded his biographical sketch of Isaiah Berlin with a caveat: "The philosophical justification of this radical pluralism and the related question of ethical and cultural relativism,

---

[15] "Of Freud's religious background not a great deal is known. There was, of course, the Catholic Nannie, and perhaps her terrifying influence contributed to his later dislike of Christian beliefs and ceremonies. His father must have been brought up as an orthodox Jew, and Freud himself was certainly conversant with all Jewish customs and festivals. His children have assured me that their grandfather had become a complete freethinker, but there is some evidence to the contrary. He was undoubtedly a liberal-minded man of progressive views and it is not likely that he kept up orthodox Jewish customs after migrating to Vienna.... Freud grew up devoid of any belief in a God or Immortality, and does not appear ever to have felt the need of it. The emotional needs that usually manifest themselves in adolescence found expression first in rather vague philosophical cogitations and, soon after, in an earnest adherence to the principles of science" (Jones, 1953, *1*, pp.19-20).

[16] Born in Riga (Latvia, Russian Empire) in 1908, Isaiah Berlin and his family emigrated to England in 1920, living for a while in Surbiton, a few miles southwest of London: "There cannot have been another Russian within miles, and there were few

remains to be fully discussed by Berlin" (p. 67). Fifteen years later, the situation had not improved by much: Alan Ryan—in his review of the biography of Berlin by Michael Ignatieff and of an anthology of essays by Berlin—was forced to admit that "Berlin's pluralism of values is hard to elucidate" (1998, p. 36).

Similar difficulties were faced by Steven Lukes, in an article, also of 1998, entitled "Berlin's dilemma: The distinction between relativism and pluralism". There, after a long introduction, Lukes raised two questions: "How exactly did Berlin distinguish the pluralism he so persistently advocated from the relativism he insistently repudiated? And did he consistently defend a plausible case for pluralism untainted by relativism?" (1998, p. 8). That the answer to these questions was not easy, is indicated by the fact that at the end of his article, Lukes had to concede that "Berlin's pluralism needs protection against the relativism of practical judgment involving real confrontations" (ib., p. 10).

Leaving aside the question of when the Berlinian pluralism is or is not relativistic, it is fascinating to see a first class *leadsman* like Berlin at work, especially when he writes, in defense of pluralism:

> I came to the conclusion that there is a plurality of ideals, as there is a plurality of cultures and of temperaments. I am not a relativist; I do not say "I like my coffee with milk and you like it without; I am in favor of kindness and you prefer concentration camps"—each of us with his own values, which cannot be overcome or integrated. This I believe to be false. But I do believe that there is a plurality of values which men can and do seek, and that these values differ. *There is not an infinity of them.... Hence the possibility of human understanding*. (Berlin, 1998, p. 57; my italics)

Jews" (Ryan, 1998, p. 30). The father, in the words of the son, "was totally emancipated, didn't eat kosher, never went to synagogue" (quoted by Ignatieff, 1998, p. 14).

Very *leadsman*: "He took three conflicting identities, Russian, Jewish, and English, and braided them together into a character at one with himself.... In intellectual terms Isaiah led a double life: the first rambling, historical, saturated in Russian, German and Jewish references, conducted in three languages at once; the second with his Oxford colleagues—precise, Anglo-Saxon, procedural and rationalistic.... In matter of faith, he remained what his friend Jerry Cohen called a 'verificationist atheist', faithful to the anti-metaphysical convictions of Oxford logical positivism of the 1930s. But he was a sceptic rather than a heretic and he expressed his radical scepticism gently" (Ignatieff, 1998, pp. 7, 81, 293).

6. Veblen's 1919 Comments on the Intellectual Pre-eminence of Jews [and ex-Jews] in Modern Europe.

So far in this section, I have discussed five eminent *leadsmen* of Jewish origin, individually, without studying them in the broader context of the great many Jews that have enriched Western civilization, and of their special relationship with the non-Jewish world surrounding them. This is a very broad subject, yet something compact was said in 1919 by the economist and social scientist Thorstein Veblen in an article, for the *Political Science Quarterly*, entitled "The intellectual pre-eminence of Jews in Modern Europe". From this article I report the following, which applies also to pre-modern times:

> It may be more to the purpose to note that this intellectual pre-eminence of the Jews has come into bearing within the gentile community of peoples, not from the outside; that the men who have been its bearers have been men immersed in this gentile culture ...It appears to be only when the gifted Jew escapes from the cultural environment created and fed by the particular genius of his own people, only when he falls into the alien lines of gentile inquiry and becomes a naturalized, though hyphenated, citizen in the gentile republic of learning, that he comes into his own as a creative leader in the world's intellectual enterprise. It is by loss of allegiance, or at the best by force of a divided allegiance to the people of his origin, that he finds himself in the vanguard of modern inquiry. (pp. 37-38)[17]

V. Napoleon

A. The Making of a *Leadsman*

In 1769 the French invaded the Italian-speaking island of Corsica, and annexed it. Nine years later, the French commander in Corsica, Count Marbeuf, who had good relations with both Carlo and Letizia

---

[17] The reader may wish to relate Veblen's remarks to what I wrote in chapter 18 of volume 1 on "The DP Origins of the Scottish Enlightenment"; in particular to my observation that "from a Scottish fundamentalist point of view, for instance from the notion that Scots should speak and write in Scots, the Scottish Enlightenment was not pure *eau d'Ecosse*." Similar comments can also be found in the previous chapter 6 "Creativity via a Detour Abroad". Both Veblen's "gentile republic of learning" and the England of my chapter 18, were "creative detours abroad".

Buonaparte, the parents of Napoleon, "aware of Carlo's financial difficulties, informed him of an arrangement whereby the children of impoverished French noblemen [by now the Buonaparte qualified as such] might receive free education. Boys destined for the army could go to military academy,... This offer was like an answer to prayer." (Cronin, 1972, p. 27)

> On Christmas Day 1778 at Marseille [the nine-year-old] Napoleone Buonaparte[18] set foot on French soil, and found himself among people whose language he could not understand [He returned home eight years later, in 1786] .... Napoleon's first task was to learn French,... He found it difficult. He was not good at memorizing and reproducing sounds, nor did he have the flexible temperament of the born linguist....
>
> There were fifty boys at Brienne when Napoleon arrived but as he went up in the school numbers increased to a hundred. Most were his social superiors. Some boys had names famous in history, others had fathers or uncles who hunted with the King, mothers who attended Court balls. In Corsica he had been near the top socially; now he suddenly found himself near the bottom. Also, he was a state-subsidized boy, and although Louis XVI had stipulated that no distinction must be made, inevitably the fee-paying boys made the others feel it. Finally, he was the only Corsican. There were other boys from overseas, including at least two English boys, but Napoleon, with his Italian accent, inevitably stood out, and for a new boy that does not pay. Alone in a strange country, far from his family, speaking a new language, still feeling awkward in his blue uniform, he certainly needed the courage his mother had wished him. (ib., pp. 30-33)

B. As *Leadsman*

Napoleon as *leadsman* is difficult to grasp because so many of his statements and decisions were predicated by urgent military and political conditions. Yet there were times when his pragmatic relativism went far

---

[18] He later changed his family name from the Italian-sounding Buonaparte (the good party) to the French-sounding Bonaparte; even before he had moved to the French "Napoléon".

beyond what other generals or politicians would have done in the same situation as is the case from the following quotes:

> [In 1798] Napoleon had read the Koran on the voyage out [to Egypt[19]], and found it "sublime".... Knowing the importance of religion in Egypt, Napoleon announced in his first proclamation: "Cadis, sheiks, imams, tell the people that we too are true Muslims. Are we not the men who have destroyed the Pope, who preached eternal war against the Muslims? Are we not those who have destroyed the Knights of Malta, because those madmen believed that they should constantly make war on your faith?"...
>
> Napoleon tried to win the support of the religious leaders. He discussed theology with the muftis and told them he admired Mahomet. To honour the Prophet's birthday, he ordered parades, gun-salutes and fireworks. One day, in euphoric mood, he boasted he would build a mosque half-a-league round where he and his entire army could worship. Then he put a request to the muftis: would they announce in the mosques that the French were truly Muslims like themselves, and advise all Egyptians to take an oath of loyalty to Napoleon's government? The muftis replied that if the French were truly Muslims, they must undergo circumcision and renounce wine. Napoleon decided that this would be carrying adaptation a shade too far. Finally they reached a compromise. Napoleon would continue to protect Islam, and the mufties issued a limited but extremely useful statement describing Napoleon as God's messenger and the friend of the Prophet. (Cronin, 1972, pp. 151-52)

VI. Shusaku Endo[20]

A. The Making of the *Leadsman*

Endo was born in Tokyo in 1923. At the age of three he was taken by his family to the city of Dairen (Manchuria). Unfortunately the parents were divorced, and the mother (to whose

---

[19] Already in 1797 "he had written 'In order to destroy England utterly, we must get possession of Egypt'" (ib., p. 144).

[20] One of the major contemporary Japanese novelists, notable for his examination of the relationships between East and West through a unique Christian perspective.

memory Endo has always been devoted) moved back to Japan with her two sons, to live with a sister in Kobe. Following her sister's example, she became a fervent, strict-observing Catholic, and then, with the encouragement of both aunt and mother, Endo himself was baptized. He was eleven years old. (Schuchert, 1978, p. 3)

Endo says that he has always felt that as a Catholic he was wearing spiritual clothes that didn't really fit him. He has often felt like abandoning them, but he hasn't "because they were the clothes given me by my mother. When I started writing novels, I made a resolution that I would make these western clothes fit me. I wouldn't change my Christian principles, but I would adapt them, by shortening them here, letting them out somewhere else. That was my motivation in starting to write." (Hebert, 1979, p. 10)

B. As *Leadsman*

Being a Catholic, he says, made him acutely aware of "the differences and distances" between the Catholic minority and other Japanese, and between his countrymen and Europeans..."I don't think I exaggerate the differences. In present day Japan, religion is totally ignored, whereas in Europe there is still a strong Christian influence." One of his most frequently recurring images is of Japan as the "swamp" in which Christianity and other western attitudes are planted, only to rot because there is something inimical in the soil and the atmosphere. The Japanese have no concept of sin, says one of his western characters. The corollary is that they are incapable of understanding the Christian concept of redemption. "Japan accepts anything from western countries, from technology to culture. But in accepting them it often neglects the background. It accepts them on the surface, without understanding the roots." (Hebert, 1979, p. 10)

Endo's foreigners, on the other hand, do not merge in with his other characters and are not regarded with affection. They reflect his own tortured conviction that the meeting of Christianity and Japan can only produce forces which will not allow a man to be at peace with himself. It is the Japanese side which wins out, partly through force majeure, partly through the 'mudswamp' quality of Japanese culture as Endo himself has described it. Endo has been a Catholic since the age of 11 and

> has spent his life setting Western spiritual values against Japanese. In his world Western missionaries to Japan are unhinged by the realization that their task is impossible. (Powell, 1983, p. 20)

Once the *leadsman* Endo felt that some sections of the Christian sea, ahead of his Japanese ship, were too cold for his shipmates, he opted for a long detour through warmer seas never before crossed by Christian vessels, yet telling his Japanese friends—in his *A Life of Jesus* (1978)—that these too were Christian waters. However such a procedure had to be explained to his Western readers who would immediately detect a Japanesation of Christianity; and Endo gave it in the preface to the American edition of his text:

> My book called *A Life of Jesus* may cause surprise for American readers when they discover an interpretation of Jesus somewhat at odds with the image they now possess. Jesus as I depict him is a person who lived for love and still more love; and yet he was put to death, for he chose to live without violent resistance. My way of depicting Jesus is rooted in my being a Japanese novelist. I wrote this book for the benefit of Japanese readers who have no Christian tradition of their own and who know almost nothing about Jesus. What is more, I was determined to highlight the particular aspect of love in his personality precisely in order to make Jesus understandable in terms of the religious psychology of my non-Christian countrymen and thus to demonstrate that Jesus is not alien to their religious sensibilities. The religious mentality of the Japanese is—just as it was at the time when the people accepted Buddhism—responsive to one who "suffers with us" and who "allows for our weakness," but their mentality has little tolerance for any kind of transcendent being who judges humans harshly, then punishes them. In brief, the Japanese tend to seek in their gods and buddhas a warm-hearted mother rather than a stern father. With this fact always in my mind I tried not so much to depict God in the father-image that tends to characterize Christianity; but rather to depict the kind-hearted maternal aspect of God revealed to us in the personality of Jesus. (1978, p.1)

Expanding on the above, the translator of the book, the Jesuit Father Schuchert, wrote:

> Endo's theory—that Japan has not accepted God because God has been presented too frequently as an authoritative father-image—merits our consideration. The Japanese have a traditional saying to the effect that the four most dreadful things on earth are "fires, earthquakes, thunderbolts, and fathers." (1978, p. 4)

In all fairness, Endo is not only a GAM *leadsman*, but also a GAM *miner* from the divorce of his parents at age ten, and a GAM *seeker/radiologist* because of early health problems as discussed in Appendix 2 of chapter 4. No doubt, such a man needs the "kind-hearted maternal aspect of God", independently of his being, or not, a Japanese thirsting for *The One* who "suffers with us" and who "allows for our weakness."

VII. In the End

According to Robert Park (1928), "It is in the mind of the marginal man that the conflicting cultures meet and fuse. It is, therefore, in the mind of the marginal man that the process of civilization is visibly going on, and it is in the mind of the marginal mind that the process of civilization may best be studied" (p. 881).

This is a high claim; still, if taken with a grain of salt, what Park said has merit because, indeed, the *leadsmen* bring new important ways of thinking, and often are highly creative because as stated correctly by Weiten (1989): "Foremost, creativity involves seeing the world in a new and different light" (1989, p. 292).

2

Nathaniel Hawthorne, GAM *Radiologist-Universalist-Miner*

1. Introduction

> In Nathaniel Hawthorne's story "The Minister's Black Veil", a New England clergyman emerges one day with his face swathed in black crepe. Ignoring the disquiet of his parishioners, the unrelenting reverend wears the ghastly cloth to his final hour—at which point he cries out: "Lo! on every visage a black veil!" The mask has served not as a concealment but as a mirror. Through a lifetime's correspondence, Hawthorne unveiled no dark, private wellspring for his sombre, soul-probing writings. As he wrote to his soon-to-be wife, Sophia Peabody, in 1842, "When people think that I am pouring myself out in a tale or essay, I am merely telling what is common to human nature, not what is peculiar to myself. I sympathize with them—not they with me." (Tursi, 2002, p. 8)

Hawthorne wrote of what he thought was common to human nature; but he did it as a somber, soul-probing *radiologist-universalist-miner*, with the richness and complexity which derives from his belonging to no less than three GAM personality families (for causes and results): *radiologists*, *universalists*, and *miners,*[21] in something like a

[21] The three GAM personality families to which Hawthorne belongs in part are described in chapter 3 of volume 1 as:

Radiologists. Cause: Painful or humiliating physical infirmities. Result: Keen eye for anything that is sick in people and society; may have some religious or mystical interests (like *seekers*), but permeated with pessimism.

Venn's diagram in which three circles (one for each of these personality families) partially overlap, thus dividing the *personality plane* in eight regions. Thus, some of Hawthorne's works are *dedicated* (the region outside of the three circles), some are predominantly *radiological* (inside only that circle, but at various *radiological intensities*), some are *radiological-universalist* (where the *radiologist* and *universalist* circles overlap), and some works are *radiological-universalist-miner* (the area in which the three circles overlap).

This description may seem unnecessarily complicated, but it can provide a new understanding of Hawthorne's works, and relate his personality to that of others whose youth, personality and works, were similarly affected. By focusing, for instance, on that part of the Venn diagram in which the ring of the *radiologists* overlaps that of the *universalists* we can compare (for causes and results) Hawthorne to Nietzsche.[22] Similarly, by focusing on the *miner ring*, we can compare some of Hawthorne's works to those by the *miners* Baudelaire and Rimbaud.

Moving to that important M of his GxAxM, Hawthorne suffered in his youth of three major misfortunes: death of father Categories A and B[23]

Universalists. Cause: Early parental death or much parental absence, coupled with love and good assistance from the remaining parent or relatives; good assistance from school and stable socio-economic status. Result: Fond of moral concepts that are logical and prescriptive; have penchant for vast, nonrelativistic, theoretical thinking; fond of higher/intellectual pursuits such as religion, philosophy, mathematics, science and law.

Miners. Cause: Lack of love (e.g., lack of love from remaining parent after one died, divorced or was pushed aside). Result: Skeptics with a tendency to pessimism and satire; sharp eye for hypocrisy (Many great poets are *miners*).

[22] Both suffered of early paternal death Category A: at age 5 for Nietzsche, and age 4 for Hawthorne; and of physical infirmity which began at the age of 9 for both.

[23] Parental death category A is the one which causes a parental vacuum, but with no additional misfortunes. Parental death category B is the one followed by problems (mainly lack of love for the children) when, for instance the remaining parent soon remarries (not to provide a substitute father or mother for the children) or has an affair; or is overwhelmed by his/her difficulties and unable to give love and attention to the children. Parental death category C is the one which starts a painful socioeconomic downfall of the family. Parental death category A is the origin of the *universalists*, category B the origin of many *miners*, while parental death category C is a the source of many *critical jesters*. (The reader will find the definitions of these terms—causes and results—in chapter 3 of volume 1, "The Fourteen GAM-Challenged Personality Families." In the case of Hawthorne, the paternal death was neither full A, nor full B as discussed later, and this fact should go a long way to explaining his rich and complex personality and opus, and why I felt the need to call him a *radiologist-universalist-miner*.

when he was four, then severe/insulting physical infirmity, from about age 9 to 10 (or 9 to 12, depending on the sources and their recollections, many years after the facts). Accordingly, I will start with Hawthorne's *radiological circle*, move to his *universalist circle*, and then to their overlap. I take this approach because, of all his personality/creativity circles, the *radiologist* has been the least studied, and yet it is capable of providing important additional understanding on both the creator and his acts of creation.

Also, what I report hereafter on Hawthorne the *radiologist*, provides evidence for the validity of the GAM theory of personality and creativity. As in the case of Nietzsche (see volume 2), when I read of Hawthorne's *radiological misfortune* of youth (discussed in the next section), I expected to find numerous and important *radiological texts* in his works, and I found them, especially in his master work, *The Scarlet Letter*.

Like Nietzsche (whose eye problems grew slowly, but never stopped), Hawthorne became fascinated (even obsessed) with diseases and cures, with patients and doctors. Then, both Nietzsche and Hawthorne (as *radiologist-universalist* because of the two major misfortunes of youth: physical infirmity and early death of the father), felt they had discovered a cure to what they saw as the fundamental psychological-physical disease afflicting humanity.

In chapter 9 of volume 2, I spoke of Doctor Nietzsche and his cure, in this chapter I will also refer to Doctor Hawthorne and his cure. On the other hand, the Nietzsche-Hawthorne comparison should not be pushed too far, because in Hawthorne there was also a major *miner component* (from paternal death category B) which is not present in Nietzsche, while the *radiologist* in Hawthorne is not as powerful as in Nietzsche.

2. *Radiological Causes*

> On November 10, 1813, Nathaniel was hit on the leg while playing "bat and ball." According to [his sister] Elizabeth, writing decades later, "no injury was discernible," yet he was incapacitated for two or three years. Apparently he refused to walk and favored the injured leg for months following the accident. Elizabeth, not the mother, wrote urgently to Uncle Robert: "I don't know Nathaniel's foot will ever get well if you don't come home. He wont walk on it & the Doctor says he must; so do come soon." Perhaps after direct orders from Uncle Robert,

> who apparently was the only one who could manage the youth, Nathaniel began to walk with the aid of crutches... The patient remained stubborn and uncooperative, and recovery was slow. Many doctors from Salem and elsewhere, including his future father-in-law, Dr. Nathaniel Peabody, examined and prescribed but produced neither cure nor diagnosis of the injury. Soon one leg was growing more slowly than the other. Only by January 20, 1815, was his mother finally confident. "Nathaniel has," she wrote, "entirely recovered the use of his foot, and walks as well as he did before he was lame, his joy was great, when he found he could walk without his crutches." She attributed the cure to a Dr. Smith of Hanover, who advised pouring cold water on his foot every morning.[24] *The disability appears to have had psychological as well as physical origins. It occurred within a year of two traumatic deaths, which reactivated Nathaniel's memory of the death of his father four years earlier. Nathaniel was moved by the almost uncontrollable grief of his Grandmother Manning, who had lost a husband and son in the same year.* (Miller, 1991, pp. 47-48, my italics)

Worth mentioning is also the following from a letter[25] that Hawthorne's sister Elizabeth wrote to her niece Una :

> Your father was lame a long time from an injury received while playing bat-and-ball. His foot pined away, and was considerably smaller than the other. He had every doctor that could be heard of; among the rest, your grandfather Peabody. But it was 'Dr. Time' who at last cured him. I remember he used to lie upon the floor and read, and that he went upon two crutches. Everybody thought that, if he lived, he would be always lame. (Hawthorne, Julian, 1884/1968, p. 100)

---

[24] Specifically, "The prescribed consisted of pouring showers of cold water from a window in the second story upon the lame foot, extended below.... Later he was able to walk with the aid of crutches. Not until after some three years, when he was twelve, was the lameness completely overcome" (Stewart, 1948, p. 4). If so, the cure may have seemed—especially to the patient—more quackery than science, and the infirmity may have lasted longer than mentioned by others: three years instead of a little more than one year.

[25] Reported by Julian Hawthorne in the biography he wrote of his father Nathaniel.

Clearly, the first of those who "thought that, if he lived, he would be always lame" was Nathaniel Hawthorne himself.[26] This thought, coupled with those on the death of his father and other relatives—and with the many failed attempts for a cure which showed how little the doctors knew—led him to have strong *radiological leanings*. Burning questions on the meaning of life—stimulated by the early death of his father—were asked together with question on the meaning of health and sickness. These questions raised by the young *universalist* became, at the same time, also *radiological questions*.

While *universalists* can, on an intellectual basis, forget the body as unimportant, and concentrate their thoughts on the mind and its involvement with truth, justice, happiness and love, *universalist-radiologists* cannot: they cannot forget their sick body, the sick body of others, the physicians who were unable to help them. Then, when the two misfortunes (early parental death, and humiliating physical infirmity) act in parallel, the infirmities of the body will be linked to the infirmities of the mind.

Hawthorne, the *universalist*, discovered what other great minds of the past had found—that pride is the gravest of all sins.[27] Then, *radiologist* as he was, Hawthorne linked pride to bodily sickness.

---

[26] It is true that the problem with his foot lasted only little more than one year (or up to three years), but Nathaniel did not known if and when he would be healed. The reader will remember the strong impact on Charles Dickens of those five months in the blacking warehouse, discussed in chapter 1 of volume 3. Dickens at that time was 12, close in age to Hawthorne who became infirm at the age of 9. In both cases, the terrible fear was that it would last for ever. Dickens never forgot, and certainly neither did Hawthorne. It shaped their personality and creativity.

[27] In classical religious terms "pride is the inordinate desire to excel which springs from a self-love that is exclusive of others.... When man isolates self from God and the rest of humanity and makes self absolute or central, either ignoring all others or using them solely toward the achievement of his own private ends, he has the vice of pride. It is in this sense that pride is traditionally regarded as one of the seven capital sins and the queen of vices. It is a capital sin because it is a source or foundation for other sins, especially such sins as presumptuousness, which inclines the proud man to attempt what is beyond his powers; inordinate ambition, which is the inclination to aim at an honor and dignity beyond one's deserts or to use sinful means in the pursuit of them; and vainglory, which is the inordinate effort to manifest one's own excellence, real or fictitious. At the root of all such sins there is an exaggerated love and concern for self that clouds one's knowledge and appreciation of the true self and the corresponding worth of others" (Parmisano, 1967, pp. 765-66).

3. *Radiological Results*

Within the GAM theory of Personality and Creativity, *radiological results* are those parts of a novel, poem, painting, conversation, allegory, philosophical-religious-social thinking, which deal with sickness: pain, wounds, symptoms, cures, doctors, hospitals. *Radiological results* remind us of our bodies, particularly of our sick bodies. They are *radiological* because they look beyond the skin, they see sickness behind the appearance of health.

Moving to Hawthorne, I report hereafter a number of *radiological results* from three of his major works: two short stories (*Egotism, or the Bosom Serpent* of 1843, and *The Birthmark* also of 1843) and his masterwork, the novel *The Scarlet Letter* of 1850. For ease of understanding I will put those words which are strongly "medical" in italics.

Egotism; Or, The Bosom Serpent (1843)

> It was not without a shudder that he found himself on the point of meeting his former acquaintance, whom he had known in the glory of youth, and whom now after an interval of five years, he was to find victim either of a *diseased fancy or a horrible physical misfortune*. (1950, p. 188)
>
> "Oh, 'tis a mere nothing! A snake! the commonest thing in the world. A snake in the bosom-that's all," answered Roderick Elliston. "But how is your own breast?" continued he looking the sculptor in the eye with the most acute and penetrating glance that it had ever been his fortune to encounter. "All pure and wholesome? No reptile there? By my faith and conscience, and by the devil within me, here is a wonder! A man without a serpent in his bosom!" (pp. 189-90)
>
> The sculptor did not pursue him. He saw that no available intercourse could be expected at such a moment, and was desirous, before another meeting, *to inquire closely into the nature of Roderick's disease and the circumstances that had reduced him to so lamentable a condition*. He succeeded in obtaining the necessary information *from an eminent medical gentleman*. (p. 190)
>
> *The symptoms caused them endless perplexity. They knew not whether ill health were robbing his spirits of elasticity, or whether a canker of the mind was gradually eating, as such cankers do, from his moral system into the physical frame,* which is but the shadow of the former. (p. 190)

What could it be that gnawed the breast of Roderick Elliston? Was it sorrow? *Was it merely the tooth of physical disease?*....There was plausible ground for each of these conjectures; but it must not be concealed that more than one elderly gentleman, the victim of good cheer and slothful habits, magisterially pronounced *the secret of the whole matter to be Dyspepsia*! (p. 191)

The empiric's cure had been a sham, the effect, it was supposed, *of some stupefying drug which more nearly caused the death of the patient* than of the odious reptile that possessed him. (p. 192)

All persons chronically diseased are egotists, whether the disease be of the mind or body; whether it be sin, sorrow, or merely the more tolerable calamity of some endless pain, or mischief among the cords of mortal life. Such individuals are made acutely conscious of a self, by the torture in which it dwells. Self, therefore, grows to be so prominent an object with them that they cannot but present it to the face of every casual passer-by. There is a pleasure—perhaps the greatest *of which the sufferer is susceptible—in displaying the wasted or ulcerated limb, or the cancer in the breast*; and the fouler the crime, with so much the more difficulty does the perpetrator prevent it from thrusting up its snake-like head to frighten the world; for it is that cancer, or that crime, which constitutes their respective individuality.... *The snake in his bosom seemed the symbol of a monstrous egotism* to which everything was referred, and which he pampered, night and day, with continual and exclusive sacrifice of devil worship.... In some of his moods, strange to say, *he prided and glorified himself on being marked out of the ordinary experience of mankind*, by possession of a double nature, and a life within a life. (p. 193)

"Could I for one moment forget myself, the serpent might not abide within me. *It is my diseased self-contemplation that has engendered and nourished him.*" (p. 200)

The Birthmark (1843)

One day, very soon after their marriage, Aylmer sat gazing at his wife with a trouble in his countenance that grew stronger until he spoke. "Georgiana," said he, "*has it never occurred to you that the mark upon your cheek might be removed?*" "No, indeed," said she, smiling; but perceiving the seriousness of his manner, she blushed deeply. "To tell you the truth it has been so often called a charm that I was simple enough to imagine it might be so." "Ah, upon another face perhaps it might," replied her husband; "but never on yours. No, dearest Georgiana, you came so nearly perfect from the hand of Nature that this slightest possible defect, which we hesitate whether to term a defect or a beauty, shocks me, as being the visible mark of earthly imperfection." "Shocks you, my husband!" cried Georgiana, deeply hurt; at first reddening with momentary anger, but then bursting into tears. "Then why did you take me from my mother's side? You cannot love what shocks you!" (1950, pp. 203-04)

In this manner, *selecting it as the symbol of his wife's liability to sin, sorrow, decay, and death*, Aylmer's sombre imagination was not long in rendering the birthmark a frightful object, causing him more trouble and horror than ever Georgiana's beauty, whether of soul or sense, had given him delight. (p. 205)

The crimson hand, which at first had been strongly visible upon the marble paleness of Georgiana's cheek, now grew more faintly outlined. She remained not less pale than ever! But the birthmark, with every breath that came and went, lost somewhat of its former distinctness. Its presence had been awful; its departure was more awful still. Watch the stain of the rainbow fading out the sky, and you will know how that mysterious symbol passed away. (p. 219)

"My poor Aylmer!" murmured she. "Poor? Nay, richest, happiest, most favored!" exclaimed he. "My peerless bride, it is successful! You are perfect!"

"My poor Aylmer," she repeated, with a more than human tenderness, "you have aimed loftily; you have done nobly. Do not repent that with so high and pure a feeling, you have rejected the best the earth could offer. Aylmer, dearest Aylmer, I am dying!" *Alas! it was too true! the fatal hand had grappled with the mystery of life, and was the bond by which an angelic spirit kept itself in union with a mortal frame.* (p. 220)

The Scarlet Letter (1850)

Trusting no man as his friend, he [the Reverend Mr. Dimmesdale] could not recognize his enemy when the latter actually appeared. He therefore still kept up a familiar intercourse with him, *daily receiving the old physician in his study; or visiting the laboratory, and, for recreation's sake, watching the processes by which weeds were converted into drugs of potency.*

One day, leaning his forehead on his hand, and his elbow on the sill of the open window, that looked towards the graveyard, he talked with Roger Chillingworth, while the old man was examining a bundle of unsightly plants. "Where," asked he, with a look askance at them,-for it was the clergyman's peculiarity that he seldom, nowadays, looked straightforth at any object, whether human or inanimate,- "*where, my kind doctor, did you gather those herbs,* with such a dark, flabby leaf?" "Even in the graveyard here at hand," answered the physician continuing his employment. "They are new to me. *I found them growing on a grave, which bore no tombstone, nor other memorial of the dead man, save these ugly weeds, that have taken upon themselves to keep him in remembrance. They grew out of his heart, and typify, it may be, some hideous secret that was buried with him,* and which he had done better to confess during his lifetime." (1935, p. 140)

There was another pause; *and the physician began anew to examine and arrange the plants which he had gathered. "You inquired of me, a little time agone," said he, at length, "my judgment as touching your health.*" "I did," answered the clergyman, "and would gladly learn it. Speak frankly, I pray you, be it for life or death." "Freely, then, and plainly,"said the physician, still busy with his plants, but keeping a wary eye on Mr. Dimmesdale, "*the disorder is a strange one; not so much in itself, nor as outwardly manifested,-in so far, at least, as the symptoms have been laid open to my observation.* Looking daily at you, my good Sir, and watching the tokens of your aspect, now for months gone by, *I should deem you a man sore sick, it may be, yet not so sick but that an instructed and watchful physician might well hope to cure you.*" (pp. 144-45)

"Be it so! But, again! He to whom only the outward and physical evil is laid open, knoweth, oftentimes, but half the evil which he is called upon to cure. *A bodily disease, which we look upon as whole and entire within itself, may, after all, be but a symptom of some ailment in the spiritual part.*" (pp. 145-46)

> "*Thus, a sickness*," continued Roger Chillingworth, going on, in an unaltered tone, without heeding the interruption,-but standing up, and confronting the emaciated and white-cheeked minister, with his low, dark, and misshapen figure,-"*a sickness, a sore place*, if we may so call it, in your spirit, hath immediately its appropriate manifestation in your bodily frame. *Would you, therefore, that our physician heal the bodily evil? How may this be, unless you first lay open to him the wound or trouble in your soul*?" "No! – not to thee! – not to an earthly physician!" cried Mr. Dimmesdale, passionately, and turning his eyes, full and bright, and with a kind of fierceness, on old Roger Chillingworth. (p. 146)[28]
>
> And thus, while standing on the scaffold, in this vain show of expiation, Mr. Dimmesdale was overcome with a great horror of mind, *as if the universe were gazing at a scarlet token on his naked breast, right over his heart. On that spot, in very truth, here was, and there had long been, the gnawing and poisonous tooth of bodily pain.* (p. 158)

Two of the most important chapters of *The Scarlet Letter*, are number IX "The Leech," and number X "The Leech and his Patient." Here Hawthorne plays constantly on the triple meaning of the word leech to describe Roger Chillingworth, a word which means: "1. [Archaic] a physician, 2. any member of a subclass (*Hirudinea*) of mostly flattened, annelid worms living in water or wet earth and having a well-developed sucker at each end: mostly are bloodsuckers, and one species (*Hirudo medicinalis*) has been used in medicine, esp. in former times, to bleed patients; and 3. a person who clings to another to gain some personal advantage; parasite." (*Webster's New Word Dictionary*, 1988, p. 770).

On the first meaning of physician, *The Oxford English Dictionary*, has this to say: "1. A physician; one who practices the healing art. Now *arch*. (chiefly *poet*.)... In the 17th c. it was applied in ordinary prose use

---

[28] In the words of Martin (1983): "Chillingworth has made the torture of Dimmesdale the very principle of his existence. The minister's perception in the forest interview with Hester seems accurate: Chillingworth is the worst sinner of all; he has violated in cold blood the sanctity of a human heart—the great sin of all Hawthorne antagonists. *He has used his skill as a physician to keep Dimmesdale alive so that he can continue to punish him spiritually and psychologically*" (p. 124).

only to veterinary practitioners, and this sense survives in some dialects.... 1807 Crabbe *Par. Reg*. III. Can this proud leech, with all his boasted skill, amend the soul or body, wit or will?... 1870 Morris *Earthly Par.* I. 1. 121 As one who lays all hopes aside, Because the leech has said his life must end" (1989, *8*, p. 791). The Reverend Mr. Dimmesdale could indeed have asked: "Can this proud leech, Chillingworth, with all his boasted skill, amend the soul or body?"

As noted by Roy Male, "Hawthorne's portrayal of Chillingworth illustrates how beautifully his imagination could weld the abstract to the concrete. *For the physician is interesting in his own right as an alchemist-psychiatrist manqué, who tries to solve the riddle of man's existence by logical or psychological analysis. As a symbol of guilt, Chillingworth is a leech, draining his patient of nerve, will, and physical energy*. But, as the whole book demonstrates, he is also the healer. Only by knowing him, confronting him face to face, is moral growth possible. Not that moral growth is guaranteed or that having this unwelcome guest is "fortunate"–it is simply inevitable in human existence" (1964, p. 96).

And yes, the leech—the animal, the *Hirudo medicinalis*—can truly heal when used, not to bleed, but to suck pus from an abscess. Gradually, the little animal grows bigger and bigger, and the abscess smaller, until—satiated—the leech detaches itself from the abscess. Chillingworth, besides draining his patient of nerve and physical energy, sucked Pastor Dimmesdale's pus: his pride, so that he could openly admit his guilt, and be saved.

One after the other, these quotes speak of the sick and suffering body, of dreadful diseases, of wrong operations, of an evil medical doctor poisoning the soul, of stypefying drugs which nearly cause the death of the patient, of wasted or ulcerated limbs. This is the human condition as seen by a GAM *radiologist* who, when he has a cure, it is a moral one, like the "Forget yourself!" of *The Bosom Serpent*.

4. *Universalist Results*

All of Hawthorne's fiction shows pride as the great besetting sin, whether it is the intellectual pride of Ethan Brand or the reforming zeal of Hollingsworth in *The Blithedale Romance* or the arrogation to oneself of godlike power, as in the case of Aylmer in "The Birthmark," or a selfish concern for one's repu-

> tation, as with Arthur Dimmesdale who played the hypocrite in his Puritan parish for seven years. Man is a fallible, imperfect being in Hawthorne's stories.... The tales and novels of Hawthorne embody these truths in enduring dramatic situations. (Stewart, 1973, p. 189)

Such strong interest in pride—the disordered love of self or success, the worse of all sins, and often the source of other sins—also says that Hawthorne is a Berlinian *hedgehogs* "those who relate everything to a single central vision, universal, one system organizing principle in terms of which alone all that they are and say has significance.... Dante, Plato, Lucretius, Pascal, Hegel, Dostoevsky, Nietzsche, Ibsen, Proust are, in varying degrees, *hedgehogs*" (Berlin, 1953/1967, pp. 1, 2). This list of eminent *hedgehogs* is important: 1. to understand the strong words that Berlin used in defining the *hedgehog*. With the great thinkers, that "single central vision" is in no way restricting, but it is instead a set of radial paths to inspect the universe); and 2. it reminds us of Nietzsche, *radiologist-universalist*, like Hawthorne for psychological causes and results.

As in the previous section, I will quote—and italicize—those texts showing Hawthorne's fascination with the evil of pride. Starting with *The Scarlet Letter*, as commented by Terence Martin (1983):

> From the beginning of her exposure to public view, *Hester bears her ordeal with haughty agony. Undeniably she flaunts the letter*;... Alone in the world with the symbol and consequence of her sin, *Hester dons an armor of pride* that is also a mantle of suffering. (p. 111)
>
> *Dimmesdale is afflicted with a devious pride*. He cannot surrender an identity which brings him the adulation of his parishioners, the respect and praise of his peers. His contortions in the guise of humility only add to the public admiration which, in turn, *feeds an ego fundamentally intent on itself.* (p. 113)
>
> Hester Prynne, as we have seen, stands in the marketplace with a "burning blush, and yet a haughty smile"; her beauty shines out and makes "a halo of her misfortune"; on the scaffold she takes refuge in public exposure. Moreover, she has embroidered her letter fancifully; *she has adorned the symbol of her sin almost as if she were parading that symbol and that sin before the public*. (p. 118)

Dimmesdale clearly suffers from an excess of self. His weakness and suffering throughout most of the romance, as I suggested earlier, have tended to blur for some readers the fact of his pride, which, like his scarlet letter, lies beneath and gives special form to his mask of saintliness. Self-condemnation, self-abnegation, and self-loathing are the stimulants of Dimmesdale's psychic life; they constitute, as well, the price he must pay if he would not abdicate the self reverenced by the public. And that self—formed out of a communal wish to admire a young, pious, and learned minister—he cannot bring himself to renounce. *That his private suffering contributes to the public mask of spirituality is a kind of masochistic dividend for him. Weak and proud, false and efficacious, the minister thus stumbles ever deeper and ever more self-consciously into thickets of hypocrisy.*[29] (pp. 122-23)

He wants to have it both ways, to leave, but to leave with his "public duty" well performed—*which is to say, with the congregation marveling at their saintly and inspired minister*. Let us note that Dimmesdale's motive here is characteristically secret. No sooner is he converted to Hester's romantic cult than he has an immediate private life outside it which gives him latitude to plan by himself, for himself. The psychological keenness informing Dimmesdale's transformation is of a high order. *No matter what "religion" the minister professes, he serves ultimately the interests of one master—himself.* Thus, in keeping with the brilliant economy of *The Scarlet Letter*, the moment at which Dimmesdale commits himself consciously to deadly liberating sin becomes the moment at which he secretly wishes to cap his public life with a final burst of eloquence on the most important occasion the Puritan community can offer. (pp. 123-24)

---

[29] As noted before, *miners* have a "sharp eye for hypocrisy." In Hawthorne's own words: "'At least, they shall say of me,' thought this exemplary man [Dimmesdale], 'that I leave no public duty unperformed, nor ill performed!' Sad, indeed, that an introspection so profound and acute as this poor minister's should be so miserably deceived! We have had, and may still have, worse things to tell about him; but none, we apprehend, so pitiably weak;... No man, for any considerable period, can wear one face to himself, and another to the multitude, without finally getting bewildered as to which may be true" (ib., pp. 230-31).

> *Chillingworth personifies the Hawthorne antagonist in all his essential pride and monomania.* (p. 126)[30]

And what, in the end, is the cure for this universal disease of mankind? This will be found at the end of *Egotism; or, The Bosom Serpent*:

> "*Could I for one moment* [says Roderick, the snake-possessed] *forget myself*, the serpent might not abide within me. *It is my diseased self-contemplation* that has engendered and nourished him."
>
> "*Then forget yourself*, my husband," said a gentle voice above him; "forget yourself in the idea of another!" ... She touched Roderick with her hand. A tremor shivered through his frame. At that moment, if report be trustworthy, the sculptor beheld a waving motion through the grass and heard a tinkling sound, as if something had plunged into the fountain.... "A tremendous Egotism, manifesting itself in your case in the form of jealousy [said the sculptor], is as fearful a fiend as ever stole into the human heart. Can a breast, where it has dwelt so long, be purified?" "Oh yes," said Rosina [Roderick's wife] with a heavenly smile. "The serpent was but a dark fantasy,... To give it its due importance we must think of it but as an anecdote *in our Eternity*." (ib., pp. 200-02)

Dr. Hawthorne's cure, then, is not the Delphian "Know thyself," but "Forget thyself" which should eliminate at one stroke—as it happened with Roderick—pride and its derivative jealousy (the same pride and jealousy which drove Chillingworth).

---

[30] But why this strong interest in the evil of pride? Why this strong link between pride and physical disease? Could it be that Hawthorne himself saw the two at work in himself? Could he have felt that what had happened to his leg was a deserved punishment for his pride?

On the other hand, the pride he noticed may have been those of the many doctors who "examined and prescribed but produced neither cure nor diagnosis of the injury." Still, among these two causes—and there may be others—I prefer the first, which does not exclude his joy at being able to walk without crutches: there are limits to masochism in every intelligent person. In essence, I feel that the pride which Hawthorne condemns is first of all his own, and this from his youth, when the miseries of life have not yet clipped of his wings.

In *The Scarlet Letter*, Hester saved her man—as Rosina did her in *Egotism*—by using similar words. Speaking to Dimmesdale: "'Thou are crushed under this seven years' weight of misery,' replied Hester, fervently resolved to buoy him up with her own energy. 'But thou shalt leave it all behind thee!... *Leave this wreck and ruin here where it hath happened. Meddle no more with it! Begin all anew.* Hast thou exhausted possibility in the failure of this one trial? Not so! The future is yet full of trial and success'" (ib., p. 212).

In *The Birthmark*, instead, that cure was not possible because Georgiana was the one directly attacked by her husband's pride: the pride of the scientist who thought he understood everything and could correct what he thought was Nature's fault:

> "Dearest Georgiana, I have spent much thought upon the subject," hastily interrupted Aylmer. "I am convinced of the perfect practicality of its removal."... "Noblest, dearest, tenderest wife," cried Aylmer, rapturously, "doubt not my power.... I feel myself fully competent to render this dear cheek as faultless as its fellow; and then, most beloved, what will be my triumph when I shall have corrected what Nature left imperfect in her fairest work! Even Pygmalion, when his sculptured woman assumed life, felt not greater ecstasy than mine will be." (pp. 207-08)

A pride so gigantic (and at the expense of one's loving wife who is reduced to the equivalent of a formless piece of marble) cannot be conquered: it will be crushed by the most tragic failure.

5. *Miner*

Miner Causes:

> Hawthorne never referred directly to his father's death, but it is written large in the fatherless universe of his fiction, from his first published work, *Fanshawe,* to the last uncompleted manuscript, *"The Dolliver Romance."* If assembled in one of Hawthorne's processionals, these orphans would number thirty or more, all of whom are as burdened as John Bunyan's Christian....The Peabodys depicted Mrs. Hathorne as a recluse who took her meals alone in her room and allowed no one to enter her almost sacred chamber.... Mrs. Hathorne's isolation

> from the world was extreme, granting all mitigating circumstances. At various times Elizabeth and Nathaniel noted without drawing somewhat obvious conclusions that their mother could not cope with feelings and demonstrations of affection. In Elizabeth's words, both the Hathornes and Mannings "possessed deep, strong, passionate affections & natures, but were very undemonstrative, with a real inability to express their deeper emotions." It was not her intention, but Elizabeth could have been describing Hawthorne's writings with the recurrent emphasis upon repression, non-communication, and the absence and fear of tactile contact and gratification. She perhaps did not perceive that the mother's failure was often her brother's subject matter.
>
> The bond of children to parent, then, was intense but somewhat chilled. The mother was unable to supply the affection and support as fully as the children expected or perhaps needed. But she set a model for their behavior. Hawthorne was in many ways his mother's son. As everybody testified and his behavior illustrated, he was reclusive and often melancholic and like his fictional characters a spectator with the coolness and distance provided by such a stance. It was one way of keeping hellfired emotions under seeming control. (Miller, 1991, pp. 28-33)

Miner Creative Results:

> Hawthorne's childscape is a universe of "wounded love," of orphans, of parents who betray their offspring through death or other form of abandonment, of quests for roots and an ancestral home or Eden, of circular journeys which return the victim to the scene of the crime, the place of birth. In age Hawthorne still wondered: "Where the child got the tenderness which a child needs to live upon is a mystery to me." He had not found the answer to the personal mystery, his own survival. (ib., p. 38)

In *The Gentle Boy* "not the first self-portrait, in Hawthorne's writings, but the most detailed recollections of his perceptions and depressions during his earliest years" (ib., p. 43), the gentle boy's parents "have sacrificed him to their religious persuasion. For years their flights from country to country denied the youth the opportunity to establish roots in any society or to form relationships. As his mother later reveals in her wild harangue to the Puritan congregation, she has placed God's will, as intuited by her inner light, above the nurturing of Ilbrahim" (ib., p. 39).

We are reminded of Baudelaire:

Ma jeunesse ne fut qu'un ténébreux orage,
Traversé çà et là par de brillants soleils;
Le tonnerre et la pluie ont fait un tel ravage
Qu'il reste en mon jardin bien peu de fruits vermeils.[31]
from *"L'Ennemi"*, in *Les Fleurs du Mal*

6. In the End

"*The Scarlet Letter*", says Benét, "treats Hawthorne's favorite themes. First, all men are guilty of secret sin. Second, greater than Dimmesdale's is Chillingworth's sin, for he has invaded the sanctity of another's soul"[32] (1987, p. 871).

That "all men are guilty of a secret sin"—and that this sin is pride—is the discovery that the *universalist* Hawthorne made. Next is the discovery by the *radiologist* Hawthorne that all men are bodily sick, medically sick, and that they do not notice or misunderstand this sickness in themselves and in the others. Third is the discovery by the *universalist-radiologist*, Hawthorne, that moral sickness (specifically pride) and bodily sickness are related, often more so when it is least apparent.

The fourth discovery is that by the *miner* who realized how the "childscape is a universe of 'wounded love,' of orphans, *of parents who betray their offsprings through death or other forms of abandonment*" (Miller, 1991, p. 38; my italics). We are reminded of the feelings of the American poet, and GAM *miner* Sylvia Plath whose father died when she was only eight:

> The manner alone of Otto's death was bound to have a traumatic effect on a girl of Sylvia's sensitivity and imagination. *Somehow she had to struggle with the fact that he had left her, that death was truly final, and that his departure was in*

---

[31] My youth was but a dark storm, Crossed here and there by brilliant suns; The thunder and the rain have made such ravages, That in my garden there remain few vermillion fruits.

[32] This last point refers to the following words by Pastor Dimmesdale: "I freely forgive you [Hester] now. May God forgive us both! We are not, Hester, the worst sinners in the world. There is one worse than even the polluted priest! That old man's revenge has been blacker than my sin. He has violated, in cold blood, the sanctity of a human heart. Thou and I, Hester, never did so!" (p. 209)

> *some way a "deliberate" act of betrayal*....It is also possible that Aurelia [Plath's mother] had perhaps unconsciously let slip some of her own feelings that Otto had left her in a financial bind, a struggling widow with two children to raise and no income available beyond what she could bring in herself by going out to work. (Butscher, 1976, p. 13, 15; my italics)

For the *universalist* (as discussed in volume 1) the missing parent is now with God, partaking in some way of the divinity, and protecting, guiding, or inspiring son or daughter. Not so with the *miner*: the memory of the lost parent is stained by the behavior of the remaining one, be it the quick remarriage of the mother of Baudelaire, or the negative comments by the mother of Sylvia Plath, or the unhealthy reclusive behavior of the mother of Hawthorne.[33]

Let us not forget that things look different to a highly sensitive youth: "Near the end of his life, Hawthorn confirmed that 'The Gentle Boy' was one of the earliest of his tales. It is his quintessential childscape, composed of his memories of loneliness, feelings of rejection and loss,... [His sister] Elizabeth insisted that his was a happy childhood—*the truth is, simply, that he viewed it and felt it that way*" (Miller, 1991, p. 38; my italics).

The biographers of the eminent Spanish novelist Pío Baroja are faced with the same problem. Seen from a distance, his youth was a normal happy one, not less than that of other children who then said to have been happy; yet Baroja did not see that way, and said more than once: "*no he tenido una infancia feliz* [I did not have a happy childhood].... *Baroja fue, sí, tratado con cariño, pero no come necesitaba él; ya en su hogar debía de sufrir de incomprensiones; se sintío abandonado, solo, más de una vez, pese, lo repetimos, a lo mucho que le quiso* [Baroja was indeed treated with love, but not the way he needed it; at home he must have suffered of incomprehension; more than once, he felt abandoned, alone, notwithstanding, we repeat it, all the love that he received]" (Juan, 1963, p. 33). On the other hand, in the case of Hawthorne, there were two real major misfortunes of youth—early death of the father and physical infirmity—and not only a problem of heightened sensitivity.

---

[33] It was "Elizabeth, not the mother, [who] wrote urgently to Uncle Robert: 'I don't know Nathaniel's foot will ever get well if you don't come home'" (Miller, 1991, p. 47).

7. Appendix: Like Tolstoy, also Hawthorne was, in part, a *Fox*

My interest in relating an author's GAM personality to his/her works—as just done for Hawthorne—should not be interpreted to say that everything of value written by these authors is directed by their major misfortune of youth. No, much of what they wrote is plain good writing to be enjoyed as good music in a concert hall. In this respect, the reader may remember, from chapter 9 of volume 1, my criticism of Isaiah Berlin's claim that Tolstoy was "by nature a *fox*, but believed in being a *hedgehog*" (1953/1967, p. 4):

> But when we come to Count Lev Nikolaevich Tolstoy, and ask this of him—ask whether he belongs to the first category or the second, whether he is a monist or a pluralist, whether his vision is of one or of many, whether he is of a single substance or compounded of heterogeneous elements, there is no clear or immediate answer. The question does not, somehow, seem wholly appropriate; it seems to breed more darkness than it dispels.... No author who has ever lived has shown such powers of insight into the variety of life—the differences, the contrasts, the collisions of persons and things and situations, each apprehended in its absolute uniqueness and conveyed with a degree of directness and a precision of concrete imagery to be found in no other writer. No one has ever excelled Tolstoy in expressing the specific flavour, the exact quality of a feeling—the degree of its 'oscillation', the ebb and flow, the minute movements (which Turgenev mocked as a mere trick on his part)—the inner and outer texture and 'feel' of a look, a thought, a pang of sentiment, no less than that of the specific pattern of a situation, or an entire period, continuous segments of lives of individuals, families, communities, entire nations. The celebrated life-likeness of every object and every person in his world derives from this astonishing capacity of presenting every ingredient of it in its fullest individual essence, in all its many dimensions, as it were. (Berlin, 1953/1967, pp. 3, 39-40)

The same can be said for Hawthorne in whose works we can find large sections of first-class *fox* material. Take for instance the first chapter of *The Scarlet Letter*, the one entitled "The Custom House".

> In my native town of Salem, at the head of what, half a century ago, in the days of old King Derby, was a bustling wharf,—but which is now burdened with decayed wooden warehouses, and exhibits few or no symptoms of commercial life; except, perhaps, a bark or brig, half-way down its melancholy length, discharging hides; or, nearer at hand, a Nova Scotia schooner, pitching out her cargo of firewood, — at the head, I say, of this dilapidated wharf, which the tide often overflows, and along which, at the base and in the rear of the row of buildings, the track of many languid years is seen in a border of unthrifty grass,—here, with a view from its front windows adown this not very enlivening prospect, and thence across the harbor, stands a spacious edifice of brick. From the loftiest point of its roof, during precisely three and a half hours of each forenoon, floats or droops, in breeze or calm, the banner of the republic; but with the thirteen stripes turned vertically, instead of horizontally, and thus indicating that a civil, and not a military post of Uncle Sam's government is here established. Its front is ornamented with a portico of half a dozen wooden pillars, supporting a balcony, beneath which a flight of wide granite steps descends towards the street. Over the entrance hovers an enormous specimen of the American eagle, with outspread wings, a shield before her breast, and, if I recollect aright, a bunch of intermingled thunderbolts and barbed arrows in each claw. With the customary infirmity of temper that characterizes this unhappy fowl, she appears, by the fierceness of her beak and eye, and the general truculency of her attitude, to threaten mischief to the inoffensive community; and especially to warn all citizens, careful of their safety, against intruding on the premises which she overshadows with her wings. (1850/1935, pp. 4-5)

Here, I could have highlighted nearly every word to show how *fox* this text is; instead let me point to the description of the American eagle, which fits perfectly with the Tolstoyan situation, *apprehended in its absolute uniqueness and conveyed with a degree of directness and a precision of concrete imagery.*

"No one has ever excelled Tolstoy in expressing the specific flavour, the exact quality of a feeling—the degree of its 'oscillation', the ebb and flow." No one may ever excelled Tolstoy in this. Yet, he was not so unique, indeed:

> The special quality of Hawthorne's achievement in *The Scarlet Letter* derives from the essential duality or ambivalence of his fictional world and its component parts.... In *The Scarlet Letter*, Hawthorne sustains a vision of the ambivalent nature of reality beyond all he had done before and in a way he was never to do it again. Hester, Dimmesdale, Chillingworth, the scarlet letter itself—all signify more than one thing; all must be considered in more than one way.... Hawthorne provides material for both portraits. He dramatizes the double (or multiple) nature of every important character, thing, and event in his romance.... As a masterpiece of literary art, however, it generates meaning by a technique of dynamic interrelationships, meaning that is for the reader in a perpetual state of becoming.... Hester believes, as lovers have always believed, in the supremacy of her passion. And it is Hawthorne's achievement to make her appeal to passion seem fresh, to lend it an emotional eloquence rendered all the more persuasive by the desert places of suffering through which she has lived in uncomplaining silence. (Martin, 1983, pp. 117-21)

My point here is not to reduce the evident merit of Tolstoy, nor to pile additional praise on Hawthorne, but to expand our understanding of the great *hedgehogs* who must be appreciated also for a large number of *fox elements*, and of the great *foxes*, like Goethe and Joyce, for their important *hedgehog* contributions, the first in *Faust*, the second in *Ulysses*. In the case of Hawthorne, to conclude with him, he must be read—and with profit—as *radiologist*, *universalist*, *miner*, and *fox*.

3

Aleksandr Pushkin, GAM *Miner* and *Trapper*, Therefore Berlinian *Arch-Fox*

1. Introduction

Pushkin (1799-1837) "Russian poet, novelist, dramatist, and short story writer; he has often been considered his country's greatest poet and the founder of modern Russian literature" (*Encyclopaedia Britannica*, 1978, 8, p. 312).

> GAM *miners* have been shaped, in their youth, by the pains of lack of love from one or both parents; many great poets are *miners*.... All *miners*, and all *trappers*, are Berlinian *foxes* (Therivel, 2001, *1*, pp. 42, 114).

A review by Clive James (2002) of T. J. Binyon's *Pushkin: A biography*, led me to check on what I had read on Pushkin many years ago, and to expand on his GxAxM of youth, his personality, and the nature of his work. In all these readings I was impressed by the strength of the *fox* element in Pushkin, and how clearly several scholars had mentioned this.

2. Pushkin's Major Misfortunes of Youth

> He was a graceless, morose infant. His face was a sallow, soft globe, with thick lips that were too bright a red and glittering,

dreamy eyes. His hair, the color of horse chestnuts, grew in tight, curly shavings from brow to nape. The nose, with flaring nostrils, was slightly flattened between well-filled cheeks. Alexander never laughed. He never ran. He let his toys drop from his hands. He dragged listlessly from room to room. The child's silence, his ponderous immobility, irritated his mother exceedingly. *She would have liked a high-spirited, mettlesome, laughing son, and preferred her daughter Olga, two years his senior, and was later to prefer her other son Leo as well. She even felt a kind of resentful loathing for him; she could not believe she could have produced such a lumpish creature.* Besides, he looked like an ill-bleached blackamoor. She wanted to shake him, wake him up, give him some energy, some muscle, a sense of repartee, a love of outdoor games and new songs and new faces. She grimly ordered him to run and leap about, strained to catch the sound of his laughter, coldly observed his lethargic breathing. Often, the harassed child took to his heels and sought refuge in some back room or inside his grandmother's workbasket. (Troyat, 1974, p. 21; my italics)

Arina's[34] room was also a safe refuge from the 'troubles' and 'unpleasantness' that beset his childhood. Very early, he discovered that he was an unloved and unwanted child. [His sister wrote]:

"Until he was six, there was nothing exceptional about him; on the contrary, he drove our mother to despair by his taciturnity and his clumsiness which was caused by his being a very fat boy. Mother used to take him for walks almost by force and make him run about and that was why he preferred to stay with our grandmother, climbing into her basket of sewing and watching her do her needlework ... It was only after the age of seven that he became very active and playful..."

In fact, his mother hated him: he reminded her too much of her father whom she detested. She did not even bother to conceal her dislike and often slapped his face. [The father] Sergey Pushkin remarked that his son Sashka reminded him of the child of an aborigine. Pushkin was a stranger in his own home. He used to appear at breakfast looking like a hunted wolf cub and kissed his mother's hand as though under compulsion. (Magarshack, 1967, p. 24; my italics)

---

34 His nanny.

> [quoted from Simmons]:
>
> On both sides of his family Pushkin could claim noble descent, and he was excessively proud of this nobility of six hundred years. On his mother's side the family line came into Russia, so to speak, by the back door. Pushkin's maternal great-grandfather, Abram Petrovich Hannibal, was an Abyssinian whom Peter the Great procured from a Turkish seraglio at the age of eight.... He prospered in Russia, gained high honors, and founded a numerous family. Pushkin liked to think of his great-grandfather as a Negro, and he gaily wrote of himself:
>
> But I, a rake on pleasure bent,
> Ugly offspring of negro descent.
>
> It would be difficult to say, however, how much, if any, negro blood ran in the veins of Abram Petrovich.... Little Sashka (Alexander) became the ugly duckling in this household. Unattractive and awkward in his movements, his dark coarse features betrayed the Abyssinian blood in him. Unsociable and rebellious, Sashka was always getting into difficulties, and punishments were frequent. His society-loving mother had more interesting things to do than bother with a moody, recalcitrant child. Her indifference eventually turned to coldness, and when the maternal spirit moved Nadeahad Osipovna, she lavished her affection on the other two children who always remained the favorites. An unloved child, Sashka was soon left to himself, a position which he seemed to prefer. (1937, p. 4-6)

This is a clear picture of a youth shaped by the lack of parental love, and who then became a GAM *miner personality* (once more, as said in table 1 of volume 1: "Many great poets are miners"). To this we must add the *genetic misfortune* of having been born both less than beautiful, and with some negroid traits ("Pushkin had black blood, but it didn't make him Harry Belafonte", James, 2002, p. 3). Something had gone wrong with him in the genetic roulette, with an "illegitimate" predominance of the physical traits of the part-Abyssinian maternal grandfather. In addition, his birth seemed to show the falsity of the mother's beauty. Indeed, Serfey's mother had been against his "marriage to the 'beautiful creole', the granddaughter of a 'dubious' and far from wealthy 'blackamoor'" (Magarshack, 1967, p. 17). Now her opposition seemed proven right by the birth of a son that could not be the offspring of "her Sergey". It is not surprising, therefore, that the mother preferred her other two children, Olga born in 1797, and Leo in 1805. Having these two finally confirmed her genetic distance from Abram Petrovich Hannibal, Aleksandr could not be but an illegitimate genetic aberration, first in the eyes of his parents and other relatives, and after a short while, in his own eyes.

With Pushkin we encounter a misfortune of youth that is both *miner* (lack of love) and *trapper* (virtual illegitimacy). Yet thanks to a high GxA[35] he managed his misfortune through intelligence, culture, wit, and poetic creativity:

> *"On veut sculpter mon buste. Mais je ne le veut pas. Ma laideur de nègre passerait á l'immortalité dans toute son immobilité sans vie....*
>
> *Il peut se faire (espoir flatteur)*
> *Qu'un ignorant, dans le futur,*
> *Montre mon illustre portrait*
> *Et dise: C'était un poète!"*[36]
>
> (quoted by Backés, 1978, p. 14)

3. *Maître Renard* Aleksandr Pushkin

With his double misfortune, which shaped him into a *miner* and *trapper*, Pushkin could hardly avoid becoming a Berlinian *fox*, indeed a *master fox*, based on the following three pieces of evidence:

---

[35] On the side of the assistances, we read that "of importance for Pushkin's literary development was the fact that both his father, Sergey Pushkin, and his uncle, Vasily Pushkin, were writers, the latter in particular enjoying quite a reputation among contemporaries. Young Pushkin was, therefore, exposed from a very early age to a literary atmosphere and to the presence of literary ambitions. The most distinguished men of letters used to gather at his father's house, and Pushkin had free access to his father's library, which contained, among other things, a good collection of French literature. By the time he was twelve years old, the future poet had read widely for his age, though without systematic guidance, had tried his hand at writing poetry (his earliest verses, which do not survive, were apparently in French), and had almost certainly conceived the ambition of making his name as a poet" (Vickery, 1970, p. 13).

[36] 'They want to sculpt my bust. But I don't want it. My negro ugliness would pass to immortality in all its lifeless immobility.... It could be (flattering hope) that some ignorant person will point one day to my famous portrait and say: 'This was a poet'.

1. [Few knew the great Russian writers better than the Latvian/Russian born Isaiah Berlin, and Berlin had no doubts]:

   "Shakespeare, Herodotus, Aristotle, Montaigne, Erasmus, Molière, Goethe, Pushkin, Blazac, Joyce are foxes.... [Pushkin was] an arch-fox, the greatest in the nineteenth century". (Berlin, 1953/1967, p. 2)

2. It is in vain that we would search in Pushkin a doctrine, some wisdom, any kind of system. The ill-informed reader is often deceived by his tales where there is never a question of God, nor of human destiny.... Rarely does the poet meditate; he does not write philosophical verses at the time when everybody tries to be an intellectual.... Even when he practices journalism, he is loath to make general considerations or to write leading articles. Everything in him leads him to inform, describe, and narrate. Moral poetry, metaphysical poetry are not for him. Many of his contemporaries love "mystical" poetry. He, instead, declared: *The closer we are to heaven, the colder it becomes*. (Backès, 1978, p. 12; italics in the original)

3. Pushkin's visual faculty was without the inbuilt dark glasses of abstraction: he never saw a type, he always saw individuals.... Binyon accepts too quickly Pushkin's written opinion that women were mentally inferiors.... His denigration was a defence mechanism, and it is probable that his coarse language was the same.[37] A pose of raw carnality staved off the unrelentingly repeated impact of the sublime. He was in love every time he lusted, and he was in love not because he saw less than other men but because he saw more. He could see everything, and he probably got sick of it. (James, 2002, p. 4)[38]

---

[37] Very *trapper*, "Pushkin was a stoat. There are less vulgar ways of putting it, but they wouldn't fit a sex drive like his. In his earlier amatory career, which appears to have got under way at about the same time as his chin grew its first whisker, he routinely referred to females, compliant or otherwise, as 'cunt'. On the eve of his marriage, he described in a letter to a similarly priapic male friend, the blissful state of wedlock as 'lawful cunt', which he further defined as 'a kind of warm cap with ear-flaps'. That was about as reverent on the subject as he ever got in ordinary speech" (James, 2002, p. 3).

[38] Toward the end, this text by James becomes of difficult interpretation, yet it reminds me of Leporello's description of Don Giovanni: "*Per piacer di porle in lista(*).... Non si picca, se sia ricca, se sia brutta, se sia bella! Purche porti la gonnella,...*

4. Conclusion

Readers may have discovered a long time ago, that I am nearly as fascinated by the *hedgehog/fox* dichotomy, as by my own classification of the *challenged personalities* into 14 families (see volume 1), because the *hedgehog/fox* partition allows for an immediate discourse with other scholars, be it on results (personality type and works), or on causes (misfortunes, or better yet, GxAxM).

In the previous chapter, I mentioned the natural and mild *fox aspects* of some—but not the most important—of the writings by Hawthorne and Tolstoy. With Pushkin, things are different: the *fox element* is primary and unmitigated; not only his writings were unmitigated, but so was his difficult private life: "Our anger [against Baron d'Anthès who killed Pushkin in duel] is wasted, because Pushkin always behaved suicidally. The duel that finished him was not his first" (ib., p. 4).

[For the sake of listing them..... He does not care if she is rich, ugly, beautiful, provided she wears a skirt,...]".

(*) "Usually, Pushkin's friends were informed by letter of any successful encounter immediately after it happened, if not while it was actually happening" (ib., p. 4), and in November, 1829—at the age of 30—he wrote down "the famous 'Don Juan' list of some thirty-seven ladies with whom he had amorous relations, which he sent to one of his correspondents. Incidentally, the word ladies is here used advisedly, since he included in the list none of the demi-mondaines with whom he had consorted" (Malevsky-Maletvitch, 1937, pp. 46-7); see also Magarshack, 1967, pp. 220-21. Again, very *trapper.*

4

Moravia, Down-to-Earth *Radiologist*

1. Introduction

For a while I thought that Alberto Moravia[39] was one of those exceptions about whom Trevor-Roper had warned me a long time ago when he wrote that "Any general theory has to take account of exceptions" (1967, p. 6). I knew of Moravia's long, severe and very painful physical infirmity of youth, but had always felt—at a distance—that he was neither a *radiologist*[40] nor a *hedgehog*, therefore an exception to be explained, given the general claims of the GAM/DP theory of personality and creativity. I was clearly both intrigued and challenged by his case.

I had seen no *radiologist* in him because he seemed to lack the required "diagnostical" approach in dealing with his subject matter. And it even seemed that Moravia himself would have agreed when he said: "At the beginning of my career as a writer I was unable to connect two

---

[39] One of the most widely translated Italian novelists, whose works have been published in more than twenty-five languages. "Moravia's work ranks on a level with that of the great Italian (and European) novelists of the 20th century: Italo Svevo, Luigi Pirandello, Grazia Deledda, Federico Tozzi, Carlo Emilio Gadda, Cesare Pavese and Italo Calvino" (Peterson, 1996, p. x).

[40] As I said in chapter 1 of volume 1., the *radiologist personalities* were challenged in their youth by the misfortune of painful or humiliating physical infirmity; later, their creativity will give evidence that they have developed a keen eye for anything that is sick in people and society.

arguments; I expressed myself only through images and dialogue, in short I had a very high capacity for representation, and almost none for reflection" (quoted by Dego, 1966, p. 16).

This *non-radiological* mental frame of youth would have been no problem if his masterworks had come later in his career based on a changed set-up, however his most famous novel, *The times of indifference,* was his very first, about which Thomas Peterson had written that "the reader is led to ask what he has learned. Are these unethical personae representative of their historical class situation? For Moravia the question is enough. By not attempting more than he does, he guarantees the authenticity of the cultural objects he depicts" (1966, p. 17). Yet, this would be like a doctor describing a patient in detail, and letting others decide which symptoms are important, and of which disease. That would not be *radiological*, which consists of the sharp identification of the sick points and their nosological labelling.

I was wrong; he was a *radiologist* after all. But before describing him as such, let's start—in good GAM fashion—with the causes, specifically with his misfortunes (severe) and assistances of youth (few). For this study, I was greatly helped by two books: *Moravia: Il bambino Alberto* [The child Alberto] by Dacia Maraini, and *Alberto Moravia: Vita, parole e idee di un romanziere* [Alberto Moravia: life, words and ideas of a novelist] by Enzo Siciliano, both books being the result of long and detailed conversations with Moravia.

2. Physical Infirmity

For about eight years, from the time he was nine years old, Moravia—born Nov. 28, 1907—was afflicted with tuberculosis of the leg bones, which prevented him from attending school regularly. Severe aftereffects of the disease remained with him until the age of twenty five. Bedridden much of the time, Moravia obtained most of his education from reading. At the age of sixteen Moravia entered the Codivilla Sanatorium at Cortina d'Ampezzo, in the Italian Alps. There he began work on a novel—*The Time of Indifference*—and he continued to work on it for three years after leaving the sanatorium in 1925.

In the words of Moravia to Enzo Siciliano:

> There is an abscessual bone tuberculosis which manifests itself in abscesses on the body which are painless. I, instead,

> suffered of dry bone tuberculosis which gives horrible pains. I suffered these pains from the age of nine to seventeen.... The disease produces acute muscular contractions so strong as to make you fall down, as it happens to epileptics. This was worse than unpleasant, and certainly had a decisive impact on my sensibility.... I was always alone, in bed, and in a state of continuous pain. (1992, p. 30)

3. Parental Assistance

There was a paucity of parental assistance, beyond providing Alberto a reasonably high socio-economic status and derived medical and cultural help.

Here is what Moravia said first about his father (a retired prosperous architect and painter), then about his mother:

A.[41] After lunch he used to sleep for a while, then he went out—and nobody ever knew where he went. He stayed away until dinner time. Then came back, had dinner and left again. This time we knew where he was going: to the *caffé* (bar/coffee house)....

D. Did your father ever bring you to see some art exhibitions?

A. No, never.

D. And to museums?

A. No.

D. Would you have liked to go? Did you love painting?

A. Very much....

A. No. My father did not take me anywhere with the exception, on Sundays to a dairy shop. My mother instead never brought me anywhere. Only a few times when she went to the dressmakers.... My mother was an absolute Madame Bovary, even in the details....

D. You began by saying that you had no love for your father, but in the end you spoke much of him and with tenderness. The best memories of your childhood seem linked to him. Instead for your mother you have given very despising judgements....

---

[41] From here onward: A. (Alberto) is Moravia, D. (Dacia) is Maraini, and E.

A. The bovarism[42]—as can be seen from the novel by Flaubert—renders the affected person less lovable because she is constantly dissatisfied, mythomaniac. (Maraini, 1986, pp. 42, 63, 64, 67)

This is definitely no great assistance, especially when compared with that received by the *radiologists* Proust and T. S. Eliot. Briefly, on Proust:

At the age of nine he was suddenly stricken with an attack of asthma and from that time on he claimed the privileges of an invalid fortunately relieved by wealthy parents from any concern with a livelihood. The doting father and mother watched over him with exaggerated care and developed in him a morbid affection" (Krutch, 1934, p. vii). "Fierce and possessive, the mutual adoration of young mother and ailing boy.... The invalid is entitled to special care and consideration. Jealous of the attention his mother paid to her husband, Proust used illness to make her more like a nurse and less like a wife (Hayman, 1990, pp. 6, 115).

And on Eliot: "From the start, Eliot's family recognized him as exceptional particularly his mother who would speak to him as an equal. He, in turn, was devoted to her" (Gordon, 1978, p. 2).

4. Depressing Solitude

Of all misfortunes suffered by Moravia, his long years of depressing solitude is the one that differentiates him most from the other *radiologists.*

A. I entered the sanatorium in March 1924, and came out in October 1925, a year and a half later. In bed, with traction on the leg: eight kilos on the foot, and two at the knee. From this traction I could not free myself for one minute....

There were three classes of rooms, and I was staying in the first class. My father had thought to do good for me in this way. But it would have been better if he had put me in the third class because in first I was alone, as a dog.... I suffered the loneliness. I envied those of the third class who were always together, playing cards, singing, laughing, moving from one bed to the other....

---

[42] The *Grand Larousse* says that "Bovarysme", from Flaubert's *Madame Bovary*, is a feeling of dissatisfaction in the affective and social domains, frequent in the evolution of certain neuroses of the woman.

D. Did this solitude affect your character?

A. I truly believe that it modified my character. It taught me not to think at all...

Because of all that solitude, I became gloomy, morbidly desperate.

D. In what consisted the morbidity of your melancholy?

A. It consisted in the fact that I henceforth deliberately searched for what was gloomy. I took delight in such matters. But because I had great vitality, these sad matters were alive in me with great strength. Those were horrible years, without affection. (Maraini, 1986, pp. 89-90, 91, 94-95, 100)

5. *Radiologist*

> [Moravia's] *Times of Indifference* is a ferocious indictment of the falseness which infects contemporary society. The youthful Moravia lays bare the corruption, baseness, and sham which characterize modern man's preoccupation with material wealth and prosperity. Moravia's young protagonists, a brother and sister, see through the hypocrisy of the adult around them.... [In the end, brother and sister] devoid of the autonomous conscience that would demand rejection of their false surroundings, capitulate and are eventually trapped by their own indifference..., while on the surface a languid yawn is the only sign. But Moravia understands that this yawn is a world-cracking scream beyond the range of hearing, like a filmed cry for help with the sound turned off. *He is as always, the therapist who dares to hear the 'dark root of the scream'*. (Ross & Freed, 1972, p. 41; my italics)

Here, indeed we have Moravia the *radiologist*, the diagnostician who dares to hear.

6. Berlinian *Hedgehog*, but Down-to-Earth

"Through his voluminous body of work and despite a dizzying array of plots and situations, Moravia displays a constant concern for the same themes: love and sexuality, and money and material survival [down-to-earth]" (Peterson, 1996, p. 3).

He exhibits this through not only the same themes, but basically through the same characters:

A. The male character of *Woman of Rome* and Marcello [of *Conformist*] are in the end the same character.

E. And so is Michele of *Time of Indifference*, and the protagonist of *Wheel of Fortune*, and *The Imbroglio*.

A. Yes, always the same character, independently of the various tone of voice I gave to each book. (Siciliano, 1982, p. 73)

As a *GAM radiologist* Moravia also had to be a *Berlinian hedgehog* ruled by his single, universal, organizing principle, and so indeed he was with his constant concern for the same themes worked through, essentially, the same characters. In addition, given that his primary concern was with such down-to-earth themes as love and sexuality, and money and material survival—especially the way Moravia handled them, it may be useful to call Moravia a *down-to-earth hedgehog*, or more precisely a *down-to-earth radiologist*. This can clearly be seen in the following comment by Luciano Rebay:

> That his writings express a tragic sense of emptiness becomes apparent rather quickly, but the fact in itself does not signify enough until one sees it as the consequence of twentieth-century man's inability to relate to reality—both to his own and to that of his time.... [His first works], which remain among his best, portray a world disintegrating, a society of individuals trapped in a circle of indifference, cynicism, cruelty, and boredom. (1970, pp. 3, 7)

7. Some Mystical and Religious Interests (not rare among the *radiologists*)

Moravia:

> We search the ultimate explanation of what happens because we notice that the reality, as it is, is absurd. This already may be a true and valid religious instance. When I was a boy, up to the age of fifteen and sixteen, I was dominated by such a strong religious feeling that it brought me unequivocal mystical crises;...[43] All this brings me to think that we are living in a time of changes, of new fact, and the new facts are always religious,

---

[43] This brings to mind something similar in the life of T. S. Eliot: "At the age of twenty-one Eliot had one of those experiences which, he said, many have had once or

> because they happen through quantitative steps. (quoted by Siciliano, 1982, pp. 246-7, 134-5)

In the end, Moravia is no radiologist like Hawthorne, Proust, Nietzsche, T. S. Eliot, and Shusaku Endo; but neither is Proust like Hawthorne, Nietzsche, T. S. Eliot, or Shusaku Endo. Yet all have in common a sharp eye for what is sick in man and society, a characteristic which I explore a little more—in accordance with T. S. Eliot's remark that: "Comparison and analysis are the chief tools of the critic" (1951, p. 32-3)—in the following three appendices, the first two of comparison and the third of analysis.

Appendix 1 – Proust and T. S. Eliot *Radiologists*

In order to compare the radiologist Moravia with other radiologists, I sought out my university's library, and there discovered two books I had missed in the past: The first is a work by Bernard Straus—professor of medicine at the Mount Sinai School of Medicine and Director of Medicine at the Beth Israel Medical Center in New York—entitled *Maladies of Marcel Proust: Doctors and Disease in his Life and Work*, whose Introduction dove-tails with what I have written on Proust in volume 1.

---

the Divine or you may call it 'temporary crystallization of the mind', he said on another occasion" (Gordon, 1978, p. 15).

I agree with Eliot that "many had once or twice" similar experiences in their lives, but would add that the majority, in all probability, were either *seeker* or *radiologist* personalities. The reader may remember that, in volume 1, I had written that many mystics are *seekers* and that the *radiologists* also may have some religious interests or mystical experiences.

"There is a great deal about Eliot's 'early environment' which we do not know.... Eliot himself did not 'wish my executors to facilitate or countenance the writing of any biography'" (Raffel, 1982/1991, p. 21). Yet, from the little that is known, it is hard not to visualize the making of a *radiologist*, as in: "He had few playmates and spent most of his time reading. He had a congenital double hernia and [his mother] Charlotte, afraid it would rupture forbade football and strenuous sports. During summers at Cape Ann, when 'the Skipper' used to give him sailing lessons, Charlotte would go along, fortified by a guard of grown-up sisters, to ensure that he did not get too wet or too hot or too tired" (Gordon, 1978, 3-4). The *radiologist* label seems even more applicable when reading the following upsetting observation by Raffel: "T. S. Eliot, a shy, introspective man who had been an even shyer and more solitary boy" (1982, p. 2).

My justification for this work—if one is needed–[wrote Straus] is the fact that illness permeates the life and work of Marcel Proust. Edmund Wilson, one of Proust's most perceptive critics, has said that all of his characters are sick with some form of the ideal. André Malraux declared that *Swann in Love* is the clinical description of a case from the beginning to its end, and remarked on the painful precision of this pathology of feeling. He added that love, jealousy, and vanity are illnesses—literally—for Proust.

Sickness of body and mind colored and influenced Proust's entire life and is mirrored in his work. His life and work cannot be fully understood without a knowledge and appreciation of the complexities of his various ailments and how they interact with one another. (1980, p. xi)

The second book, *Theory and Personality: The Significance of T. S. Eliot's Criticism* by Brian Lee, also provided evidence of *radiological* thinking:

Eliot's metaphors....are frequently surgical-medical....

[In Eliot's words]: "More important than the lack of balance is the lack of detailed analysis ... Wyndham misses the cardinal point in criticizing the Elizabethans: we cannot grasp them ... without some understanding of the *pathology* of rhetoric [which] was *endemic*, it pervades the whole *organism*; the *healthy* as well as the *morbid tissues* were built up on it .... [we must] *diagnose* the rhetoric." (*The Sacred Wood*, pp. 30-1)

"One must draw attention to Eliot's habitual use of medical-physical metaphors to lend again that curious sense of objectivity to his judgements. In the passage above we have seen diagnosis, pathology and faculty. In addition Eliot constantly sees the activity of thinking as like the mechanical operation of an organism". (1979, pp. 5, 99-100; Lee's italics)

Appendix 2 – Endo *Miner-Leadsman-Radiologist*

Shusaku Endo—the author of *Silence*, discussed at length in volume 3—is a GAM *miner-leadsman-radiologist*. *Miner* because of the divorce of his parents at age 10; *leadsman* due to his conversion to Catholicism at age 11 and his long stay in France; and—important for this chapter—*radiologist* for his early health problems which became severe, from 1951 onward, requiring long years of hospitalization first in France then in Japan, as well as major operations.

In his own words:

> [In 1952] After two and a half years in Lyons, I went to Paris. I was constantly feverish, so I went to see a doctor, who diagnosed my problems as a lung disease and ordered me into the hospital immediately.... I spent the winter of my last year in France in a Paris hospital.... [Back in Japan in 1953] for over a year I lay in bed unable to do anything..... [In 1958] the doctor told me that my lungs had suffered a relapse. I spent the next three years in a hospital bed, and in 1961 underwent three major operations.... During the [last one], my heart stopped for several seconds. The doctors thought I was dead. But the devil's own luck is with me, and I have managed to survive. I had one entire lung removed during the operations. (1979, pp. 8-9)
>
> Even now I am attracted to hospitals, and several of my novels are set in them. There people must cast off all the decorations of society and battle face-to-face with disease. There the company president and the politician must don pajamas and sleeping gowns. No social privilege is of any avail in combating the realities of disease. An operation is painful to each individual, and everyone hates injections. No accumulation of wealth or wielding of power hold sway over the fear of pain and death" (ib., pp. 7-8).

The *radiologist* in Endo is not limited to his interest in hospitals and medical diseases, but like Proust (see chapter 3 of volume 1), he uses his X-rays to uncover severe ethical diseases, such as in *When I Whistle* of 1979, in which two doctors, with the interested consent of the hospital's director—experiment with a new drug on a patient without her consent; worse they administer the new drug based on data provided by the pharmaceutical company owned by one of their fathers. And the patient dies of complications which additional non-invasive tests would have forestalled.

Appendix 3 – Moravia, the *Non-Medical Radiologist*

But why are there no doctors and medicines in Moravia's novels? The answer is that Moravia's feeling of a tragic sense of emptiness, a fundamental inability to relate to reality, a world disintegrating, a society of individuals trapped in a circle of indifference, could not be expressed in medical terms or in relation to the medical profession because both are strongly related to reality. There is indeed little that is more real, for the individual man or woman, than modern medical science and the medical profession: individual diseases can be clearly identified, and hopefully cured with drugs, operations, special diets and physical rehabilitation. We all want to be cured quickly, and nice words do not count. In this—for us, for our children, or other beloved relatives or friends—we are all realists from Missouri where "frothy eloquence does not convince nor satisfy: you have got to show results."

Some doctors can be criticized, as Hawthorne and Endo did, but implied is the notion that there are good doctors who know what they are doing and who can cure. This distinction of doctors into good and bad, of good and bad therapies, in turn, would soften Moravia's portray of a disintegrating world, a society of individuals trapped in a circle of indifference, cynicism, cruelty, and boredom.

In a certain sense, the world of medicine was a "luxury" that the well-assisted Hawthorne, Proust, Eliot and Endo, could allow themselves. Moravia's X-rays, instead, are not directed at uncovering pride, as done by Hawthorne, or the lies and hypocrisy of individual persons as done by Proust and Endo, but at more down-to-earth needs, those lowest in the Maslowian hierarchy of needs. As observed by Dego:

> Money and sex, the two constant poles between which bourgeois and neo-capitalist society move, are then, in the writings of Moravia the only basic criteria for every interpretation of existence. In a world populated by men who tend to reduce the reasons for their existence to the most rudimentary skeleton, the only realities that we cannot reduce further are sex and money. (1966, p. 50)

It is true that Moravia wrote a short story, *A Patient's Winter*, whose action takes place in a hospital. However in it, "again we experience the depressing atmosphere of boredom that hangs over the existence of the tennis club members, and of Carla, Mariagrazia, and Leo [of *The Time of Indifference*]. Here too boredom inevitably leads to the 'abyss.' The protagonist is Girolamo, a boy suffering from tuberculosis of the knee bone. A patient at a sanatorium in the mountains (the setting recalls Moravia's Codivilla clinic),... [At the end of the story] Girolamo finds himself again utterly alone, ashamed, frustrated, despairing even that he will ever recover from illness" (Rebay, 1970, p. 16). Again, doctors and medicines were not—could not have been—Moravia's concern, but the probing of man's inability to relate to reality, man's tragic sense of emptiness.

5

Deliberately More *Hedgehog* or *Fox*

1. Introduction

At a certain point in their lives—in late adolescence or early maturity—those with a high GxAxM discover that they are either *hedgehogs* or *foxes*[44]. They then embrace what nature and nurture made them, and set themselves deliberately to the task of being more of what they are, or at minimum they feel consciously comfortable at being the *hedgehog* or *fox* that life made them.

This said, it is fascinating to distinguish the deliberate self-developmental approach of the *hedgehogs* from that of the *foxes*: the first being more rational, more isolated in an ivory-tower, continuous and consistent

---

[44] See in chapters 9 and 10 of volume 1, "GAM's *Hedgehogs* and GAM's *Foxes*" and "*Maître Renard* James Joyce", and in chapters 5 and 7 of volume 3, "Like Goethe also Calvino was a Berlinian *Fox* by being a GAM *Alchemist*", and "Why Berlinian *Fox* or *Hedgehog*?". For those who do not have these two volumes handy, Isaiah Berlin defines the difference between the two as: "[hedgehogs] relate everything to a single central vision, one system less or more coherent or articulate, in terms of which they understand, think and feel—a single, universal, organizing principle, in terms of which alone all that they are and say has significance—and, on the other side, [the foxes] those who pursue many ends, often unrelated and even contradictory, connected, if at all, only in some de facto way, for some psychological or physiological cause, related by no moral or aesthetic principle; these last lead lives, perform acts, and entertain ideas that are centrifugal rather than centripetal, their thought is scattered or diffused, moving on many levels, seizing upon the essence of a variety of experiences and objects for what they are in themselves, without, consciously or unconsciously, seeking to fit them into, or exclude them from, any one unchanging, all embracing, sometimes self-contradictory and incomplete, at times fanatical, unitary inner vision" (1953/1967, pp. 1-2).

along the years; the second being more heart-felt, immediate, related to people newly met and original experiences. For the first approach, of the *hedgehogs*, let me quote hereafter on Solzhenitsyn; for the second approach, of the *foxes*, on Goethe and Picasso. I will then conclude with what happened to the *fox* Goethe when, later in life, he tried to be a *hedgehog*.

2. Solzhenitsyn: *Hedgehog*

Solzhenitsyn's father, Isai, died three months before the birth of his son. His mother did not remarry or have affairs: she devoted herself completely to her son.

> Isai's death had a profound effect on the boy's development; his father remained a permanent symbol in his thoughts, perhaps even more commanding than it would have been had the living man supervised his upbringing. It was a symbol of innocence, passion, and pursuit of excellence—of a clean young life cut short before it fulfilled its promise...."My mother....never married again, fearing possible sternness toward me by a stepfather. When I grew up and was able to judge, I came to the conclusion that this sacrifice had been unjustified: in my opinion sternness encountered at an early age is ordinarily only helpful for a boy."[45] Here is an unmistakable indication of longing for the source of external authority that he lacked, and that had to be found within himself. The desired authority might even be stern—for *surovost,* Solzhenitsyn's word, has a strong implication of welcome sternness, limited by and applied for the sake of justice. *In this sense, he was to become his own father, limiting his boyhood pleasures, disciplining himself with uncommon exactitude, subjugating his entire life to a goal. If there is a central element in his personality, it is this severe self-discipline which sets standards at least as high as could be*

---

[45] Solzhenitsyn's mother may indeed have mentioned her fear of possible sternness by a stepfather, but the real danger was another: at minimum the distinct possibility that there might have been stepsiblings (his mother was in her early twenties when Isai died) who would have been preferred by the stepfather (in addition to probable sibling antagonism). In such a case, instead of becoming a *universalist/ hedgehog*, Solzhenitsyn would probably have evolved into a *miner/fox* like Baudelaire and Sartre.

> *expected from the most demanding father.* These standards no doubt contributed to his later strivings toward ethical maximalism as well as his perfectionism in writing. Moreover, a feeling of incompleteness stimulates some fatherless boys to compensation, which can manifest itself in greater self-assertion and independence. Such a young boy learns to fend for himself rather than await help—and insofar as this analysis pertains to Solzhenitsyn, the self-fathering was also decisive in the development of independence of mind. (Burg & Feifer, 1972, pp. 24-25; my italics)

3. Goethe: *Fox*[46]

Goethe's misfortune of father failure has been discussed previously, starting with chapter 3 of volume 1. Hereafter I will concentrate on his *fox* self-development: life, works, and their strong interaction. The contrast with the prior *hedgehog* developmental method of Solzhenitsyn is striking.

As discussed by Jeanne Ancelet-Hustache:

> There is no literary work, like his, in which one can less dissociate the writer from the man.... From his first adventure to his last desperate and renouncing love, he has sung what has moved him: "All my poems are poems of circumstance stimulated by the reality at their base." He gave duration to the instant, the external circumstance was the shock which brought into action the affective powers. (1976, p. 5)

Then, as discussed by Goethe himself in his autobiography of 1811-22:

> And thus began that tendency from which I could not deviate my whole life through; namely, the tendency to turn into an image, into a poem, everything that delighted or troubled me, or otherwise occupied me, and to come to some certain understanding with myself upon it, that I might both rectify my con-

---

[46] Please refer to chapters 3 of volume 1, 5 of volume 1, and 1 of volume 3 for information on the youth of Goethe and Picasso, especially on their misfortune of father failure.

ceptions of external things, and set my mind at rest about them. The faculty of doing this was necessary to no one more than to me, for my natural disposition whirled me constantly from one extreme to the other. All, therefore, that has been confessed by me, consists of fragments of a great confession, and this little book is an attempt which I have ventured on to render it complete. (p. 240)

At the time when I was pained by my grief at Frederica's situation, I again, after my old fashion, sought aid from poetry. I again continued the poetical confession which I had commenced, that by this self-tormenting penance I might be worthy of an internal absolution. The two Marias in *Götz von Berlichingen* and *Clavigo,* and the two bad characters who play the part of their lovers, may have been the results of such penitent reflections. (p. 453)

The resolution to preserve my internal nature according to its peculiarities, and to let external nature influence me according to its qualities, impelled me to the strange element in which *Werther* is designed and written. I sought to free myself internally from all that was foreign to me, to regard the external with love, and to allow all beings, from man downwards, as low as they were comprehensible, to act upon me, each after its own kind. (p. 470)

I had freed myself from that stormy element, upon which, through my own fault and that of others, through a mode of life both accidental and chosen, through design and thoughtless precipitation, through obstinacy and pliability, I had been driven about in the most violent manner. I felt, as if after a general confession, once more happy and free, and justified in beginning a new life.

The old nostrum had been of excellent service to me on this occasion. But while I felt myself eased and enlightened by having turned reality into poetry, my friends were led astray by my works for they thought that poetry ought to be turned into reality, that such a moral was to be initiated, and that at any rate one ought to shoot oneself [like Werther]. What had first happened here among a few, afterwards took place among the larger public, and this little book, which had been so beneficial to me, was decried as extremely injurious. (1897, pp. 511-12)

4. Picasso: *Fox*

Also with Picasso, the misfortune of father failure has been discussed previously, already in chapter 3 of volume 1, then in chapter 1 of volume 3. Hereafter I will concentrate on his *fox* type of self-leadership. As discussed by John Richardson:

> "I paint the way some people write their autobiography," Picasso told Françoise Gilot. Likewise Dora Maar—another of the artist's mistresses—said that the transformations in Picasso's style reflected transformations in his private life. When the woman changed, everything changed: not just the art, but the house Picasso lived in, the poet he would have around, the circle of friends, and the dog. (1984, p. 21)

Marilyn McCully (2002) said similar things:

> Picasso never completely dissociated himself from the women who had shared his life once a new lover occupied his attention. This is evident in his work, in which one mistress often turns into another, for instance in a private sketchbook (number 99, 1929) Picasso's portrait drawings betray his double life, for the pictures of his then secret mistress evolve into horrific images of screaming Olgas. (p. 862)

Implied, in the description by Berlin in the first footnote of this chapter, whatever the *foxes* mastered they quickly left, going on to the challenge of new problems. As discussed by Michel Leiris (1954):

> Almost all the themes used by Picasso in his successive periods have linked up intimately with his personal life: details of personal experience, people for whom he had developed a sensual or sentimental attachment, pathetic or picturesque figures that had caught his youthful eye. Then came big compositions on the epic scale, beginning with *Guernica,* and characters stemming from the myths of classical antiquity or a folklore of his own invention. All these motifs, not only those relating to things actually seen but also those which conjure up some more or less imaginary world, have a precise connection with the artist's physical or emotional condition at the time; in fact Picasso's art is

autobiographical through and through. Indeed so close is the association between the subjects of his paintings and the man himself that their evolution seems to run parallel with his own. Moreover, far from being milestones as it were in his career as an artist, which once attained are left behind for good and all, all these various themes and *trouvailles* are incorporated in his artistic baggage, intermingle and give rise to new creations. (p. 7)

And by Tim Hilton (2002):

As we know, there has never been another artist with Picasso's ability to change his manner, or to work simultaneously in more than one manner. His supporters usually attributed such changes to genius. His detractors (quite numerous and vocal until the mid-1950s) saw the changes as proof of charlatanism or trickery. On the whole, modern Picasso experts have associated changes of style with the artist's varying circumstances. They point to his friendships and loves, or his travels, to his interest in ballet, bullfighting, or poetry. (p. 19)

And by Picasso himself, as quoted by Françoise Gilot (1964):

Juan Gris said, "I take a cylinder and I make a bottle out of it," reversing, in a sense, Cézanne's remark. His idea was that beginning with an ideal plastic form—a cylinder—one can make a portion of reality—the bottle—enter into that form. Gris's method was a grammarian's. I suppose you might call mine an entirely romantic one. I start with a head and wind up with an egg. Or even if I start with an egg and wind up with a head, I'm always on the way between the two and I'm never happy with either one or the other. What interests me is to set up what you might call the *rapports de grand écart*—the most unexpected relationship possible between the things I want to speak about, because there is a certain difficulty in establishing the relationships in just that way, and in that difficulty there is an interest, and in that interest there's a certain tension and for me that tension is a lot more important than the stable equilibrium of harmony, which doesn't interest me at all. Reality must be torn apart in every sense of the word. What people forget is that

> everything is unique. Nature never produces the same thing twice. Hence my stress on seeking the *rapports de grand écart*: a small head on a large body; a large head on a small body. I want to draw the mind in a direction it's not used to and wake it up. I want to help the viewer discover something he wouldn't have discovered without me. That's why I stress the dissimilarity, for example, between the left eye and the right eye. A painter shouldn't make them so similar. They're just not that way. So my purpose is to set things in movement, to provoke this movement by contradictory tensions, opposing forces, and in that tension or opposition, to find the moment which seems most interesting to me. (pp. 59-60)

5. Not "Cobbler Stick to your Last", but "Stick to your Basic Lifestyle"

Things went wrong with Goethe when he decided to act at the same time as *fox* and as *hedgehog*, and develop a scientific theory of colors which would prove Newton wrong:

> As an old man he could still say to his famulus, Riemer: "Above all no professionalism. That is against my nature. I want to do everything I can as though it were a game, just as it occurs to me and for so long as the inclination lasts. In my youth I used to play like this unconsciously, now I want to continue to do it consciously for the rest of my life." This is a somewhat dangerous principle to follow when one is trying to establish a scientific theory of colour, and is abusing all physicists as fool, which is precisely what Goethe was doing at the time he said this. (Friedenthal, 1965, p. 24)
>
> *The urge to discover, which he sees at work all round him, is infectious, and he sees no reason why, with his 'good brain', he too should not make one of the great discoveries of the century*. He is on the track of one. The world of colour has never yet been adequately described or understood; he is about to reveal it....The style is quite different from that of his metamorphosis of plants. He writes with a poetic verve that sometimes sounds as though he were telling fairy tales. Italy, with the harmony of its pure deep blue sky, is a fairy tale; regretfully he says he must leave the lovely picture of this southern

> landscape and take a calmer view. Calm, however, he is not; hidden anger is stirring within him. *He is dealing with an abstract theory and the experiments are difficult. He has been told that it is impossible to understand and solve optical problems without a knowledge of mathematics. To this view he opposes his 'pure experiences'*; the word 'pure' recurs again and again, in the sense that theologians speak of their own doctrine as pure while all others are impure. (ib., p. 285, 287; my italics)

And in the words of Hume Brown:

> During the first months of 1810 he saw through the press his *Farbenlehre*, the work which he considered the greatest achievement of his life, deliberately placing it above *Faust* as a product of his genius.... A hasty experiment made in May, 1791, convinced him that the Newtonian theory of light was founded on an initial error, and henceforward the subject had engrossed his thoughts and entailed an amount of toil beyond every other.... There is both a pathos and an irony in the fact that a mind like Goethe's should have given the best years of his life to the pursuit of an illusion; yet if the main thesis of the *Farbenlehre* is mistaken, the work remains an extraordinary memorial of his capacity for toil and of the range of his powers. (1920/1971, pp. 571-72)

Yes, *Farbenlehre* (830 pages in my Artemis/dtv edition of 1977) may, indeed, be a "memorial of his capacity for toil and of the range of his powers," but it was also a betrayal of his stated "tendency to turn into an image, into a poem, everything that delighted or troubled me, or otherwise occupied me, and to come to some certain understanding with myself upon it, that I might both rectify my conceptions of external things, and set my mind at rest about them."

With this, am I saying, with the great painter Apelles: "Let the cobbler stick to his last"? No, because my point is not the profession, but the philosophy of life and related methodologies. Goethe could have been a scientist, but of the Leonardo da Vinci *fox* type, not of the Newton hedgehog *type*: "[Leonardo] had a scientist's vast curiosity about nature's secrets,... [But he] always seemed to go on to other things before he took the final step of bringing his projects to concrete, functioning reality" (Wallace, 1967, p. 110). Better to be called "capricious and fickle," and be criticized for being unreliable and dilatory, as Leonardo was because of his incessant drive to do new things, than building such a dubious memorial to one's capacity for toil and range of powers as Goethe did. All this discussion would have little importance if it did not confirm the need to make a jewel out of one's basic philosophy of life as originally shaped by one's specific GxA or GxAxM of youth. In the end, the rule is: "Be true to yourself, and stick to your "personality last"."

Interestingly, Goethe's *fox-hedgehog* failure with his *Farbenlehre* parallels that of his alter ego, the *fox-hedgehog* doctor Faust, who—at the beginning of *Faust I*, says: "Philosophy have I digested, the whole of law and medicine, from each its secret I have wrested, theology, alas, thrown in. Poor fool, with all this sweated lore, I stand no wiser than I was before . . . Meanwhile I live penury, no honour falls to me, no dog would linger on like this, and so I turn to the abyss of necromancy." (Night. Faust's study).

One may ask the reason for this failure, and a normal explanation would be that he had dispersed his efforts badly: medicine like his father, but then also philosophy, law, and theology! This is the perfect recipe for the making of an inept and incompetent medical doctor, the ideal combination for failure.

All this is important because of the contrast it provides between the successful creativity of *Maître Renard* Goethe, in his poems and novels, and Goethe's failure when he acted the *fox-hedgehog*. The temptation was strong: Why not also be a great scientist given that he was so great in other fields? Why not, indeed? But he cannot be a great scientist when proclaiming, as previously reported: "Above all no professionalism. That is against my nature. I want to do everything I can as though it were a game, just as it occurs to me and for so long as the inclination lasts." This is too *fox* even for a sorcerer's apprentice.

But this is not the whole story. The reader may remember how in chapter 6 of volume 1, I discussed Goethe's *Theory of Colors* as an example of Eccentric Science which is not true science. The eccentric–shaped by unpleasant parental domination (father or mother may have been the source of unpleasant critical comments and arguments)–has the tendency, as a youth, to refuse any parental or pedagogic involvement or counseling, and as an adult reacts, to the slightest pressure, by telling people: "Go fly a kite; I'll do it my way." This in turn prevents the eccentrics from establishing contacts with the best minds of the present (and the past) in their field of interest.

Expanding on what I said on Goethe in volume 1, let me report two brief comments on the cause (the father), and the result ("Go fly a kite; I do it my way") of his personality.

> *Il parle presque toujours de son père avec une sourde rancune, en dépit de l'objectivité formelle et du ton respectueux. Il lui a à peine pardonné l'éducation austère qu'il a reçue, moins encore l'esprit de pédagogue.... En l'absence du fils, la situation familiale ne s'est pas améliorée. Les défauts du conseiller sont devenus intolérables et il s'exaspère maintenant de voir que les années d'études de Leipzig ont abouti à un tel échec.*[47] (Ancelet-Hustache, 1976, pp. 19, 33)

> The fight [against Newton] begins with his very first conversation with a professional, a physicist, who cautiously points out to the *Herr Geheimrat* that Newton did not in the least maintain that, looking through a prism against a white wall, one would be able to observe the division of light into separate colours. On the contrary, he had postulated quite definite conditions for his experiment: the light had to pass through a very fine aperture, the prisms had to be placed in a certain way, and a whole series of strictly formulated precautions had to be observed. Goethe grimly enumerates these conditions under eleven headings and remarks naïvely: one only has to alter these conditions, and make the aperture large, 'and the beloved spectrum cannot and will not appear!' He dismisses the whole thing as 'hocus-pocus'. The very simple and fundamental requisite for a scientific experiment, that it must be subjected to certain clearly defined conditions which cannot be arbitrarily varied, he sees merely as a convention of the scientists' guild, and with that he refuses to have anything whatever to do. (Friedenthal, 1965, pp. 285-86)

---

[47] He nearly always speaks of his father with an underhanded grudge, notwithstanding a sense of formal objectivity and a respectful tone. He hardly pardoned him the austere education he received, and even less his pedagogical attitude.... In the absence of the son [Goethe], the family situation has not improved. The defects of the councillor [the father] have become intolerable and he becomes exasperated in seeing that [the son's] years of study in Leipzig have produced such a failure.

As usual, there are good and bad refusals of conventions, however a *fox* who wants to play the *hedgehog* had better understand the ways of *hedgehogs* before discarding them, otherwise he risks being out-foxed.

6

The Berlinian *Arch-Fox* Isaiah Berlin

1. Introduction

Few cases are more interesting than that of Isaiah Berlin[48]—an *arch-fox* who wanted, later in life, to be a *hedgehog,*[49] and—in the interim—claiming that the *hedgehog* Tolstoy was by nature a *fox*. In his

---

[48] British historian and writer (1909-1997), noted for his writings on political philosophy. For additional information see chapter 1 of this volume.

[49] In Berlin's own words "There exists a great chasm between those on one side [the *hedgehogs*], who relate everything to a single central vision, one system less or more coherent or articulate, in terms of which they understand, think and feel—a single, universal, organizing principle in terms of which alone all that they are and say has significance—and on the other side, those [the *foxes*] who pursue many ends, often unrelated and even contradictory, connected, if at all, only in some de facto way,... moving on many levels, seizing upon the essence of a vast variety of experiences and objects for what they are in themselves, without, consciously or unconsciously, seeking to fit them into, or exclude them from any one unchanging, all-embracing, sometimes self-contradictory and incomplete, at times fanatical, unitary vision.... Dante belongs to the first category, Shakespeare to the second; Plato, Lucretius, Pascal, Hegel, Dostoevsky, Nietzsche, Ibsen, Proust are, in varying degrees, hedgehogs; Herodotus, Aristotle, Montaigne, Erasmus, Molière, Goethe, Pushkin, Balzac, Joyce are foxes" (1953/1967, pp. 1-2).

More on this subject can be found in my chapters 9 and 10 of volume 1: "GAM's *Hedgehogs* and GAM's *Foxes*", and "Maître Renard James Joyce"; and in chapters 5 and 7 of volume 3: "Calvino, like Goethe,was a Berlinian *Fox*, by being a GAM *Alchemist*", and "Why Berlinian *Fox* or *Hedgehog*?"

desire to be what he was not, particularly late in life, Berlin resembles Goethe.

Berlin's case is also interesting for the evidence it provides in support of the GAM theory, both at the level of causes and results, and in confirmation by outside observers (I saw him as *fox*; others as *arch-fox*).

Lastly, Berlin's case is fascinating for the subject matter of his studies (e.g., the beauty and plusses of plularism; the feasibility or infeasibility of scientific history) most of which have a direct appeal to students of the impact of united or divided power .

2. *Fox*

During my work on the section on Isaiah Berlin in chapter 1 of this volume, I came across a 1969 essay by him entitled "The concept of scientific history", which I read with much interest and admiration for its breadth of culture and depth of understanding. Midway through my reading, I realized that this essay was the work of a Berlinian *fox*, as were his writings on pluralism and on Tolstoy.

His was not the strong attitude of "those who pursue many ends, often unrelated and even contradictory, connected, if at all, in some de facto way" (Berlin, 1953/1967, p. 1), but the *fox* attitude of those who pursue many ends in parallel, loving their multiplicity while looking at each from a unique point of view; and opposing those who instead try to relate everything to a single, universal, organizing principle.

Following are some excerpts from his *The Concept of Scientific History*, which I consider *fox* passages.

> Yet the truth about history—perhaps the most important truth of all—is that general history is precisely this amalgam, *a rich brew composed of apparently disparate ingredients*. (p. 46; my italics in this and the following quotes)

> We wish, ideally at any rate, to be presented, if not with a total experience—which is a logical as well as a practical impossibility—at least with something full enough and concrete enough to meet our conception of public life (itself an abstraction, but not a deductive schema, not an artificially constructed model), *seen from as many points of view and at as many levels as possible*, including as many components, factors, as-

> pects as the widest and deepest knowledge, the greatest analytical power, insight, imagination can present. (p. 47)

> *A concentrated interest in particular events or persons or situations as such, and not as instances of a generalization*, is a prerequisite of that historical sense. [To this, Berlin added the following footnote: "There are really only two ways of acquiring knowledge of human affairs" says Ranke: "through the perception of the particular, or through abstraction...the former [is the method] of history. There is no other way.... Two qualities, I think, are required for the making of the true historian: *first he must feel a participation and pleasure in the particular for itself ... Just as one takes delight in flowers without thinking to what genus of Linnaeus ... they belong ... without thinking how the whole manifests itself in the particular*."] (p. 54)

> [History's real] subject-matter involves a 'thick' texture of criss-crossing, constantly changing and melting conscious and unconscious beliefs and assumptions some of which it is impossible and all of which it is difficult to formulate, on which, nevertheless, our rational views and rational acts are founded, and, indeed, which they exhibit or articulate. (p. 55)

3. *Arch-Fox*

My next GAM task was to look at Berlin's youth, and see if I could find an explanation for his *fox* vision of life—and I found what I was expecting. But before presenting the result of this search, let me report something—in two parts—which gives evidence for the validity of the GAM theory, and for which I thank Michael Ignatieff.

The first part has to do with Berlin's initial thoughts on the division of great thinkers in *foxes* and *hedgehogs:*

> [For Berlin,] Goethe and Pushkin were foxes; Dostoyevsky and Tolstoy were hedgehogs. The distinction dramatized the ultimate incompatibility between types of artistic personality and the moralities of creative life. In 1951 the distinction recurred to

> him when he was working on a study of Leo Tolstoy's sense of history: it seems to capture the fissure between Tolstoy's fox-like gift as a novelist for conveying the fine detail of human life and his hedgehog-like search for an overarching theory of human existence. (Ignatieff, 1998, p. 173)

Here Tolstoy is seen as a true hedgehog, at least in part, (see my pages on Tolstoy in chapter 9 of volume 1), and not as "a natural fox who tried to become a hedgehog"[50] as Berlin claimed in his 1953 book.[51] I suspect that in a second phase, Berlin decided that he himself was like Tolstoy, or that Tolstoy was like him, Berlin: a natural fox who tried to become a hedgehog.

The second and important part deals specifically with Berlin:

> This [in Tolstoy] was also a fissure within himself [Berlin]. *Most of his friends saw him as an arch-fox—nimble, cunning, quick-witted, darting from subject to subject, eluding pursuit. Yet he was also the type of fox who longs to be a hedgehog*—to known one thing, to feel one thing more truly than anything else. It would take more than a decade for him to discover what this was. (ib., p. 173; my italics)

So, Berlin was not only a *fox*, but an *arch-fox.*

4. Father Failure and the Making of a *Fox*

In the search for what made Berlin a fox, I discovered a certain amount of the misfortune of father failure (even if mainly in the eyes of the mother and son). The misfortune was not strong enough to make him an *alchemist* like Goethe and Calvino, but sufficient to make him a *leadsman*[52] *of the fox variety.*

From Ignatieff's biography of Berlin, let me report those sections which point to Berlin's less than positive vision of his father, Mendel. For

---

50 In Annan's compact formulation, 1998, p. xiv.

51 "The hypothesis I wish to offer is that Tolstoy was by nature a fox, but believed in being a hedgehog; that his gifts and achievements are one thing, and his beliefs, and consequently his interpretation of his own achievement another" (Berlin, 1953/1967, pp. 3-4).

52 See chapter 1 of this volume for the misfortune of uprootedness/rootlessness.

simplicity I will follow the sequence of events and comments as given by Ignatieff.

> In Berlin's papers there is a manuscript, eighty-six pages long, written in a spidery hand in an accountant's duplicate book, dated March 1946. It is his father's call to his son to renew the connection to the family's Riga past, to the Berlins, Volschonoks and Schneersons, the rabbis, scholars and merchant princes of Jewish Russia. I asked whether Isaiah was aware of his father's memoir—he had no recollection of it whatsoever.... At my urging, Isaiah read his father's memoir. *All very unreliable, he said sternly, as if marking a deficient undergraduate essay. "Pure sentimental return to roots," he insisted.* "He worked up all this Jewish feeling in old age." In 1946, when Mendel composed the memoir, he was sixty-three, nearing the end of his professional life; he had been in exile in London for twenty-five years; the Soviet conquest of Latvia meant that he would never see Riga again; the Holocaust had turned his past to dust. It was hardly surprising that the roots he had sunk in England should suddenly have seemed insubstantial and that he should have attempted to salvage what he could of his deeper allegiances.
>
> His son would have none of this. "He broke away from it. He was totally emancipated, didn't eat kosher, never went to synagogue." When I pointed out that his father taxed him gently in the memoir for not knowing his Talmud, Isaiah shot back, "He didn't, either." (pp. 12-14; my italics)
>
> Mendel rarely dined out, because he did not like the rough conviviality of the London timber trade; he would rather spend the evening with his wife, although, as Isaiah began to understand, *he bored her.... He did not like arguments and discussions; she thrived on them. She read the latest fiction, Knut Hamsun, D.H. Lawrence, a bien pensant Austrian feminist named Baroness Zutner, and thirsted for intellectual stimulation; he did not.* "If you put a book in his hand, he read it." Otherwise, he preferred the operettas of Offenbach. He had a clear, bright, clerical mind and would have made, his son believed, a better civil servant than a businessman. He lacked the daring and dynamism for big business gambles. *In later life, he could not help feeling sorry for him. 'He never lived his life.*

> *And if you haven't lived your life, what have you had?'*
>
> She, on the other hand, did live her life—through her son. In all his recollections there was never a note of disparagement or pity towards her; rather a kind of awe at the force of her love, the size of her frustration and vitality. "*She resented being married to him. Felt he was dull, depended on her. She wanted to be loved, she wanted to be lifted, nothing ever happened, so all her love was turned on me.*" (pp. 37-38; my italics)

Without the added misfortune of rootlessness which made Berlin into a *leadsman* (see chapter 1), Berlin would have probably become an *alchemist* like Goethe and Italo Calvino: yet something of the *alchemist* is in him (and also of the *fox*): "nimble, cunning, quick-witted, darting from subject to subject, eluding pursuit." Expanding on this subject, there is an interesting little *alchemist* story about Berlin which, especially in its last part, brings to mind Goethe:

> One day Isaiah told "a great big bold gratuitous lie". He set off from the hotel, telling his parents he was going to Quebec Street [i.e., to the crammer school in Quebec Street, where he prepared for common entrance to the London public schools]. Instead, he walked up and down Kensington High Street, looking in the shop windows, until it was time to go home. He returned to find that his father had rung the school and discovered that he had not showed up that day. All his life Isaiah remembered his father's dressing-down, and even more Mendel's tearful, uncomprehending disillusion. Though it passed, it left the child aware of just how-deeply his parents idealized their only son. (ib., p. 37)

In his memoirs, Goethe had written "[My father] valued my natural endowments all the more because they were lacking to him, as he had acquired everything by incalculable diligence, persistent application, and repetition. He often assured me, in season and out of season, in jest and in earnest, that with my talents he would have behaved himself quite differently and not have such a frivolous use of them" (1811-22/1949, p. 21).

5. From *Fox* to *Hedgehog*

The comparison with Goethe can go one step further. Long before Berlin, also Goethe wanted to be a *hedgehog*, and tried hard in his *Theory of Colors* of 1805-1810, of which he was more proud than of his poetic works (see chapter 5 of volume 1). Yet the fox in him was too powerful, and too often distracted the hedgehog: "As an old man he could still say to his *famulus*, Riemer: 'Above all no professionalism. That is against my nature, I want to do everything I can as though it were a game, just as it occurs to me and for as long as the inclination lasts.'... This is a somewhat dangerous [*fox*] principle to follow when one is trying to establish a scientific theory of colour, and is abusing all physicians as fools, which is precisely what Goethe was doing at the time he said this" (Friedenthal, 1965, p. 24).

With Berlin things were different, because of the strong *leadsman* component of his personality. Yet, also in him, as seen by Ignatieff, there was an evolution from *fox* to *hedgehog*, even if late (and probably incomplete):

> He had never had a very strong sense of his own intellectual agenda. Indeed, it was only when others began to take up his ideas in the late 1970s, and the 1980s that he began to understand the extent of their hidden coherence. Now (in 1988, at the age of 79) he could look back and see the path he had been following. For the Agnelli Prize (of 1988), he attempted to express the one big thing that the hedgehog had known all along.... [The resulting essay] "The Pursuit of the Ideal" made plain that all the disparate studies of Machiavelli, Vico, Herder, Herzen and Tolstoy were not disparate exercises but a sustained philosophical enquiry into the history of human values. (Ignatieff, 1998, p. 284)

7

How the Mind of Young Abraham Lincoln Was Shaped by the Creativogenic Misfortune of *Relative Father Failure*

Several times, in previous chapters (e.g., 1:2; 1:10; 3:5), I have discussed the creativogenicity of the misfortune of father failure when interacting with a high GxAxDP.

In this chapter, I want to go one step further and distinguish between the misfortune of *relative* father failure (as was the case with Abraham Lincoln) and the misfortune of *real* father failure (as was the case, for instance, with Picasso). I will also contrast these cases of early discovery of father failure (when the son was in his early teens or younger) with a case of sudden and late discovery (at the age of seventeen): that of Biff, the son of Willy Loman, the main character of Arthur Miller's *Death of a Salesman.*[53]

Part of the discussion will deal with labelling "misfortune" a condition which fosters creativity and eminence. However, this misfortune and

---

[53] My apologies for using a fictional case, but very few books have been written about people who accomplished little. On the other hand, Arthur Miller is a superb writer with a strong grasp of psychology as noted by Richard Evans— professor of psychology at the University of Houston, at the beginning of his book *Psychology and Arthur Miller*—when addressing the writer: "Mr. Miller, there is no doubt that some of the greatest insights into human personality have come not from the discipline of psychology, but from the humanities—art, literature, drama. Many of us would regard your insights concerning man in your many plays as a superb example of this kind of

others listed previously become "lucky events" only in those rare instances in which the misfortunes interact with a high GxAxDP. In addition, there is no guarantee that the high potential for creativity of youth will be actualized in the following decades. Again, the sad story of Biff, even if fictional, is instructive, and so are the depressing real conclusions—discussed shortly—of the study by Elsa Ferri on the misfortune of early paternal death in the absence of a high GxA.

Major misfortunes are objective, one can even say measurable, and so are intelligence, talents, assistances and a broad condition of division of power. I will discuss each of Lincoln's four factors in his GxAxMxDP while, in parallel, adding comparisons to Picasso, Dickens, Mozart, Piaget, and Biff Loman.

Relative father failure follows the birth and growth of the son, and it is a failure by contrast, mainly in the eyes of the son who is more intelligent, educated, with more drive and creativity; the father loses by comparison. This was the case with both, Abraham Lincoln and Italo Calvino.[54]

Real father failure, instead, is independent of the birth of the son or daughter, and often precedes it. This failure is objective (not only in the eyes of his child), and is described in nearly all biographies. Real paternal character and/or business failure shaped Cervantes, Goethe, Mozart, Dickens, Einstein, Picasso, and James Joyce.[55]

Relative Father Failure: the Case of Abraham Lincoln

In a compact way we read from David Donald (1996) that

> Abraham's pulling away from his father was something more significant than a teenage rebellion. Abraham had made a quiet reassessment of the life that [his father] Thomas lived. He kept his judgement to himself, but years later it crept into his scornful statements that his father "grew up, literally without education," that he "never did more in the way of writing than to bunglingly sign his own name," and that he chose to settle in a region where

---

[54] The Italian novelist Italo Calvino (1923-1985) is discussed in my chapter 5 of volume 3 (2003) "Like Goethe also Calvino was a Berlinian *Fox* by being a GAM *Alchemist.*"

[55] Joyce was discussed in detail in my chapter 10 of volume 1 (2001) "Maître Renard James Joyce".

"there was absolutely nothing to excite ambition for education." To Abraham Lincoln that was a damning verdict. In all of his published writings, and, indeed even in reports of hundreds of stories and conversations, he had not one favorable word to say about his father. (p. 33)

However, as William Miller (2002) pointed out:

> [The father] did not deserve to have his name accompanied by some derogatory comment,... He did take the initiative to bring his family through two long and complicated moves, the second involving an entourage of thirteen people. He did take the initiative to return to Kentucky to find a woman he had known, Sarah Johnston, to be a wife and mother to his children....
>
> *What happened to poor Tom Lincoln was that, somehow, he had the bad luck to have this impossibly bright and strong-willed son. Had that not happened, his deficiencies, such as they may have been, would not have been spread across the pages of a thousand books for all the world to read. He appears to have been a limited man, not a bad man. But clearly there was a mismatch between father and son.*
>
> *Abraham Lincoln, discovering intellectual abilities in himself and discovering books, came to value education, intellectual development, exceedingly highly*—and it is difficult for any of us to value something highly, and strive for it mightily, and not then to be disdainful of those, perhaps especially those, who are close to us, who seem not to value it at all.
>
> The critic of Thomas Lincoln who is most convincing, partly by what he does not say and do, is Abraham, his son. He may have been unfair to his father, as ambitious and able sons sometimes are to not very talented and not very successful fathers, but his lack of filial piety is striking. If you are looking for some fault in Abraham Lincoln to offset his too consistent goodness, I suppose you might name his insufficient obedience to the commandment to honor his father. (pp. 60-61; my italics)

For Lincoln, the paternal failure was real, because this was the father he was dealing with every day, the father whose scripts he refused more and more, replacing them with his own: "He did not yet know who he was, or where he was heading, *but he was sure he did not want to be another Thomas Lincoln*" (Donald, 1996, p. 37; my italics). This last comment points to the essence of the creativogenicity of parental failure (when interacting with a high GxAxDP).

This refusal of the paternal scripts can be better understood when compared with similar ones by Picasso and Dickens:

"Picasso found it necessary to denigrate his father and even to render an unflattering portrait of him" (Gardner, 1993, p. 144).

> Around this time Pablo [aged 12 to 13] must have realized that, compared with other bourgeois families, his parents were poor and déclassé, and that his beloved father was a pathetically bad teacher and painter. This realization would have dealt Pablo's pride a grievous blow. His response was embodied in a determination to exorcise the stigma of paternal failure by a triumphant display of his own gifts, something that can only have increased his guilt toward his father. (Richardson, 1991, p. 48)

> The time that Charles [Dickens] had spent in the warehouse amounted to no more than five months, but the lonely anguish he endured had made it seem an eternity of suffering. The wound was so deep that its emotional scar remained with Dickens forever. All the remainder of his life he lay under the double shadow of the Marshalsea [where his father had been gaoled for debt] and the workroom dungeon where he had toiled in despair.The experience was crucial for Dickens's entire future course. It is hardly fanciful to say that in the blacking warehouse that unhappy child died, and into his frail body entered the spirit of a man of relentless determination. Deep within him, he resolved that he should never again be so victimized. He would toil, he would fall prey to none of his father's financial imprudence, he would let nothing stand between him and ambition. He would batter his way out of all the goals that confine the human spirit. (Johnson, 1969, p. 45)

From the pen of an eminent psychologist, Piaget, we have a first-hand description of both the misfortune and its influence:

> My mother was very intelligent, energetic, and fundamentally a very kind person; her rather neurotic temperament, however, made our family life somewhat troublesome. One of the direct consequences of this situation was that I started to forego playing for serious work very early; this I obviously did as much to imitate my father as to take refuge in both a private and non-fictitious world. Indeed, I have always detested any departure from reality, an attitude which I relate to the second important influential factor of my early life, viz., my mother's poor mental health; it was the disturbing factor which at the beginning of my studies in psychology made me intensely interested in questions of psychoanalysis and pathological psychology. (1952, pp. 237-8)

Here we have a compact description of a reduced learning of normal scripts (few scripts from mother, father who concentrated on his work, and peers), and, instead, the building of highly personal scripts (from thoughts on the meaning and cause of poor mental health). These personal scripts, clearly, built on the foundation of a high G, and with the help of high A. On this last, Piaget wrote that his father had devoted "his writing mostly to medieval literature, and to a lesser extent, to the history of Neuchâtel. He is a man of a painstaking and critical mind, who dislikes hastily improvised generalizations, and is not afraid of starting a fight when he finds historic truth twisted to fit respectable traditions. Among many other things he taught me the value of systematic work, even in small matters" (ib., p. 237).

Finally, with Biff Loman, we have a case of a too late discovery of paternal failure:

Biff is the first son of Willy Loman, the main character of Arthur Miller's *Death of a Salesman* (1949). We hear of Biff early in the play when his father complains about him to his wife, Linda, and she tries to defend him:

Willy: In the beginning, when he was young, I thought, well, a young man, it's good for him to tramp around, take a lot of different jobs. But its more than ten years now and he has yet to make thirty-five dollars a week!

Linda: He is finding himself, Willy.

Willy: Not finding yourself at the age of thirty-four is a disgrace! (1976, pp. 15-16)

By this time, Biff had discovered his father's character and professional failure, but this had been of no help to him because the discovery had come late, at age seventeen, and only by chance, not in his early teens as was the case with Goethe, Lincoln, and Picasso, to name only a few.

> Willy's frustrated "success dream"—he has insisted that "appearances" and being "well liked" are keys to social advancement—is complicated by the fact that years ago Biff discovered Willy in a hotel with a strange woman. Since that time, Willy has been a "fake" to his son *and an excuse for Biff's own failures*. (Murray, 1967, pp. 22-23; my italics)

As summarized by Santosh Bhatia:

> The hollowness and falsehood is revealed to Biff when he discovers his father *whom he always looked upon as an ideal*, with a half-naked woman in a room of the Boston hotel. Biff looks at her "open mouthed and horrified." Willy tries to convince him by making all kinds of flimsy excuses: "this is Miss Francis, Biff, she's a buyer. They're painting her room. Go back, Miss Francis, go back." *But Biff's faith in his father is shattered*. In reply to Willy's "I'll see Birnbaum first thing in the morning," Biff says, "He would not listen to you."[56] He leaves crying, and calls Willy: "You fake! You phony little fake! You fake." *Biff loses faith in his father. He comes back to his*

[56] Biff had gone to Boston to ask his father's help in convincing professor Birnbaum, back in New York, to give him four more points in the math finals, without which Biff would not have enough credits to graduate.

> *room the same day, burns his sneakers, and thus puts an end to a bright career in sport.*[57] (1985, pp. 53-4; my italics)

Yet, as stressed many times, the misfortune is only one of the four multiplicative factors. It is Mx(GxAxDP), not only M!

Starting with Lincoln's superb genetic endowment G, and in addition to what was said above:

> Abraham Lincoln believed that his mother was illegitimate... He and his law partner, William H. Herndon, were about to try a case in Menard County Court that involved a question of hereditary traits, and Lincoln observed that illegitimate children were "oftentimes sturdier and brighter than those born in lawful wedlock." To prove his point he mentioned his mother, who he said was "the illegitimate daughter of Lucy Hanks and a well-bred Virginia farmer or planter." *From "this broad-minded, unknown Virginian" Lincoln believed he inherited the traits that distinguished him from the other members of his family: ambition, mental alertness, and the power of analysis.* (Donald, 1996, pp. 19-20)

> *He developed confidence in himself, that he could take up a subject, read the books about it, and acquire a mastery of it sufficient to his purpose*—as he would do repeatedly throughout his life. And he developed a confidence with respect to others, that he need not defer to those with more formal learning. (Miller, 2002, p. 47)

Moving to assistances, the first came from his mother:

> According to tradition, [Lincoln's mother] was able to read... On the rare occasions in later years when he mentioned her, he

---

[57] Obviously, there are many other differences. At no moment does Arthur Miller describe Biff as particularly intelligent or endowed with special talents, at no moment is there mention of important assistances: neither by the father, nor by the mother. On the contrary, later in the play, Biff complains to his father for the wrong assistance he received: "And I never got anywhere because you blew me so full of hot air I could never stand taking orders from anybody! That's whose fault it is." (1949/1976, p. 131).

> referred to his "*angel mother*," partly in recognition of her loving affection, but partly to distinguish her from his stepmother, who was very much alive. (Donald, 1996, p. 23)

Assistance from books was fundamental, starting with those that his stepmother had brought with her from Kentucky: "One was her family Bible. Abraham read it at times, she remembered, 'though not as much as said: he sought more congenial books—suitable for his age.' *The Pilgrim's Progress* was one of them, and the biblical cadences of Lincoln's later speeches owed much to John Bunyan. Another of Sarah Bush Lincoln's books was *Aesop's Fables*, which it was said Abraham read so many times that he could write it out from memory" (Donald, 1986, pp. 30-1).

Later, in a speech of 1859, Lincoln spoke about the democratizing effect brought by readily available books: "At length printing came. It gave ten thousand copies of any written matter, quite as cheaply as ten were given before; and consequently a thousand minds were brought into the fields where there was but one before." And as commented by Miller (2002): "Surely one might say that he is himself one of those thousand minds of obscure rank in a remote place who were now 'brought into the fields' by the production of vastly increased numbers of books—enough copies of the King James Version of the Bible so that there could be a copy even in the bookless Lincoln cabins in Kentucky" (p. 46).

Equally important was the great multiplying factor of living under DP:

> For all the poverty and obscurity and unpromising character of his origins, Lincoln was in some other ways fortunate.... [He was born a] citizen of a budding new power among the nations of the earth, *a repository across the water of the liberal idealism of Europe, of which rising nation he himself would one day give incomparable definition...* Young Lincoln accomplished his stunning ascent primarily through his extraordinary efforts, as we might say, *but he might not have undertaken those efforts, or persisted in them, if the ethos of the new country had not encouraged his ambition. The young Frenchman Tocqueville was describing the "individualism"*

> *of American democracy just at the time Lincoln was exhibiting one form of it in Indiana and Illinois. The institutions of the New World, such as they were, encouraged the picture of individuals with possibilities not bound by social ties and limits.* (Miller, 2002, pp. 21, 23; my italics)[58]

In essence, Lincoln's strong *ambition, mental alertness, and power of analysis* cannot be understood without the interaction of his powerful genetic endowment (especially for intelligence and stamina) with the many assistances he received (from mother, stepmother, father, books), and the DP of the American intellectual climate. However, without paternal failure there probably would have been no great ambition. As noted above, "[the young Lincoln] did not yet know who he was, or where he was heading, but he was sure he did not want to be another Thomas Lincoln" (Donald, 1996, p. 37).

Discussion

The misfortune of *relative father failure* is objective in its impact on personality, so objective that many biographers of Lincoln have accepted at face value Abraham's low esteem of his father. Equally objective—evident, and measurable if needed—are all other fourteen major misfortunes of youth discussed by my GAM theory such as early parental death, physical infirmity, parental domination, lack of parental love, and suffered illegitimate birth.

However, it is only when interacting with high GxAxDP that these misfortunes become a *positive event*. Indeed, rarely does anybody see their misfortune of youth as a positive event, certainly not, for instance,

---

[58] Similar observations on the importance of growing up in a DP climate were made in my 1999b article, in a section entitled "Mozart and Salieri: Children of the Enlightenment." There I wrote: "We are therefore squarely within the Enlightenment, which gave us the Declaration of Independence, the American Constitution, and two revolutions: the American and the French. We are in a period of rapid increase in the division of power.... Mozart's two greatest operas, the revolutionary *Don Giovanni* and *Le Nozze di Figaro*, are revenges against the aristocratic world,... His self-confident, courageous, persistent, *challenged personality* (and the Vienna of the Enlightenment) help us also to understand his liberal-reformist Masonic music (*The Magic Flute*, the cantatas K.619 and K.623) in honor of the freemasonry's humanitarian and enlightened aspiration for equality, liberty, and tolerance" (p. 73).

those with a low SES studied by Elsa Ferri (1976) for the misfortune of early paternal death.[59]

The distinction between *real* and *relative* father failure is important for the light it throws on the relationship between parents and children. Normally a parent will welcome a brilliant child with his/her high G, and yet in some cases—as that of Lincoln— that high G works against the parent, while working creatively for both the youth and society.

---

[59] In Ferri's concluding words: "The results of the analyses carried out showed that, overall, children in one-parent families had a lower level of attainment in school and were less adjusted than their peers from unbroken homes.... Bringing up children singlehanded is an arduous task, both physically and mentally. Help is needed, not only in providing for the family's material welfare which is so gravely threatened by the loss of a parent but also in offering guidance, reassurance and moral support to unsupported parents in their lonely role of bringing up children without another adult to share the responsibility. If such is not forthcoming, the strains and pressures on some lone parents may become so intolerable that they are finally forced to relinquish their burden, resulting in perhaps the worst of all possible outcomes—a no-parent family" (pp. 147, 149).

8

Further Evidence for GAM: Tedlow's *Giants of Enterprise*

1. Introduction

Richard S. Tedlow—Professor of Business Administration at the Harvard Business School—has recently written a fascinating book entitled *Giants of Enterprise: Seven Business Innovators and the Empires they Built.* The seven giants are: Andrew Carnegie, George Eastman, Henry Ford, Thomas J. Watson Sr., Charles Revson (the founder of Revlon), Sam Walton, Robert Noyce.[60]

Given that the book was written without knowledge of the GAM Theory of Personality and Creativity, if many of its seven giants were *challenged personalities* with their high GxAxM in which the M of the misfortunes can be clearly and immediately recognized, then this high incidence of *challenged personalities* is proof that the GAM theory is

---

[60] Carnegie (1835-1919) was the industrialist who led the enormous expansion of the steel industry in the late 19th century. He was also one of the most important philanthropists of his era; Eastman (1854-1932) was the founder of the Eastman Kodak Company, whose introduction of the first Kodak camera helped to promote large-scale amateur photography; Ford (1863-1947) was the founder of the Ford Company. He revolutionized factory production with his assembly-line methods; Watson (1874-1956) was the industrialist who built International Business Machines Corporation (IBM) into the largest manufacturer of electric typewriters and data-processing equipment in the world; Revson (1906-1975) was the businessman who turned a $300 investment into the largest retail cosmetic and fragrance manufacturing firm; Walton (1818-1992) was the retail magnate who founded Wal-Mart Stores, the largest retail-sales chain in the United States; Noyce (1927-1990) was the cofounder of Intel and the inventor of the silicon integrated circuit, which gave Silicon Valley its name.

correct, and specifically that the theory is correct in saying that the M of a major misfortune(s) of youth is one of the needed personal ingredients for greatness (together with a high GxA), and in saying that the pinnacles of creativity are *challenged personalities*, and this in every field.

Basically:

1. Carnegie was affected by the misfortune of father failure;
2. Eastman suffered from the death of his father at age seven;
3. Ford suffered from the death of his mother at age thirteen;
4. Watson was affected by father failure;
5. Revson is one of the two for which no misfortune of youth is mentioned;
6. Walton was affected by father failure; and
7. Noyce is the second for which no misfortune of youth is mentioned.[61]

The fact that five of seven have suffered of a major misfortune of youth—of which two suffered of early parental death (compared to the average of 8% for the general population experiencing early parental death by age 16)—is evidence for the validity of the GAM theory.

Additional evidence for GAM—to be derived from Tedlow's book—is that several personality traits match those foreseen by the theory in function of the nature of each misfortune; for instance *fox* traits (see chapter 9 of volume 1) in the *alchemist* Carnegie, challenged by the misfortune of father failure, *hedgehog* traits in the *architect* Ford challenged by the misfortune of early death of his mother. The same is true for the fact that the greatest of the seven—Carnegie, Ford, and Eastman—are those for which the misfortune is most evident.

2. Carnegie – *Alchemist* and *Berlinian Fox*

Tedlow began the detailed analysis of his seven giants with the following brief story on Carnegie:

---

[61] In terms of eminence, each of the first six are discussed in individual sections of the *2002 Encyclopaedia Britannica*, with the largest space devoted to Carnegie and Ford; only Noyce has not received that accolade, yet.

> Sometime between 1886 and the turn of the century, Andrew Carnegie was visiting his widowed sister-in-law Lucy at the estate his late brother had bought in Dungeness, Florida, just off the coast of the border between Florida and Georgia. Andrew and Lucy got along well. He had named his first blast furnace for the production of pig iron after her (it was the custom in the iron trade to name furnaces after women); and she had named one of her sons (she and Tom were the parents of nine children) Andrew. During this visit, Lucy was complaining to her bemused brother-in-law that her son Andrew neglected to write her from the college which he was attending. Without missing a beat, Carnegie told her that he could induce a response from the young collegian by return mail. He was sure enough that he bet her $10, a wager to which she immediately agreed. So Andrew drafted what has been described as a "nice, newsy" letter to his nephew and namesake. He added a postscript that he was enclosing a check for $10 as a gift. He then deliberately left the check out of the envelope. In no time, young Andrew Carnegie wrote back to his uncle, gratified by the gift but distressed that it had not been enclosed. Carnegie triumphantly presented the letter to Lucy; she paid him the $10 she had lost on the bet, and Carnegie sent that $10 off to his nephew. (p. 19)

In this story I see a personality at work that is close to the *alchemists-foxes* Goethe and Picasso (Goethe who wrote *Reynard the Fox*, as discussed in chapter 5 of volume 3; Picasso who created the *Bull's Head* of 1943 from a saddle and handlebars of a bicycle). Readers of Goethe, and viewers of Picasso, must have felt like Lucy: losing (or winning), while still admiring the creativity of the daring fox.

On the misfortune of father failure (to be compared with that of Goethe and Picasso, discussed in previous volumes), I report hereafter some of the things that Tedlow said on Will Carnegie, the father of Andrew:

> After Andrew was born, Will's business was brisk briefly; but the inexorable and foreseeable results of steam power soon made themselves felt throughout the world of handloom weavers. Will Carnegie began a dreadful economic descent. It lasted for more than a decade in Scotland and then continued in the United States. He cleaved to the old ways.... Harold Livesay, in

> his brief but engrossing Carnegie biography, asserted that Will Carnegie's "wife and elder son clearly thought him a failure, their love and respect gradually changing to shame and contempt."... . Andrew was a ball of fire who seemed not at all to mind being the head of the house. It was he, after all, who was to care for his father, not the other way around.... Will Carnegie was not a fool; and although there is no evidence that his family belittled him for it, his failure in the New World was hardly a state secret. Will was a failure, and he knew it. Andrew, whose long life was (with a few exceptions) one success after another, did not understand failure. (pp. 25-26, 30)[62]

3. Ford – *Architect* and *Berlinian Hedgehog*

I have already discussed Ford in chapter 8, volume 3, "GAM *Architects*", where I explained the strong relation between GAM *universalists* and GAM *architect.*

Architects, in GAM terminology (see chapter 3 of volume1), are those who have been shaped by the same misfortune as the *universalists* (early parental death or much parental absence coupled with love and good assistance from the remaining parent or relatives; good assistance from school and a stable socioeconomic status) but have been raised (or involved from an early age on) in an aristocratic, upper class, military, professional, or business environment. They show vision and great enterprise, give priority to the institution they are part of or have created. (Many great statesmen, and many founders of long lived family enterprises were *architects.*) Here and there, one feels in the *architects* what would have been otherwise a *universalist vision* but bent down by state or professional demands. In the end their view of the world approaches the detachment of the *universalist* in that the state or the enterprise has priority over everything else: family, friends, happiness. As reported by Tedlow: "Henry Ford's mother died on March 29, 1876. Years later, Ford said *the house became like a watch without a mainspring*, a touching expression of grief even if slightly jarring in its use of a mechanical, inanimate simile" (p. 142; my italics). However, this statement about the watch without a mainspring can also bring to mind similar comments by a number of *universalist* philosophers and math-

---

[62] Here, and elsewhere in this and the next chapter, all quotes not otherwise marked, are from the book by Tedlow.

ematicians, like Descartes (his mother died when he was 1) and Leibniz (his father died when he was 6), who were fascinated by the metaphor of God as the *Great Clockmaker*: Once set in motion by Him, the clock of the universe ticks with a life of its own. "In Leibniz's sarcastic caricature, Newton's God is a clumsy watchmaker, 'obliged to clean' his creation 'now and then by an extraordinary concourse, and even to mend it.' On the contrary, Leibniz insisted, 'the same force and vigour remains always in the world, and only passes from one part of matter to another, agreeably to the laws of nature, and the beautiful preestablished order'" (Gay, 1969, p. 143).

The *universalist-architect* and *hedgehog* Ford is also evident from Tedlow's book:

> In the autumn of 1916, John and Horace Dodge, minority shareholders, brought suit against Henry Ford to compel him to distribute more of the Ford Motor Company's profits as dividends. What follows is an excerpt from Ford's testimony on the witness stand. He was being questioned at this point by the lawyer for the plaintiffs:
>
> Q. You say you do not think it is right to make so much profits? What is this business being continued for, and why is it being enlarged?
> A. To do as much [good] as possible for everybody concerned.
> Q. What do you mean by "doing as much good as possible"?
> A. To make money and use it, give employment, and send out the car where the people can use it. (pp. 119-20)

Clearly, being a GAM *universalist-architect* does not mean being a generous and friendly saint, yet it means to have, from time to time, new high ideas of a kind not expected in that line of business. (More on this point will be found in the next chapter).

4. Eastman – *Miner*

> In 1858, [Eastman's father] sold the Waterville business; two years later, after some difficulty, he also sold the Waterville home. The family was back in Rochester to stay. "Then," writes

> Brayer, "total calamity." Eastman died, reportedly of a "*brain disorder*," on May 2, 1862....When G. W. Eastman died, his son was seven years old. The downward glide in social and economic standing which commenced in 1857 turned into an abrupt tailspin in 1862. Stuck in Rochester, Maria [his mother] took in boarders to make ends meet.... *His father left the family not with assets but with debts, a fact which the son is said to have resented.* (pp. 74-75; my italics)

The causes of the *miner personality* in him are: early death of the father, reportedly by brain disorder which may have affected his previous dealings with business and family, SES downfall of the family not attributed to the evils of society but to the father himself. These are not the conditions which shape an *universalist* or an *architect* who looks back with fond memories to the deceased father or mother, but conditions which form a *miner* with a grudge, a lack of love for the deceased, and a reduced enthusiasm for life itself.

Miner may be some of the resulting personality characteristics, as "most of his life he was described shy" (p. 75), and the painfully brief note he left explaining his suicide: "'To my friends: My work is done—Why wait?'" (p. 72). His suicide at the age of 77 can be explained by his health problems and by his personal solitude (he never married,[63] he did not have children); yet that suicide note brings to mind the darkness in which the professional miners work in the entrails of the earth.

5. High G

It is often the high genetic endowment (G) of the son or daughter which says that the father is a failure. To less endowed children the father would be average:

---

[63] "GE was profoundly unmarried. His primary love relationship was with his work. His passion for Kodak became a major reason for keeping marriage at a distance, as he himself frequently explained. Perhaps, too, the memory of a father who struggled to establish himself in two occupations while shuttling his family between two locations, and who then died at an early age deeply in debt, was the example that made him focus solely on vocational goals. And by the time he had enough money to place himself and his mother forever out of the reach of poverty, the thought did nag him that women were mainly interested in him for his wealth and position" (Brayer, 1996, p. 402).

> With the odds against her [the mother of Carnegie], she never gave up. As for her eldest son [Andrew], he was smart, shrewd, charming, self-confident. (p. 26)

> On his own [around 1850, at 15], with no lessons, he learned telegraphy. He mastered it; he was quicker than anyone else.... Carnegie [is] said to be the third person in the United States to learn to take messages by sound rather than having to transcribe the rolls of dots and dashes which Samuel F. B. Morse had assumed would be necessary.... (pp. 28, 36)

Similar comments can be found for Watson and Walton:

> At some point—it is not clear when, but apparently in his late teens—Watson took the measure of his father. "I was absolutely sure that I was a smarter man than my father," he said years later. "I was positive of that. I felt I could prove it, if I were called upon to prove it." By his early twenties, Watson was for all intents and purposes the head of his household. When he went to Buffalo in the mid-1890s, his family followed. (p. 217)

"Walton believed that 'Dad never had the kind of ambition or confidence to build much of a business on his own, and he didn't believe in taking on debt'....Walton seems to have learned more from his family about how not to behave than the opposite. Don't squabble like your parents. Think big, unlike your father" (pp. 320-21).

In the GAM formula GxAxM, a very high G can compensate for a moderate A and M. This may have been the case with Ford, in which much of the G was Gi/Ta/St (the Gi of intelligence, the Ta of mechanical talent, and the St of stamina) especially the last two. That high G was the "supercomputer" that magnified a thousand times certain inputs and neglected others.

Worth remembering is that the misfortune of death of one parent works on many different levels, one of them being the changes it introduces in the remaining parent who will not be as "normal" as otherwise, for instance more often sad and bitter, overburdened by more tasks, less available to the children (or exceptionally over-protective, or excessively involved in their lives). The high Gi/Ta/St of Ford can be seen at work in the following from Tedlow:

> At any rate, during the 1890s, Ford, like Eastman early in his career, held down two jobs: salaried master mechanic by day and freelance experimenter and inventor by night. By 1896, the work was overpowering in its grip. Ford felt that he was on to something new, different, and better than others who were experimenting with the horseless carriage. By May of 1896, his friends "often wondered when Henry Ford slept." [His wife] Clara feared he would have a nervous breakdown, but characteristically kept her fears to herself. The result of all this was the emergence, between two and four o'clock on the morning of June 4, of the quadricycle. Henry Ford had done what not one top automobile executive in the world could do today. He had built a complete car with his bare hands. (pp. 148-49)

Having paid tribute to both G and M, it behooves us—before moving to the next chapter fully devoted to Ford—to pay homage to the A of assistances:

> The influence of Mary Ford on her son, by his own evidence, profoundly affected his whole career and work. "I have tried to live my life as my mother would have wished," he told Edgar Guest. "I believe I have done, as far as I could just what she hoped for me."[64] The influence of his father was undoubtedly exerted with no less benefit in a stabilizing direction. Henry as boy and youth was highly charged with energy, and increasingly drawn toward things mechanical. His parents guided him to a working adjustment between his environment and his passionate interest in tools and wheels. (Nevins, 1954, p. 51)

---

64 Very *universalist* comments.

9

Let the Cobbler Stick to His Last! Wrong Advice, to a *Universalist-Architect* like Henry Ford!

Thanks to the book by Tedlow, discussed in the previous chapter, I also discovered two major aspects of Henry Ford's personality and life which can be explained by the GAM theory.

1. Ford and the Normal Inconsistencies of a Subliterate and Underdeveloped *Universalist-Architect.*

Inconsistencies

Tedlow remarked that,

> a good example of the problem Ford's personality poses is his pacifism. This is a belief that he trumpeted with increasing bombast as World War I dragged into 1915.... Ford himself publicly stated he would burn his factory to the ground rather than supply war material.... In late November, Ford chartered his famous "Peace Ship," the *Oscar II,* to sail to Europe to bring peace to the warring nations. On November 22, Ford visited the White House, where he found President Wilson cordial but noncommittal. Shortly before setting sail, Ford announced that the soldiers in the trenches should call a general strike.... In the Nevins

> and Hill study, chapter 2 of the volume on this period is entitled "Peace Crusade." The reader turns the page to see chapter 3, which is entitled "Production for War"... On February 5, two days after President Wilson broke diplomatic relations with Germany, Ford said, "I cannot believe that war will come"; but if it did he promised to "place our factory at the disposal of the United States government and will operate without one cent of profit.".... If one is to stick with historical narrative, one finds a man who, for an extended period of time, was a belligerent pacifist—if that is not an oxymoron—willing to see his factory burned down rather than accept an order for a bullet. Then, suddenly, having virulently opposed a reasonable plan for preparedness which had been backed by the nation's chief executive, Ford changed his mind completely and for no apparent reason. "I am a pacifist first but perhaps militarism can be crushed only with militarism. In that case, I am in on it to the finish." Why the change?... Was Ford really a pacifist? when he said he was, did he mean it?... As Anne Jardim has pointed out, "for all his protestations, Ford was not a pacifist in any real sense: his were not the beliefs of a Quaker." She explains what role his supposed belief played: "*Pacifism for him was a means of expressing emotion unencumbered by conviction....* The belligerence, earlier turned against war and the makers of war, was now expressed in waging it." (2001, pp. 130-35; italics added by Tedlow)

I disagree with Jardim, because what Ford believed for a while, and tried to achieve—world peace—was of high universal quality, higher and more universal than say the construction of Versailles by Louis XIV. One may disagree with Ford for abruptly changing his positions, but one cannot label his efforts "means of expressing emotion unencumbered by conviction." By being a *challenged personality* with few normal scripts because of the early death of his mother, his pacifism was not a script engraved in his mind from early youth as is the case with the Quakers[65];

---

[65] The initial problem is the M of GxAxM which reduces normal script adoption, not being subliterate. Illiterate and subliterate people nearly always have the strongest conviction, based on strongest scripts: few ideas, strong as rock, unchangeable. The *challenged personality* has strong ideas, the ones they developed on their own, and yet these can be changed by new inputs and new rumination. In other words the *challenged personality* is a combination of both stubbornness and flexibility.

and by being *subliterate*[66], his pacifism was not a script he had gradually developed—at the light of his misfortune of youth—from books and related ruminations. The ruminations were probably there, but not the extensive reading, for the simple reason that he read little.

The reader will remember—from the very first chapter of volume 1 of *The GAM/DP Theory of Personality and Creativity*—that the key characteristic of the *challenged personalities* is their paucity of normal

---

[66] "Henry attended rural schools only to the age of 15, when he found employment as a machinist's apprentice in Detroit. In his spare time he repaired watches and clocks to improve his knowledge of mechanical things, an interest that never waned. Even after he had become the master of an industrial empire, he delighted in disassembling the watches of his friends or joining the mechanics in his plant in a greasy repair job" (Rogers, 1978, p. 520).

Ford's lack of education was made evident in the summer of 1919 when cross-examined by the lawyers of the *Chicago Tribune* which had been sued by Ford for libel. In the words of Robert Lacey: "If Henry Ford had been cross-examined on his knowledge of birds, he would have been on strong ground, but the lawyers of the *Chicago Tribune* chose instead to probe his knowledge of American history:

Q: Have there been any revolutions in this country?

A: Yes.

Q: When?

A: In 1812.

Q: One in 1812, eh? Any other time?

A: I don't know of any others.

Q: Do you know that this country was born in a revolution?

A: Yes, in 1776.

Q: Did you forget that revolution?

A: I guess so...

Q: Do you know what forced us into the Revolutionary War?

A: No, I do not....

Q: Do you know of any great traitors?

A: No.

Q: Did you ever hear of Benedict Arnold?

A: I have heard the name.

Q: Who was he?

A: I have forgotten just who he is. He is a writer, I think.

Asked to define 'a large mobile army,' he came up with 'a large army mobilized,' and as a meaning for 'ballyhoo' he offered, 'Oh, a blackguard or something of that nature'" (1978, pp. 200-01).

The word *subliterate* for Ford was used by John Kenneth Galbraith in his 1986 review of Robert Lacey's *Ford: The man and the machine*.

scripts, the result of their major misfortune(s) of youth: "Lack of regular scripts forces challenged youths to build their own scripts, guided in this by their long ruminations on the causes of their specific misfortunes. And they do it, more than others, from books ..." (*1*, p. 26). But this was not the case with Ford: books were not his friends, either early or later in life. On the other hand if he had not been the first class mechanic he was, we would not have had him and his Ford Model T. Ford is an exception, and a fascinating exception, among the challenged personalities I have studied.

Pacifism was an idea that Ford's *universalist mind* could easily love[67] and his *architect mind* and money could attempt to bring to fruition (e.g., his "Peace Ship"). However, lacking deep roots, lacking the immediacy and connectedness that he had when dealing with automobiles, those great ideas could not survive for long the heat of wide spread criticism:

> The [Peace Ship] *Oscar II* departed from Hoboken, New Jersey, on December 5, 1915 - a fool's errand if ever there was one. When Ford arrived in Oslo on December 18, he said nothing for five days. He then called a press conference to explain that his new tractor was unpatented and that armaments makers could realize greater profits on tractors than on the instruments of war. Ford betrayed not the slightest inkling of the causes or potential consequences of the greatest armed conflict in history. Observed one attendee at the press conference, "He must be a very great man who permits himself to utter such foolishness." (p. 131)

"The ridicule that surrounded the departure of the Peace Ship continued undiminished during the voyage. Reporters on board described the delegates as "a bunch of nuts" (Jardim, 1970, p. 128).

Ford had good intentions, but not the intellectual stamina to steer his peace initiative to a valid conclusion, nor the intellectual stamina to hear for long that what he was doing was "a fool's errand if ever there was

---

[67] "Ford, we are informed, 'grew up in an era marked by an increasing devotion to peace.' Leading American statesmen had been crusading for peace for more than a decade prior to World War I. 'Pacifism in the United States ... was thus not only respectable but little short of triumphant.' Ford had not embarked upon a 'wild, perverse crusade.' To the contrary, 'millions in spirit marched with him'" (ib., p. 133).

one," and that "He must be a very great man who permits himself to utter such foolishness [read: Ford is a complete fool]". On top, it must have become evident to him in Oslo that he knew abysmally little about Austria, France, Germany, Great Britain, Italy, Russia, Serbia, and why they were at war.

He would have been a complete fool indeed if he had continued to let his noble rethorical aims guide him with something he knew nothing about. On the other hand he would not have been the *challenged personality* that he was if he had simply watered down his pacifism. He had to adopt another universal idea. So, "on February 5, two days after President Wilson broke diplomatic relations with Germany, Ford said, "I cannot believe that war will come"; but if it did he promised to "place our factory at the disposal of the United States government and will operate without one cent of profit."

This last statement, about not making one cent profit, was certainly said in good faith by Ford, but this, too, was not rooted in a clear understanding of the relationship of the individual entrepreneur to the national effort (as defined by the President), and with time it was eroded by the realities of industrial life, month after month. Hypocrisy? No, simply lack of follow-up by somebody lacking strongly established scripts, lacking the intellectual training which would weigh an important idea five times before proffering it.

However, lacking long-term consistency in high universal thoughts and plans is not the same as "*expressing emotion unencumbered by conviction.*" Ford's wonderful, universalist convictions were indeed strong when dealing with automobiles:

> Prior to Ford, the automobile had been predominantly a high-priced luxury item assembled in small numbers by skilled craftsmen. But Henry Ford developed a vision of what the car should be. That vision is expressed in the first paragraph of this chapter ... "I will build a car for the great multitude," Henry Ford said early in his career, "constructed of the best materials by the best men to be hired after the simplest designs that modern engineering can devise... so low in price that no man making a good salary will be unable to own one — and enjoy with his family the blessing of hours of pleasure in God's great open spaces." This elegant, eloquent passage ranks among the finest statements of corporate purpose in American business history. (ib., pp. 153, 119)

> Ford, on the other hand, held fast to his dream of "a car for the common man." And "dream" is precisely what it was, because it was very difficult to see at the time how Ford's vision could be turned into a profitable business reality.... We would deprive ourselves of the lessons Henry Ford's life has to teach if we allowed his later failings to obscure his vision and his talent. (ib., pp. 153-54)

We would also deprive ourselves of other lessons, if we would label his—albeit inconsistent—high universal thoughts and plans[68] *emotions unencumbered by conviction.*[69] This definitely is not what can be read from Nevins (1954): "There can be no question, for example, of his whole-hearted enthusiasm for the five-dollar minimum and the work of the sociological department in 1914-15. He was a man of generous impulses. His idealism was as yet a cardinal characteristic" (p. 580).

Ford had the high aspirations of a *universalist*—e.g. pacifism, operating a business for the purpose of doing "as much [good] as possible for everybody concerned," announcing, in 1914 when workers in United States manufacturing industries were making about $11 a week on the average, that every Ford worker would start receiving a minimum wage of $5 a day, and announcing that workers would also participate in a profit-sharing plan. Even if he did not have the education of such *universalists* as Dante, C. S. Lewis, Solzhenitsyn, Tolstoy (see volume 1), like them he saw the world as something imperfect, but also as something that could be perfected, and perfected on the basis of a vision of how things should be.

---

[68] Not all his plans were as highly universal as the "Peace Ship". Sadly—outside of automobiles—he was as uncritical in his adoption of good as of bad ideas and action plans (particularly bad was his anti-Semitism, as discussed, for instance, by Tedlow, pp. 136-138).

Yet, good or bad, his ideas and plans had a broad range, not the down-to-earth owning and running a stable of race horses or race cars, nor the construction of a huge baronial castle in northeast Scotland or a Renaissance castle in Southern California(*).

(*) "[Ford mansion] was on the banks of the River Rouge, not far from the place of his birth or from his gigantic plant. Alone among the great men of the automobile industry, Ford always remained deeply rooted in Detroit. The same is true of his family to this day" (p. 175).

[69] Every challenged personality has strong emotions: see section "Passion, Drive, Self-confidence" of chapter 1 of volume 1.

2. A Different Story: Ford as Victim of the Law of Power of Lord Acton

Tedlow has devoted a whole section of his book to "the derangement of power" in which he expanded on Lord Acton's dictum by stressing that power not only corrupts but also deranges: "Corruption and derangement are not mutually exclusive, but derangement brings an added dimension to an understanding of the impact of power" (p. 60). I agree, indeed, Ford who was one of my three examples, together with Napoleon and Arthur Koestler, in my chapter of volume 1 "Lord Acton's Law of Power, and its Four Corollaries" of volume 1.

This said, we must not confuse Ford's universalist inconsistencies with the derangement he suffered from the effects of the Law of Lord Acton. Ford would have remained inconsistent in his high ideas even if he had not acquired much power. On the other hand, many people were corrupted by power without showing inconsistent high ideas, without ever having had generous aspirations.

So we have two stories: the first on his inconsistencies resulting from his being a subliterate *universalist*, the second on his being corrupted and deranged by power which is rather common. Remember the comments by Octave Gélinier reported in the French business magazine *Entreprise*: "Success, for many, brings with it the supreme luxury of despising the facts".

3. Let the Cobbler Stick to his Last! Wrong Advice, Especially to a *Universalist-Architect* like Ford!

In 1923, the Rev. Dr. Samuel S. Marquis,[70] who knew Ford well for more than twenty years (for a time Ford was his parishioner, and for a

---

[70] Nevins (1958) calls Marquis "The able and enlightened Episcopalian clergyman who succeeded Lee as head of the (sociological) department" (p. 555). Lewis (1976) wrote that Marquis's *Henry Ford: An interpretation*" stands even today as one of the finest and most dispassionate character studies of Ford ever written" (p. 215), and Jardim (1970) wrote that: "the Reverend Samuel Marquis, an Episcopalian clergyman who joined the company in 1915 to run the Sociological Department (in effect personal and labor relations) and who left at bitter odds with Ford in 1921, later published a perceptive account of Ford as he knew him" (p. 32).

time his employer: Marquis serving as head of the Sociological Department)[71], published a book: *Henry Ford: An interpretation.* In it he said:

> In my opinion he could realize his supreme ambition if he were to follow the example of a good shoemaker and stick to his last, that is, to the human and production problems in industry, and leave national, international and racial problems alone. It is human to grow weary of achievement in one direction. Like Alexander, we tearfully long for adventures in other worlds instead of trying to bring a little nearer to perfection the world we have conquered.... On the return journey from Europe above referred to I found it necessary to make a very frank criticism of certain ideas advanced by Mr. Ford. It was to the effect that if he stuck to the things he knew, and let those alone about which his training had not qualified him to venture an opinion, he would avoid placing himself in a foolish position. The criticism stuck. I have heard him refer to it many times since. The last time he mentioned the matter in my presence he added, "And I have come to the conclusion that the best friend one has is the man who tells him the truth." (pp. 8-9, 13-14)

Allan Nevins, the author of the extensive study *Ford: The Times, the Man, the Company*, agreed with Marquis:

> It was his capacity for rash acts, however, which was most alarming. The peace ship venture, the abrupt dropping of Couzens, with his invaluable abilities, and other acts which followed these, disclosed a combination of ignorance, impulsiveness, and bad judgment that would have wrecked a man of less appealing personality, less faith in his own instincts, and less dazzling constructive achievements, past and potential. His motive in the peace ship venture was highly commendable; his method was appall-

---

[71] Its purpose was "to administer the five-dollar plan and ensure that the higher wage would serve constructive and not pernicious ends" (Nevins, 1954, p. 551).

"During the time I [Marquis] was with the company he [Ford] gave to his employees, in addition to generous wage, more than a hundred million dollars out of his profits, all of which he could have retained as his own, and which the average man would have put into his pocket.... He did what no other man had ever been able to do,—touched the hearts of the people through their pocketbooks" (1923, pp. 34, 46).

> ing. When Dr. Marquis expostulated, he resorted to a childish logic. "It is right, is it not, to try to stop war?" "Yes." "Well, you have told me what is right cannot fail." Ford would have been scornful of an artist or writer who meddled with the machine shop at Highland Park. But being an intuitive thinker, with a trust in his own star- or comet - he could not comprehend why men were scornful of his meddling with complex national, international, and racial problems. *He could not understand Dr. Marquis's advice that he stick to production problems, labor management, and public relations in the automotive industry*. (1954, p. 579; my italics)

I disagree with Marquis (and Nevins): It was wrong to tell Ford: *Let the cobbler stick to his last!* A man with Ford's personality could not be limited to production problems, labor management, and public relations in the automotive industry. And not only a man like Ford, but everybody with an ounce of aspiration beyond "the last."

What Marquis should have done was to outline to Ford a number of major causes tailored to Ford's *universalist personality*, not to his mechanical aptitudes. Marquis should have presented the beauty of each cause, the related high moral challenge, the chances for success and glory; he should have outlined—even in a very preliminary form—the ways to build a team of dedicated collaborators for each task, the sequence of steps in accomplishing the goal, etc. And even if none of the ideas presented would have been accepted, Marquis would have trained Ford in considering alternative possibilities and in choosing one or two that made sense.

Instead, with that *Let the cobbler stick to his last!*, Marquis simply pushed down Jack in the box and closed it; but little was needed to open the lid, and the compressed Jack was out again with a vengeance.

Marquis propounded the old social rule (see chapter 15 of volume 3) to "Maintain proper station." He, Doctor of Divinity, understood the higher things, Ford—rich as he was—understood the lower things, and had to stay with them. What I am now saying may be unfair to Marquis who meant well, yet from his speech and subsequent comments, I do not find a trace of an attempt to help Ford to leave his "last", to stay out of the mechanical box. Clearly, in order to work, any major non-automotive task had to be tailored to Ford's *universalist aspirations*.

What we know, instead, is that Marquis contented himself with passing to Ford that demeaning injunction of *Let the cobbler stick to his last!* Worse, Marquis "rubbed it in", by adding—at the end of the last paragraph quoted above: "I hope he will receive the critical portion of these pages [those on the cobbler] in the same spirit. They are meant to help, for I would like that halo stick" (p. 14).

Did Marquis really expect Ford to accept meekly that same unpleasant sermon, this time in print? No, those criticisms were not accepted gladly. And there was action: "The volumes by Marquis and Pipp were, in fact, never available to the public in large numbers, for the Ford organization was fairly successful in buying up and suppressing them" (Lewis, 1976, p. 215). Yes, Pastor Marquis meant to help, but according to his own scheme which left little room for Ford to expand beyond being a mechanic, even if of great quality and with huge industrial success.[72]

The present disquisition would be of little value—what happened happened—if this same problem, that of leaving one's last from time to time, did not concern every human being: each—at some point in life—aspiring to be something more than a mechanic of one type or another, but without going to Faustian excesses. And that aspiration was, indeed, very strong in Ford, who left upon Marquis "the impression that his chief ambition is to be known as a thinker of an original kind. He has the not uncommon conviction among mortals that he has a real message for the world, a real service to render mankind. .... [He said:] 'I want to live a life, to make the world a little better for having lived in it'" (1923, p. 8). Very *universalist,* indeed!

---

[72] It is true, as said by Nevins, that "much of both his originality and his soundness of instinct in mechanical matters came from the immense amount of time and effort he spent in working, in watching others work under his direction, and in thinking about his work. In building up the company he had labored all the time.... His intense concentration on engineering, on design, and on manufacturing processes does much to explain why he so far outstripped his competitors. When he moved outside the field of mechanics and production, however, his concentration was slight; he neither thought through nor felt through any political or social question. He discarded the experience of the past - a little later he was to call history bunk. Yet his cocksureness increased. He not merely delivered hasty judgments, but delivered them with absolute finality. (1954, pp. 579-80)

On the other hand, if he would have been helped very early to develop his *universalist aspirations*, he may have learned not to deliver hasty judgements with absolute finality. Once again, what happened happened, yet we must be careful not to work backward from later developments to prove the proposition that Ford had to stay with his cars and plants, without doing anything socially advanced.

The question, then, is when in life should one start thinking at being more than an expert in production problems, labor management, and public relations in the automotive industry, even at being more than a giant of enterprise? And the answer is: "early." And "early" means long before reaching the zenith in one's profession. If Marquis had, of his own initiative, quite early, suggested to Ford, that he gradually begin to think of doing higher things, and suggested three or four possibilities, then Marquis might have acquired a valid influence on Ford.

What had to be avoided was to create in Ford the feeling that everybody wanted to close the lid on him, to keep him chained to his last. Once that mistrust was created, he chose, alone, his big projects and the way to implement them.

We will never know how things might have evolved differently with Ford, but the world has progressed to the point in which a large number of *cobblers* do not need to stick all the time to their last, and may instead aspire at doing something of general and generous value.

PART II

DP

10

DP Christian Iconography Versus UP Christian Iconography

I. Introduction

The DP theory, which I have presented throughout the previous three volumes states that the prevalent culture of a given society, including its religiosity, will be different in a climate of division of power or of unity of power. If so, the iconography of Jesus—the subject of this chapter—should reflect the alternating conditions of DP or UP in the history of Christianity, with representations which correspond closely to the power conditions of the time. And indeed it was so, as discussed for the following phases:

Phase 1 – Early Christian, 3rd century to 5th century of high division of power.

Phase 2 – Byzantine I (powerful), 5th century to 8th century, under the high unity of power of the Byzantine emperors.

Phase 3 – Byzantine II (bleeding), 8th century to 13th century, under heretical division of power and major losses of power.

Phase 4 – Communal, 13th and 14th centuries of division of power, the long term consequence of the humiliations of Emperor Henry IV at Canossa in 1077, and of Emperor Frederick I Barbarossa at Legnano in 1176 and at Venice in 1177. Phase 4 ends with the new power of the *signori*, with the Renaissance, and with the steady growth of royal power.

Later Phases — Domination of Artistic Styles (e.g., Renaissance, Mannerism, Baroque) under an alternation of unity of power and division of power.

Now — Present, Under DP, and *Jesus of the People*

Clearly, there are many more phases. Already between Phase 1 and Phase 3, there are the Pre-Carolingian, Carolingian, Romanesque, Early Gothic, Refined Gothic, and Late Gothic forms of art. Some of this is discussed in the next chapter on the two types of Western Christian imperial iconography in the period 800 to 1100.

For each phase, I will start with a number of reproductions representative of the iconography of the epoch.

II. Phase 1 – DP – Early Christianity, and Jesus as the Young Good Shepherd

This first phase of the representation of Jesus is probably the least known, and yet the most important because it reflects the beginning of the new religion, which conquered the hearts of the peoples of the Roman Empire. As discussed shortly, the contrast with the next phase is shocking; but only shocking in hindsight and when compressed in time. It was far less shocking to those who lived in Phase 2.

1. Iconography

Figure 1 *Jesus as the Good Shepherd*, late 3rd century; wall-painting, Catacomb of Priscilla, Cubicolo dei Bottari, Rome.

Figure 2 *Jesus entering Jerusalem*, detail, 4th century; sarcophagus, Museo Lateranense, Vatican.

André Grabar wrote, in his discussion of "Christian Painting and Sculpture before the Peace of the Church [of 313]":

> These paintings represent figures and events whose choice and combinations show that all are intended to convey a similar

> message. *They are images of God's loving-kindness toward those who put their trust in Him.* (1968, pp. 67, my italics)

Similar comments were made by H. W. Janson on a painted ceiling of the Catacomb of Sts. Peter and Marcellino: "In the central medallion we see a youthful shepherd, with a sheep on his shoulder....; he stands for Christ the Saviour, the Good Shepherd who gives His life for His sheep" (1991, p. 256).

2. Division of Power

There was a major division of power (especially at the level of the scripts) both within the Roman Empire and within Palestine, at the time of the birth of Christianity.

Rome

> The formation of an all-embracing empire promoted the decay of the local religious cults of the several states and cities which were brought within the inclusive political unity. To be sure, many were maintained as a matter of custom or civic pride, but the heart had largely gone out of them.... Some cults were maintained by the state. These included the gods of Rome and of the cities of the Empire. The Roman Empire was in part a congeries of city states. Many of these cities had existed before the formation of the Empire and had been autonomous. *Each had felt itself dependent upon the favour of its gods and had seen to it that the worship of its official divinities was maintained.* As we have suggested, the state religions were no longer believed in as strongly as formerly. However, the continuance of their rites was believed to be necessary for the welfare of society. They were, accordingly, kept up, often with great pomp. (Latourette, 1953, pp. 22, 23; my italics)

> Rome's own republican religion was primitive and jejune. It was a religion of State, concerned with civil virtues and outward observance. It was administered by paid government functionaries and its purpose and style were indistinguishable from those of the state. It did not touch the heart or impose burdens on man's credulity.... The State's compulsory but marginal civic

> creed thus left ample freedom for the psyche within the empire. Every man could have and practice a second religion if he chose. Or, put it another way, the mandatory civic cult made possible freedom of worship.... [New gods and goddesses came to Rome from all parts, from Babylon, Persia, Egypt, etc.] It became less likely that an educated man would support the cult of his parents, let alone his grand parents; or even that he would fail to change his cult once, perhaps twice, in his life. (Johnson, 1976, pp. 6-8)

It would be difficult to find a stronger division of power in religious matters, at both the level of the state and of the competing cults.

In Palestine

> Judaism itself was not the monolithic religion it appeared to be to those outside its bounds. In Palestine at the time of Jesus, Sadducees, Pharisees, Herodians, and Essenes represented different interpretations of Judaism and jostled for predominance. Between them and the millions of Jews who remained in Babylonia or had settled in the Greek-speaking East there were great differences involving language, religious ideas, and attitudes toward pagan society. Yet for all these differences, there remained a certain unity of observance and the passionately held conviction that the Jewish way of life was inspired and directed by God. (Frend, 1984, p. 12)

Expanding on the above, inside Judaism there was a major religious-cultural-political division among several antagonistic groups, principal among them:

- the high priesthood which was in the hands of Sadducee aristocrats who supported and upheld the Roman occupation;
- the Pharisees, who held no brief for the Romans, and who were noted for their piety, earnest prayer, and their careful observance of the Sabbath. Preserving this religious heritage, the Pharisees strove to imbue the daily lives of the Jewish people with a sense of God;
- the Essenes clustered in monastic communities. With few exceptions, they shunned Temple worship, content to live

> ascetic lives of manual labor in seclusion. "The Essenes, in their concern for the purity of the Temple, stumbled on a theological concept which made the Temple [in Jerusalem] no longer significant as a physical and geographic fact, and thus opened the way to the universalist principle.... The Temple is not the building, but the worshippers, that is, the Church. Once this concept is married to the quite different, but contemporary, concept of the Pharisee synagogue, that is, a building which may be sited anywhere in the world where the faithful gather to worship and hear scriptures explained, then we are very close to the primitive idea of the Christian community." (Johnson, 1976, pp. 16-17)

In this matter, Jesus was even more revolutionary, when he said to his disciples "For where two or three are gathered together in my name, there am I in the midst of them" (Matthew 18:19). Here we have the highest religious division of power: no buildings, no priests, but only two or three gathered in his name.

I cannot stress enough the importance of these changes which liberated people from the *unity of power of the place*, be it of birth or residence, especially of the city-state and its main god and temple (e.g., Athena and her Parthenon). The contrast is evident with the attempts by the Roman emperors to expand their control over other nations by making their people "Roman citizen," who venerate the one single emperor, looking to him for directions and instructions: "From the rule of Julius Caesar onward, initially in the East and then creeping through the West, the goddess Roma (representing Rome itself) and the emperor too received all the gestures of devotion of ancient religion (temples, their images reverenced, sacrifices, hymns), expressing loyalty to Rome and genuine gratitude for the benefits of Roman rule.... When cities or districts sought to express their devotion to the Roman emperor, they often sought his permission to erect a temple dedicated to him, or to place an image of the emperor alongside the image of the goddess Roma in the temples erected to her, thereby making the temple a holy place for both Roma and the divine emperor" (Hurtado, 2000, pp. 15, 19-20).

First, the synagogues, then the Christian communities which followed Jesus's instructions of "for where two or three are gathered together in my name, there am I in the midst of them", totally upset the old *unity of power of the place,* be it of the Temple in Jerusalem, of local pagan temples, or of the goddess Roma and her Roman emperor.

Under these conditions of division of power in the Roman Empire, especially in Palestine, and under the new freedom from the *unity of power of the place*, first the Jews, then the Romans and Hellenized people of the empire welcomed a monotheistic salvationist religion—with clear moral rules and emphasis on brotherly help—integrated within a historical explanation going as far back as the creation of the world; a religion uninterested in political power, in coercion; a religion which believed in "Render unto Caesar the things which are Caesar's, and unto God the things that are God's" (Matthew, 22: 21-22).

III. Phase 2 – Byzantine I (powerful) and Jesus as Emperor and Pantocrator

> The basic principles of Byzantine imperial ideology were formulated in the 330s by the court prelate Eusebius of Caesarea, during the reign of the first Christian emperor, Constantine I; and they prevailed virtually unchanged until the last emperor, Constantine XI, fell fighting the Turks on the walls of the city in 1453. The Eusebian theory can be summed up in the phrase "As in heaven, so on earth". The emperor is to be seen as the living icon of Christ, God's viceregent on earth. The terrestrial rule of the emperor reproduces God's rule in heaven. In Eusebius' words, "Crowned in the image of the heavenly kingship, gazing upwards, he steers and guides humans on earth according to the pattern of his prototype." Just as God regulates the cosmic order, so the emperor regulates the social order. God by his divine activity brings all creation into ordered harmony under his absolute rule; and in similar ways the emperor brings all humankind into ordered harmony under his own absolute rule, within the framework of a universal Christian state. (Ware, 1993, p. 134)

1. Iconography

Figure 3 *Jesus as the Good Shepherd*, ca. 430-450; mosaic, Mausoleum of Galla Placidia, Ravenna.

"In the centre is the figure of the Good Shepherd whom the ancient painters had so often frescoed in the Catacombs, but while in the Catacomb painting the Good Shepherd had almost been represented as a countryman with short tunic and stick, here he wears a purple mantle covering a tunic of gold and carrying a tall cross in his left hand" (Bovini, 1991, p. 17).

We have here a first evidence of the transformation of Jesus, from the son of a carpenter into a princely figure donning the imperial purple with golden bands, and holding in his hand a cross of gold which has no relation to the heavy cross of wood of the Crucifixion. It is now the imperial purple and the golden cross which identify the importance of Jesus.

Figure 4 *Jesus resurrecting Lazarus*, early 6th century; mosaic, St. Apollinare Nuovo, Ravenna.

Here too, Jesus wears the purple ornated with several bands of gold.

Figure 5 *Jesus enthroned flanked by Angels*, early 6th century; mosaic, St. Apollinare Nuovo, Ravenna.

Imperialist escalation: not only does Jesus wear the purple with golden bands, but he is seated on a throne, and holds a scepter.

Figure 6 *Jesus between San Vitale and Bishop Ecclesius*, early-mid 6th century; mosaic, St. Vitale, Ravenna.

Another imperialist view of Jesus, who now, in purple, sits on the orb of the world.

Figure 7 *Emperor Justinian with his retinue*, detail, early-mid 6th century; mosaic, St. Vitale Ravenna.

Figure 8 *Empress Theodora with her retinue*, early-mid 6th century; mosaic, St. Vitale Ravenna.

The three mosaics of Figures 6, 7, and 8 are placed close to each other in the apse: Jesus in the center, to his right the emperor, to his left the empress. All three wear the same imperial purple.

In discussing Figures 7 and 8, H. W. Janson wrote that:

> Every hint of movement or change is carefully excluded—the dimensions of time and earthly space have given way to an eternal present amid golden translucency of Heaven, and the solemn, frontal images seem to present a celestial rather than a

> secular court. This union of political and spiritual authority accurately reflects the 'divine kingship' of the Byzantine Emperor. *We are, in fact, invited to see Justinian and Theodora as analogous to Christ and the Virgin*: on the hem of Theodora's mantle is conspicuous embroidery showing the three Magi carrying their gifts to Mary and the newborn King; and Justinian is flanked by twelve companions—the Imperial equivalent of the twelve apostles (six are soldiers crowded behind a shield with the monogram of Christ). (1986, p. 218; my italics)

Figure 9 *Jesus Pantocrator*, ca. 1100; mosaic in the dome of the Monastery Church, Daphni, near Athens.

The caption for this figure, in David Talbot Rice's *Art of the Byzantine Era*, reads: "Mosaic in the dome, Daphni. This strange, awesome figure belongs to the oriental conception of Christ as the stern judge—the Jehovah of the Old Testament; it is far removed from Western conception of Him as suffering mortal" (1963, p. 91).

The **Pantocrator** (in Greek: Ruler of the Universe), was a representation of Christ as the almighty, combining in one person the Creator and the Savior. In its definitive form "the Pantocrator is exclusive to Byzantine art. Here Christ, bearded, mature, with an exceedingly stern expression, is shown in rigid frontality, usually in bust, raising his right hand in a gesture of benediction or admonition and holding a book of the Gospel in his left" (*Encyclopaedia Britannica*, 1978, *vii*, p. 726).

Indicative of the radical changes, are the following comments by H.A. Drake, in his recent *Constantine and the Bishops: The Politics of Intolerance*:

> Constantine's distributions [of privileges on the imperial post] to the bishops clearly marked them out as favored members of the new order, significantly enhancing their prestige. From this time bishops begin to adopt the ceremonies and even raiment of the late imperial court. Even the image of Christ was not immune, as Christ the youthful shepherd of the third-century catacomb painting yields to Christ the Lawgiver and Christ the Judge, sitting on a jeweled throne and wearing imperial regalia, flanked by angels or apostles arrayed like so many imperial counselors,

> in churches built from the fourth century onward. (2000, pp. 317-18)

In addition, Drake said in the captions to his figures 5 and 6 [my figures 1 and 8]:

> Christian art reflects the changing fortunes of the church. A third-century fresco [Fig. 1] from the Catacomb of Priscilla in Rome shows Christ as he was commonly represented prior to Constantine's conversion, in the guise of a youthful shepherd. A mosaic from the fifth-century Church of San Apollinare Nuovo [Fig. 8] in Ravenna shows him endowed with the dignities of a Roman emperor, seated majestically amid angelic counselors. (2000, p. 318)

2. Unity of Power

In H. W. Janson's *History of the Art* we find the following fascinating distinction between the Byzantine world of the unity of power and a Western world moving toward the division of power:

> The division of the Roman Empire soon led to a religious split as well. At the time of Constantine, the bishop of Rome, deriving his authority from St. Peter, was the acknowledged head, the pope, of the Christian Church. His claim to pre-eminence, however, soon came to be disputed by the patriarch of Constantinople, differences in doctrine began to develop, and eventually the division of Christendom into a Western, or Catholic, and an Eastern, or Orthodox, Church, became all but final. The differences between them went very deep; *Roman Catholicism maintained its independence from Imperial or any other state authority and became an international institution reflecting its character as the Universal Church, while the Orthodox was based on the union of spiritual and secular authority in the person of the Emperor, who appointed the patriarch. It thus remained dependent on the power of the State, exacting a double allegiance from the faithful and sharing the vicissitudes of political power. We will recognize this pattern as the Christian adaptation of a very ancient*

> *heritage, the divine kingship of Egypt and the Near East; if the Byzantine emperors, unlike their pagan predecessors, could no longer claim the status of gods, they retained an equally unique and exalted role by placing themselves at the head of the Church as well as of the State.* (1986, p. 198; my italics)

The movement, away from the Christianity of the Gospels and the letters of St. Paul, began with the conversion of Constantine (312-313) who would not "have been able to see the point of Jesus's Crucifixion, at least without some mental acrobatics. It was embarrassing, to say the least, that God's own son had been subjected to this humiliating end, and contemporary art avoids the whole question, because it seemed to involve such ignominy; and for this reason Constantine abolished the crucifixion as punishment.... He hardly ever mentioned the theology which proclaimed and explained the Crucifixion, *and he saw the Cross not so much as an emblem of suffering as a magic totem confirming his own victoriousness.* This sort of philosophy is excellently illustrated by sarcophagi of the time, which combine the Cross with the *labarum*[73] and laurel-wreath—with which it was also combined on imperial standards—so that it becomes a Cross not of humiliation but of *triumph*" (Grant, 2000, p. 149; my italics).

Very soon, there was again, and stronger than before, the unity power of state and church:

> [Emperor Constantine] completely controlled the bishops himself. His dominance over them, embodied in the statement 'my will must be considered binding'—which some earlier Christians would never have accepted—was a prime example of that monarchical control of the church described as Caesaropapism. Athanasius declared that the Fathers never sought imperial sanction, but in Constantine's time they had to, and they did. And it was he himself who chose every bishop when a vacancy arose.... For the God he believed in was a God of power, who had given him victory, and he would have had little sympathy with the idea

[73] Sacred military standard of the Christian Roman emperors, first used by Constantine. It consisted of a long gilded pike, from whose crossbar hung a jewelled square of purple cloth; atop the spear, and essential, a large golden wreath enclosed the sacred monogram of Christ: the Chi-Rho.

> that Christianity meant love, or charity, or humility,... Furthermore, he was utterly confident that he himself was the man of God, God's servant and representative who was constantly in touch with him and was told by him what to do. (ib., pp. 159, 222)

> In 341 pagan sacrifices were forbidden; in 353 the cults of idols were declared illegal and the temples were closed; and in the reign of Theodosius the Great (379-395) Christianity was declared the official religion of the state and the faith required of Byzantine citizens. This adoption of a compulsory state religion was in part the result of a growing need for a strong unifying force in an empire that embraced a diversity of people and was besieged by dissensions from within and by barbarian invasion from without. In pagan Rome such unifying force had been to some extent provided by the practice of emperor worship. Prostration in the presence of an emperor and the burning of incense before his statue were acts of political allegiance to the divine head of state. Could not Christianity now be used to foster unity in the same way? The church readily agreed, rejoicing in its new status and tempted by the vision of a Christian society in which men were brought to salvation by law as well as by grace. In submitting to this temptation, the ministers of the Church were of course forgetting that such a close identification of Church and state was a reversion to pre-Christian practice. (Sherrard, 1966, p. 94)

3. Prior Incorporation of Imperial Scripts Within Christianity

In truth, the changes from Phase 1 to Phase 2 had already begun long before the conversion of Constantine, when Christianity began to be adopted by a large number of highly educated Romans who could not avoid integrating the new religion with many of their standard Roman scripts.

Already Tertullian (b. 155/160, d. after 220; an important early Christian theologian) was highly critical of Marcion whose God was the Pauline God of love who would not use fear to compel obedience: "This reliance on love alone as the mechanism underpinning ethics was the main burden of Tertullian's complaint against Marcion and his sympathizers. For them, he sneered, 'God is purely and simply good. He indeed forbids all sin, but only in word ... for your fear he does not want ... they have no fear of their God at all.... Do you say that he whom you call *Lord* ought not to be feared, whilst the very title you give him indicate a power which *must* be feared?'" (Johnson, 1976, p. 47). The roots of Christ Pantocrator are already here: Christ is Lord, he must be feared, as one fears the Emperor.

Two hundred years later, an *insular* and pessimistic St. Augustine (354-430), from the central area of the Roman empire, began a bitter fight against the *proto-visitor* Pelagius (c. 354-after 418) from Britain probably: "Augustine saw in Pelagius a form of arrogance, a rebellion against an inscrutable Deity by an undue stress on man's power.... What Augustine wanted [from Pelagius and his followers] was what he had already obtained in the case of the Donatists, absolute condemnation followed by total submission—monitored by State enforcement. He did not want discussion" (ib., pp. 119, 120). In this, Augustine already behaved like a Byzantine emperor demanding first an absolute condemnation, then total submission enforced by the State.

In other words, the fault was not only in Constantine and his successors, but in the culture of the times. The message of the Gospel was not sufficient to eradicate the old scripts. There would always be Roman thinkers like Tertullian and Augustine who would transform the early Christian message into the commands of a Roman Pantocrator.

What was needed, for a radical change, and this is the main point of the DP theory, was a rare combination of factors: Germanic invasions and derived new culture/scripts, isolated position of the popes in Rome, war against simony, the minority of Henry IV, the personality of Gregory VII, etc., which started the *War of Supremacy* (1075-1313) between popes and emperors, as discussed in chapters 11 and 23 of volume 1.

4. Phase 2 in Full

"In the thirteenth century, the emperor finally acquired the specific iconographic attributes of an angel, namely the wings.... Angels, of course, were provided with wings to represent their powers of heavenly ascent and descent" (Maguire, 1997, pp. 252-3). A century later, in a further escalation of imperial power, Jesus was painted in full imperial apparel, as in the frescoes of Hagios Athanasios at Kastoria, where "we see Christ wearing a *loros*[74] and a domed crown with *prependoulia*[75], attended by the Virgin, also with an imperial crown, and by saints dressed as contemporary courtiers" (ib., p. 258).

More degrading for Christianity was the fact that, "in Byzantine texts, also, angels and archangels do not appear on earth dressed as emperors but as court eunuchs. For example, a text of the Miracles of St. Michael contained in an eleventh-century manuscript speaks of the archangel making a nocturnal appearance in a church while clad in the garments of *praipositos*, that is the chief of the eunuchs. In the legendary *Narratio de S. Sophia*, the angel who guards the church of St. Sophia is described as 'a eunuch dressed in bright clothing, beautiful in appearance, as if in truth sent from the palace'" (ib., pp. 255-56). The submission to the emperor was now total.

5. Full Obedience to the Religious Superior

Not only was the iconography a reflection of the imperial absolutism, but also the behavior of the leaders of the religious communities who demanded complete obedience without doubt or question.[76]

For instance, the Benedictine abbots, everywhere, trace their authority to the rule written in 535-540 by St. Benedict. In this rule, the abbot, elected for life by his monks, has supreme power, and under all normal circumstances is accountable to no one. In the words of Dom David Knowles, Monk of Downside Abbey:

> The keystone of Saint Benedict's monastery is the Abbot, and the descriptions of his office and duties in the Rule have just been recognized as masterpieces of spiritual wisdom. *The Abbot stands in the place of Christ; the monks are his, that he may lead them to God*; he is their shepherd, as Christ is of all men; it is from him that God, the Master of the estate, will demand an account of the souls he has under his hand; every fault of theirs will at the last judgement be charged to him, and upon him will lie the burden of the proof before he can be absolved

[74] The imperial insignium par excellence: an all covering costume, full of precious stones and pearls.

[75] Strings of pearls, hanging down on both cheeks.

[76] The obedience to a religious superior will be discussed three times in this chapter: from total obedience, here in Phase 2 (under strong UP), to moderate obedience in Phase 3 (of high DP), up to minimal obedience in Phase N (of even higher DP).

from having caused their ruin. *On the other hand, the obedience of the monks must be absolute;they have renounced their own wills;that is why they have come into the monastery;they can rely on their Abbot as upon their divine Lord.* (1930, p. 39, my italics)

Even today, in the Greek monasteries of Mount Athos, the situation has not changed much, and the old Byzantine unity of power continues to rule undiminished. There, "all the monasteries are cenobitic, meaning that each monk is part of a community of men who live, eat and work together, and accept the absolute authority of an abbot" (Basbanes, 1999, p. 95).

IV. Phase 3 – Byzantine II (bleeding)[77, 78]

1. The UP/DP Evolution of the Representation of Jesus on the Cross

The crucifixion was not portrayed in Christian art during the first three centuries. Hence, his suffering and death are seldom represented, or merely hinted at. Indeed, crucifixion remained a form of savage execution up to the time of Constantine. Till then, and for two more centuries, Christian artists were unwilling to depict Jesus in such a degrading condition. From the 5th century on, Christ is represented on the cross, attached to it by four nails, his two arms outstretched with regal grandeur, his eyes open, and with no trace of suffering, as though to proclaim his victory over death. He is *Christus Victor* (Victorious Christ), or *Christus Triumphans* (Triumphant Christ). This representation parallels those of *Christus Imperator* and *Christ Pantocrator*, discussed previously, and was the favored one when the emperor in Constantinople was seen as the alter-Christ on earth, when Christ was supposed to bestow all his blessings on the imperial cause. The crucifixion of Jesus could not be negated, but it had to be represented in a way that commanded respect, not pity, as proper for an alter-Emperor.

Then, in the 9th century, the Byzantine Church introduced more realistic representations of the crucified Christ: first that of *Christus*

---

[77] The bleeding refers to the damages and sufferings of the Iconoclastic Age, 726-843, and to the related losses of territories to Islam

[78] For simplicity, I do not follow the standard division of Byzantine Art in: First Phase (330-726); Iconoclastic Age (726-843); Second Golden Age (843-1204); Final Phase (1204-1453).

Dolor (Sad Christ), with his head tilted to the side, his arms drooped, and his eyes in the process of closing; then that of *Christus Patiens* (Patient or Suffering Christ) with his head sagging down on his chest, his arms hung heavily with pain, and his legs twisted in agony by the use of only one nail instead to the previous two. The final stage of this development was that of *Christus Mortuus* (Dead Christ), in which all the aspects of physical death and tremendous suffering were represented.[79] (Often these three types are subsumed under the title of *Christus Patiens).*

"The shift from depicting a living to a dead Christ in Byzantine Crucifixions dates from the second half of the 8th century. This development was connected with the Monophysite heresy[80] of the 7th century, when Christ's human nature was denied. To counter this, Christ's mortal nature was emphasized by focusing on his physical death. Opponents of Iconoclasm in the 8th-century Byzantium defended images of Christ on the cross as representing Christ Incarnate" (Oakes, 1996, p. 210).

This same point was highlighted by Osborne: "In the 9th century, the Byzantine Church introduced a more realistic type of the crucified Christ, wearing only the *perizonium*, with eyes closed in death and blood flowing from His side. This version, stressing the human vulnerability of Christ, emphasized the reality of His incarnation and it seems to have been evolved by the opponents of the Iconoclasts, since the doctrine of the incarnation played an important part in the justification of religious images" (1970, p. 291).

The *Iconoclastic Controversy* of the 8th and 9th centuries was a major division within the Byzantine empire (and therefore, a major form of DP) between those who objected to icon worship on one side and those who insisted on the symbolic nature of the images and on the dignity of created matters on the other.

In addition (and greatly expanding the DP) the origin of the Eastern Iconoclasm is to be found in a series of major territorial losses by the Byzantines at the hands of Islamic warriors.

> Since the early seventh century, the cult of the icons had continued to penetrate all levels of Byzantine society, compelling a devotion in emperors and patriarchs, bishops and craftsmen,

---

79 This representation of *Christus Mortuus* was later pushed to its extreme by the German painter Matthias Grünewald (c. 1470-1528), in his various Crucifixions of Basel, Isenheim, Haarlem, and Karlsruhe.

80 Heresies are always major DP phenomena, at least at the level of the scripts.

> workmen and soldiers, monks and peasants. The growth of icon veneration was stimulated by their 'official' use as military ensign and protective devises, during the 626 siege for example.... But during the 720s, as the enemy repeatedly attacked without hindrance, captured fortified settlements, and carried off prisoners, Byzantine morale sank.... If God continued to favour the Arabs, the Byzantines were surely being punished for some dreadful sins. While they may not have been able to identify these offenses, they may have begun to doubt their own faith in the divine protection previously attributed to holy objects. In particular, if they had sought refuge in a walled town where the invocation of spiritual aid through icons had *not* guaranteed safety, they must have cursed their icons... As soon as the icons failed to guarantee a successful defence, they were despised rather than revered and were held responsible for misleading people. In this respect, excessive icon worship may have been identified by iconoclasts as something likely to deceive the simple minded. (Herrin, 1987, pp. 307, 330, 338)

If Christ was to be King, if he was *Victor*, it had to be as originally presented in the Gospels (and not later in Byzantium) when Jesus told Pilate: "My kingdom is not of this world" (John, 18:36); or when the *penitent thief*, crucified with him on Golgotha, said to Jesus "Lord, remember me when Thou comest into Thy kingdom. And Jesus said to him, Verily I say unto thee, Today shalt thou be with me in paradise" (Luke, 23:43). Never had the Jesus of the Gospels promised safety from the enemies: definitely not for those invoking protection from icons. He himself had suffered death on the cross. So, as soon as Christians once more faced persecution, as soon as they and their emperors faced tragic losses, the crucified Jesus began looking like a crucified person: *Dolor, Patiens, Mortuus*, not *Victor*, not *Triumphans*.

In the West, the representation of Jesus on the cross also changed from *Victor* to *Patients* but for a different reason: the evolution from UP to DP reason. "By the 11th century, the dead Christ on the cross was universal in Byzantine art, but it was not introduced in the West before the 13th century, except for rare examples painted under Byzantine influence.... In the 13th century, a still more realistic conception of the crucified Christ was created in Italy, under the influence of the preaching of St. Francis of Assisi, which showed Christ no longer indifferent to death but suffering and 'degraded' by His ordeal as in Cimabue's *Crucifixion* (1260) in the Upper Church of Assisi"[81] (Osborne, 1970, pp. 291-92).

Indeed, St. Francis of Assisi is one of the greatest exponents of the DP that followed the imperial defeat of 1176.[82] In turn, the suffering and degradation of Jesus of Cimabue's *Crucifixion* are the visual equivalent of the emotions and vivid piety of *The Lament of Our Lady*, to be discussed hereafter. Once again, Christianity was crying, as the *penitent thief* "Lord, remember me when Thou comest into Thy kingdom," and not "God save the Emperor."

2. Another Revolutionary Change

The DP brought by foreign invasions fostered other astonishing changes in the Byzantine iconography: again away from the almighty imperialism.

> [A major modification of Byzantine art grew] in the provinces and Patriarchates under the rule of Moslems, and subsequently of crusaders. For the common people, the Byzantine Christ Pantocrator had become altogether too fearful and remote; and his mother, the popular medium of intercession, too much a queen enthroned. The Cappadocian[83] school of illustrators, recruited from the monks and so, largely, from the people themselves, also responded to this other demand for increased humanization, particularly by the enlargement of the cycles of the Passion and Crucifixion, with all their dolorous incidents from

---

[81] Sadly, no good picture is possible: "The natural damage brought by age has now been compounded by the far more disastrous chemical degeneration of the white lead paint, which in some parts of the cycle has caused an inversion in the relationship between light and dark colours, altering irreversibly the chromatic values of the painting" (Chiellini, 1998, p. 17).

[82] In less than 40 years there was a very substantial expansion of the division of power: 1176, defeat of Emperor Frederick Barbarossa by the Lombard League at Legnano; 1177, humiliation of the Emperor before Pope Alexander III at Venice, and Peace of Venice between the Empire and the Church; 1183, Peace of Constance between the Emperor and the Lombard League (the Magna Carta of the Italian Communes); 1198, surrender of Conrad of Urslingen to Innocent III; 1209, Francis of Assisi obtained the approval of the "Franciscan Rule" from Innocent III; 1208 Innocent III laid the interdict on England and excommunicated King John; 1212 John accepted Langton as Archbishop of Canterbury on the pope's term; and 1215, Magna Carta.

[83] Cappadocia: an ancient district in east central Anatolia, now part of Turkey.

> the capture in the garden to the mother grieving over the body of her son. (Byron & Rice, 1930/1968, pp. 17-18)[84]

We are back to a strong division of power (actual power division between the Orthodox religious leaders and the military power of the Muslim conquerors; and scripts division of power: Christian and Islamic). Here, there is no place for a Byzantine emperor who has lost his power completely, and therefore there is no place for a representation of Jesus as Emperor or Pantocrator. Emperor and Pantocrator have become meaningless words. The consequence is a different type of religiosity, in which the Christians suffer as their Leader did on the Cross; they suffer as Maria suffered "grieving over the body of her son."[85] However, this situation did not last: Cappadocia gradually became Muslim, and the Cappadocians became Turks.

IV. Phase 4 – DP/Communes, and Jesus as Elder Brother and Teacher

1. Iconography

Figure 10 Jesus and the Samaritan Woman; detail, tempera. Duccio. Museum Thyssen-Bornemiza Collection, Madrid.

With Duccio, we are back to the friendly generous themes of Phase 1, even if Jesus is stern, and still wears noble-royal attire with a gold border or band added to his robe.

---

[84] Steven Runciman had similar comments even if geared more towards explaining a change of technique: "In the old days, when the Empire was the undoubted microcosm of Heaven on earth, the glowing, mysterious splendour of the mosaic was suitable for the Sacred Palace and the shrines of the capital. But, now, with growing disillusion and fear, men wanted something more human and intimate. For some time past scenes of Christ's life and of his mother on earth had been popular, to illustrate the Liturgy and the Gospels. For such human scenes the fresco-painter [as had been done in Cappadocia] had advantages over the mosaic-designer" (1975, p. 149).

[85] Gradually, these motives "spread, not only throughout the Greek lands, but to Italy, carried thither by Byzantine artists of Sicily and Calabria, by the traffic in Byzantine icons, and by returning pilgrims from the Holy Land (Byron and Rice, 1930/1968, p. 18). Yet, this adoption of the Cappadocian ideas would probably not have taken place, or on a smaller scale, if the situation in Italy had not been so severe, as discussed later in relation to the poems of Jacopone da Todi.

Figure 11 *Jesus Pantocrator*; fresco. Giotto, 1266-1327. Scrovegni Chapel, Padua.

A gentle Pantocrator!

Figure 12 *The kiss of Judas*; detail, fresco. Giotto, 'Scrovegni Chapel, Padua.

Jesus is now presented in his full humanity: betrayed by one of his twelve apostles. Yet, Jesus does not show anger, or hate, but, perdominantly, serious curiosity.

Figure 13 *St. Francis of Assisi preaching to the brids*; detail, fresco. Giotto, Upper Church, Assisi.

With these paintings we are firmly on *visitor ground*, as highlighted by Ernst Kitzinger in referring to "Giotto's new image of man dignified, autonomous, creating, and completely dominating of his spacial environment" (1956/1976, pp. 376-77).

Figure 14 *Jesus paying the tribute money, by instructing St. Peter*; fresco. Masaccio, Florentine, 1401-1428. Brancacci Chapel, S. Maria del Carmine, Florence.

This fresco illustrates the story in the Gospel of Matthew (17:24-27) in which Jesus consents to pay the tax collectors, but with miraculous money. Speaking to Peter: "Lest we should offend them, go thou to the sea, and cast an hook, and take up the fish that first cometh up; and when thou hast opened his mouth, thou shalt find a piece of money: that take, and give unto them for me and thee." This time, Jesus' robe is no longer bordered in gold. He, Jesus, obeys the lay authorities. He is no Pantocrator, above everything and everybody.

Indicative of the new DP times are: the friendly Pantocrator of Giotto, not awe inspiring, neither judging nor condemning; the betrayal by Judas, showing the human limitations of Jesus; the gentle communal Francis, son of a cloth merchant, talking to the birds, and not hunting them for the repast of his master; and the communal Jesus willing to pay taxes to the "commune." How different, one may say revolutionary, is this art in comparison with the Byzantine art!

2. The New Man: St. Francis of Assisi (1181-1226)

In the words of Chesterton:

> St. Francis anticipated all that is most liberal and sympathetic in the modern mood; the love of nature, the love of animals; the sense of social compassion; the sense of the spiritual dangers of prosperity and even of property. (1923, p. 9)

> There is not a trace in the poetry of this first Italian poet of all that pagan mythology which lingered long after paganism. *The first Italian poet seems the only man in the world who has never heard of Virgil.*[86] This was exactly right in the special sense on which he is the first Italian poet. It is quite right that he should call a nightingale a nightingale, and not have its song spoilt or saddened by the terrible tales of Itylus or Procne. *In short, it is really quite right and quite desirable that St. Francis should never have heard of Virgil.* (ib., pp. 182-83; my italics)

The poem in question is *The Canticle of the Sun*, of which I report the first and last parts from an old anonymous translation:

> O most high, almighty, *good Lord God*, to Thee belong praise, glory, honour and blessing.
>
> Praised be my Lord God with all his creatures, and especially our *brother the sun*, who brings us the day and who brings us the light, fair is he and shines with a great splendour. O Lord, he signifies to us Thee.

---

[86] The reader may wish to relate this Chesterton quote, to what I wrote in chapter 2 of volume 2 "But Weren't also the Classics at Fault? On the *Griselda* by Boccaccio, Petrarch, and Chaucer." By way of contrast, Peter Gay correctly stressed how the ancients "had been tough-minded, worldly, philosophical" (1967, p. 75), and E. A. Judge had remarked that for the ancients "pity was a defect of character unworthy of the wise and excusable only in those who have not yet grown up. It was an impulsive response based on ignorance. Plato had removed the problem of beggars from his ideal state by dumping them over its borders" (quoted by Stark, 1997, p. 212). Francis and his first companions were mendicants, i.e. beggars whom the tough-minded Greeks and Romans would have dumped over any border.

> Praised be my Lord for our *sister the moon*, and for *the stars*, which he has set clear and lovely in the sky.
>
> Praised be my Lord for our *sister water*, who is very serviceable unto us and humble and precious and clean.
>
> Praised be my Lord for our *brother fire*, through whom Thou givest us light in the darkness; and he is bright and pleasant and very mighty and strong.
>
> Praised be my Lord for our *mother the earth*, which does sustain us and keep us, and bringeth forth divers fruit, and flowers of many colours, and grass.
>
> ....
>
> Praised be my Lord for our *sister death of the body*.
>
> Blessed are they who are found walking by Thy most holy will.
>
> Praise ye and bless *ye the Lord*, and give thanks unto Him, and serve Him with great humility. AMEN

Such a poem would have been inconceivable in both the previous Phase 1 (when people soon expected the end of the world) and Phase 2 when the strong unity of power of the Roman or Byzantine emperors left little else to be praised. Bertrand Russell, in his *A History of Western Philosophy*, also had the highest praise for St. Francis:

> In the matter of saintliness, Francis has had equals; what makes him unique among saints is his spontaneous happiness, his universal love, and his gifts as a poet. His goodness appears always devoid of effort, as though it had no dross to overcome. He loved all living things, not only as a Christian or a benevolent man, but as a poet. His hymn to the sun, written shortly before his death, might *almost* have been written by Akhnaton the sun-worshipper, but not quite—Christianity informs it, though not very obviously. (1945, p. 450)

The profound revolution consisted of the fact that Francis of Assisi was not a mighty pharaoh, like Akhnaton. He was not even an aristocrat, but a common man, the son of a cloth merchant. In turn, the link between St. Francis and Giotto's rendition of his *Sermon to the Birds* was highlighted by Holland Smith, at the very beginning of his biography of St. Francis:

> "My brothers, praise God who created you, and dressed you with feathers, and gave you wings to fly with, and endowed you with the pureness of the air, and gives you the food you need."
>
> St. Francis's preaching to the birds has been one of the favourite subjects for the Christian artist since Giotto painted the first cycle of the scenes from the saint's legend in the Upper Church of Assisi. The story is capable of touching even those rationally convinced that if Francis ever did preach to the birds, the birds did not sit and listen, for all that the legends assure us that they did. (1972, p. 1)

In terms of division of power, it is fascinating to read how in his introductory pages Holland Smith described the demise of the Germanic imperial control over Italy:

> [In 1198], when Francis of Assisi was between seventeen and eighteen years of age, Conrad of Urslingen, Lord of the Duchy of Spoleto—of which Assisi was part—had to relinquish power, and at Narni, he had to kneel in submission before the legates of the Lord Pope Innocent III and surrender into their hands the fief he had once received as reward for valour from his sworn liege-lord, the late Christian of Buch, Archbishop of Mainz and commanding general in Italy to the Emperor Frederick Barbarossa; the Duke of Spoleto would kneel to the representative of the Servant of the Servants of God. These were humiliating days for every German in Italy[87].... *It is impossible to catch the whole spirit of an age, or even a year, in a single, simple*

[87] The same would soon also apply to King John of England: "In 1213 [two years before Runnymede], Philip Augustus accepted the Pope's offer when he deposed John Lackland and invited the King of France to take possession of the vacant realm. John ran a great risk of being dethroned but was advised that Innocent III despite his obligations to Philip Augustus was ready to pardon him *on condition of a complete humiliation. He accordingly humiliated himself* and, on 16th May, 1218, he agreed to pay to the Holy See an annual tribute of a thousand marks sterling as the dues of a vassal for the kingdoms of England and Ireland" (Petit-Dutaillis, 1964, p. 329; my italics).

I focus on these humiliations of mighty kings and emperors, at the hand of a basically unarmed chief of a religion, to highlight a kind of division of power which the

> *image. But the picture of the knightly Duke Conrad riding to surrender his land and people to the pope's commissioners in 1198 reflects much of the truth about the world in which John Bernardone, Saint Francis of Assisi, came to maturity.* (1972, p. 2, my italics)

In parallel, as described in chapter 11 of volume 1: "Thanks to the continuous fights between popes and emperors from 1075 to 1330, the cities could gradually free themselves from the domination of their masters, the German emperors. In the absence of kings and emperors these cities could develop their 'modern' commercial and political institutions: 'Europe's towns were marked by an unparalleled freedom. They had developed as autonomous worlds and according to their own propensities.'"

This was totally new, as stressed by Judith Herrin: "The later separation of church and state would come to constitute a more fundamental break from ancient traditions *in that it opened the way to the parcellized sovereignty of feudalism so antithetical to Roman imperialism.... The interaction of vicars of St. Peter with claimants to the imperial title was to become one of the determinants of western European development thereafter*" (Herrin, 1987, pp. 305, 446, my italics).

In addition there was a division of power within the Church itself that St. Francis had originated, and which further explains the change in poetry and iconography:

> In its origins, his movement was a reaction against the oversophistication of monastic capitalism,[88] the quasi-aristocratic life of the high clergy and, at the same time, the Christ-less-ness of ordinary Christians of the middle classes now becoming powerful. Having tasted poverty, it is doubtful whether he had any romantic illusions about the godliness of the poor, but his deliberate courting of poverty was a romantic rather than a pragmatic

---

rest of the world had neither known nor dreamt of before. This, in turn, is reflected in the evolution of the visual arts: the religious (the subject of this chapter) and the lay of which the *Good Government*, of 1330-1340 in the Palazzo Publico of Siena, by Ambrogio Lorenzetti is the best example.

[88] "The first missionaries whom Saint Francis sent to distant places were taken for heretics, because they practiced poverty instead of (like the monks) only taking a vow which no one regarded seriously" (Bertrand Russell, 1945, p. 449).

> gesture. He was an extremist and, in his own way, a rebel. He carried his personal protest against the corruption of riches to extreme limits. (Holland Smith, 1972, p. 65)

The power acquired through the acceptance of poverty is enormous: there is little one can take away from such persons, there is little that one can give them.

3. Less Obedience to the Religious Superior

The DP age of Phase 3 also saw a major reduction in the obedience owed to the *fathers superior* of the new religious orders (Franciscans, founded in 1208; Dominicans founded in 1215): "For the medieval Dominicans, obedience did not involve a surrender of the individual friar's freedom of judgement [as was the case before with the Benedictines]. Furthermore, unlike the other religious communities, Dominicans were not bound in conscience to obey Dominican law" (Wittberg, 1994, p. 126).

St. Francis, too, was revolutionary in this respect: In one of the brief *Admonitions* in his *Opuscola* he said: "'If a superior should order something contrary to the spiritual good, though the friars ought not to obey him, let them at no time separate themselves from him.' Brothers unhappy with the orders they received were to find a way of disobeying without rebelling" (Holland Smith, 1972, p. 149).

4. However not all was Rosy in Phase 4

It would be wrong to only highlight the serene and friendly developments so far discussed, indeed:

> It is a mistake to believe everyone had his or her place in society. On the contrary: between the orderly guilds of the towns and the feudal hierarchy of the countryside there was an immense chaos teeming with the displaced, the dispossessed, the chronic sick or crippled, beggars, lepers, runaway serfs. Perhaps a third of the population did not fit into official categories, but formed the raw materials of huge crowds which formed and dispersed mysteriously and rapidly.... Religious hysteria expressed itself in almost every imaginable form of outrageous behaviour. Self-flagellation, for instance,... (Johnson, 1976, pp. 256, 259)

On a separate level, under Pope Innocent IV (r. 1254-1261),

> the papacy took a downward turn from the high peak attained under Innocent III and Gregory IX. The entire energies of the Church had been devoted to political objectives. It had slipped from the position of impartial judge of disputes in Christendom as a whole to that of one of two contestants fighting furiously for power in the Italian peninsula. Absorbed in diplomatic, juridical. and financial activities, the Curia neglected its religious functions. Even the friars, bursting with spiritual zeal, had been diverted to mundane ends, and the huge material resources of the Church had been squandered for political and military purposes. The moral decline of the papacy and its instability as an institution are illustrated by the short pontificates of the second half of the thirteenth century, during which no fewer than thirteen Popes came and went. Without regaining moral stature they lost their grip on political affairs and became virtually appendages of French royal hegemony. (Cheetham, 1992, p. 139)

The impact of these tragedies is reflected in several DP poems of the epoch, in particular in the *Pianto della Madonna* (The Lament of the Madonna), of ca. 1280, by the Franciscan friar Jacopone da Todi[89] of which I report a few sections, followed by my translation:

Nunzio: Donna del Paradiso
lo tuo figlio è priso, Jesu Cristo beato.
Accurre, donna, e vide che la gente l'allide;
credo che lo s'occide, tanto l'on flagellato
Mess.: Lady of Paradise

[89] A near contemporary of Dante, he practiced law until his sudden conversion following the death of his wife, about 1268. He became a Franciscan tertiary and, ten years later, a lay brother. The importance he attached to the ideal of poverty made him sympathize with the Franciscan Spirituals, and brought him into conflict with Pope Boniface VIII [the same who caused Dante's exile from Florence]. In 1297, he signed the manifesto which declared Boniface's election invalid. Boniface retaliated by first excommunicating and then imprisoning him from 1298 to 1303 when the new Pope Benedict XI released him. He died in 1306. (*Encyclopaedia Britannica*, 1973, *12*, p. 839).

your son has been taken, blessed Jesus Christ.
Hurry, woman[90], and see how they are beating him;
they are killing him, so hard have they have flogged him

Maria: Como esser porria , che non fe' mai follia

Mary: How can this be, he never did wrong.....

Nunzio: Succurri, donna, aiuta, ch'al tuo figlio se sputa,
e la gente lo muta; onlo dato a Pilato.

Mess.: Hurry, woman, help, they are spitting on your son,
they are changing him [for Barabbas], and have given him to Pilate.

Maria: O Pilato, lo figlio mio non tormentare,
ch'io te posso mostrare como a torto è accusato.

Mary: Oh Pilate, do not torment my son,
I can prove to you how wrongly he is accused. .....

Popolo: De spine se coroni, che rege s'è chiamato.

People: Crown him with thorns, he who proclaimed himself king.

Maria: O figlio, figlio, figlio, figlio, amoroso giglio,
figlio chi dà consiglio al cor mio angustiato?

Mary: Oh son, my son, my son, my son, loving lily,
son who can counsel my anguished heart?

(Vicinelli, 1943, pp. 160-61)

Here, Mary is a suffering woman and mother, no longer the Byzantine Empress Mary, mother of the Pantocrator. And Jesus, too, is the original "man of sorrow". He does not wear a purple mantle covering a tunic of gold, nor does he hold a cross of gold in his hand, as shown in Figure 3. Because of this, I labelled this a DP poem. The reader is expected to have compassion for Jesus, not fear, definitely not think of him as Pantocrator. This, in turn, explains a new emphasis, stronger than before, on vivid representations of a suffering Jesus on the cross.

---

[90] Quite early in Italian there developed the distinction between *Madonna* (My Lady) and *donna* (woman), even if *donna* stemming from the Latin *domina* (the feminine of *dominus*, lord)–precedes historically *Madonna* which is the result of attaching the French *ma* (my) to *donna* (see Pestelli's 1961 *Dizionario delle parole antiche*, pp. 17-20).

VI. Later Phases—Domination of the Artistic Styles

Starting with the Renaissance, until shortly before our times, religious iconography is dominated by artistic styles (e.g., Renaissance, Mannerism, Baroque) which evolved—more than in the past—for purely artistic reasons guided by technical or culturally distant interests such as perspective, light, movement, colors, the recovery of the classical proportions and a brotherhood with great Greek sculptors, and the discovery of Platonic geometry in all things. With these priorities, artists took such liberties with religious subjects, that it is difficult to distinguish between DP iconography and UP iconography, and between religious art and art based on religious subjects. While this is nothing new, it became much stronger from the Renaissance onward.

For instance, it is difficult to pray in front of Michelangelo's famous *Pietà* of 1499-1500 (located in the first chapel on the right of St. Peter's Basilica), as soon as one discovers that Michelangelo had sculpted "ostentatiously, upon the ribbon that holds the Madonna's gown in position: *Michaelangus Bonarotus Florent. Faciebat*. Not in a discreet place, but across the Holy Virgin's breast. He was proud of his Pietà" (Brion, 1940, p. 90). Also his Maria—a beautiful young girl, who could not, by her age, have been the mother of a man of 33—is not made to evoke religious feelings. In the words of Howard Hibbard: "The lovely Florentine face of the Virgin is without legible expression yet is full of finely-incised detail. Although the group was severely criticized in later years for lack of religious feeling, the Pietà became one the images closest to Michelangelo's own spiritual aspirations" (1974, p. 46). Of Michelangelo's personal spiritual aspirations, no doubt; yet those who criticised the Pietà for lack of religiosity— as understood by the community of the faithfuls—were probably right.

Similarly, while most Catholics will be able to pray in front of a Byzantine icon of the Madonna, or even of a mosaic showing Jesus Pantocrator, they may have problems praying in front of Michelangelo's athletic *Doni Madonna* of 1503-04, in the Uffizi in Florence: "A Mary of heroic proportions" says Hibbard (ib., p. 67). In addition, not very understandable to the faithful are the five stark nude young male figures in the background: "What incomprehensible caprice caused him to fill the background with young men who have no connection with the Holy Family.... Michelangelo did not make angels of these incongruous spectators. They have no wings; they are naked youths of the sort found in Greek bas-reliefs... These intruders seem quite unconcerned with the Holy Family. They are looking toward the center of the picture, behind

Saint Joseph's back, perhaps at a runner who just came up, or at a group engaged in some philosophical discussion. Nothing could be less spiritual than the composition. There is a look of physical effort about the Holy Family, and the young athletes contribute a sense of open air, of the stadium" (Brion, 1940, p. 115).

Very strange, also, from the general religious point of view, is another large sculpture of Michelangelo, his *Risen Christ* of 1518-21 for the Dominican church of Santa Maria sopra Minerva, in which Michelangelo's passion for the male nude went beyond the limits: It is "a free-standing figure, with a twisting *contrapposto* pose that seems more suitable to a pagan statue. Indeed, the nature of this Christ is ambiguous and the reasons for its total nudity are obscure. He holds instruments of his Passion, including an undersized symbolic cross.... Just as in the *Dying Slave*, we see here Michelangelo's unabated love, even hunger, for the beauty of the nude figure, which from the rear could be in every sense a pagan work. It is ultimately this conflict between pagan nude and the Man of Sorrows of Christian iconography that has placed Michelangelo's *Christ* in a special limbo, separate from all his other works" (Hibbard, 1974. pp. 167-69).

Also Caravaggio's *Rest on the Flight to Egypt* of 1594-96, now in the Galleria Doria Pamphili, in Rome, shows this same total priority for the arts. Not Jesus, nor the Madonna, nor Joseph, is the well-lit central figure in the foreground, but a beautiful quasi-naked angel (a lovely young man or girl, with wings) seen from the back, who plays the violin—probably a lullaby for Baby Jesus—while Joseph holds the score.

These works are neither UP religious iconography nor DP religious iconography: *they are ART, and as such they are DP in a different sense, of a work which is at the same time religious and not religious, affirming the religious tradition and negating it by giving priority to artistic considerations.*

VII. A Later DP Exception to the Domination of the Artistic Styles

There were, however, exceptions, between the Renaissance and present times, to the domination of the artistic styles. There were truly religious iconographic innovations: more popular, less artistic, and, again, the reflection of major power changes. One of these exceptions is the representation of the *Sacred Heart of Jesus*: a form of devotion to Jesus consisting of the veneration of his physical heart. The Heart, alone, or painted on his white robe, was popularly depicted with a wound, encircled by a crown of thorns, and a small cross, the whole radiating light. The caption for the figure of the *Sacred Heart of Jesus*, included in the 1995 *HarperCollins Encyclopedia of Catholicism* says: "A type of depiction of the Sacred Heart of Jesus that was common in pre-Vatican II Catholicism. It is one of many mass-produced devotional items in an art style in which Jesus is portrayed as effeminate, if not sexless" (p. 1150). In a similar tone, the McGraw-Hill *New Catholic Encyclopedia*, laments that "Statues and painting of disputable taste, often of vulgar sentiment repulsive to educated sensibilities, proliferated from the beginning of the 19th century. Occasional efforts have been made in reaction to these depictions of the Sacred Heart, but they have not effected *a widespread and deeply meaningful image*" (1967, xii, p. 820, my italics).

"The iconography of the Sacred Heart, properly speaking, does not appear in Christian art until the end of the 17th century following the apparition [of Christ] to St. Margaret Mary Alacoque" (ib., p. 820). "These experiences, although subject to human error, were valuable for reminding contemporaries and others of truths forgotten or neglected. In 17th-century France both irreligion and Jansenism had in different ways obscured the doctrine of Christ's love for all men, sinners included" (*The Oxford Dictionary of Saints*, 1997, p. 10).

Once more, in a situation of strong religious DP (Catholicism under severe attack by irreligion and Jansenism), the Catholic iconography turned away from the Baroque triumphalist representations of Jesus.[91] Instead it created small and intimate images of him, which highlighted his redeeming pains, his redemptive love. These are images which evoked in

---

[91] Examples of this exuberant triumphant image of Jesus can be seen on the ceiling of many Baroque cupolas, starting with that of the Jesuit *Church of Gesù* in Rome with its *The Triumph of the Name of Jesus*, which was painted circa 1670 by Giovanni Battista Gaulli, in which "sculpture and painting are played against the vaulted surface of the nave in a deliberate desire to mislead the spectator as to the physical dimensions of the space in the building, and to create the illusion of a boundless vista opening toward Heaven" (Gardner, 1959, p. 405).

Very correctly, Gardner speaks of spectators, not of the faithful, because "*le baroque veut la magnificence, le surhumain, la sensation, l'étonnement* (the baroque wants magnificence, the superhuman, sensation, astonishment)" (Simpson, 1962, p. 17). However, the *surhumain* of Jesus Emperor or Pantocrator or of boundless vistas opening toward Heaven, works only as long as the Byzantine or Papal UP triumphs; when they fail, the true religious iconography turns small and intimate: Jesus is again the sufferer, the crucified, the kind brother.

the faithful a dedication to Jesus, the gift of the self in response to Jesus' love, a sense of reparation for individual and communal sins, as sharing in Jesus' redeeming sacrifice of atonement. Later, in the 18th and 19th centuries, the DP (actual power, and scripts) was even stronger, thereby explaining the above mentioned widespread diffusion of the deeply inspiring statues and paintings of the Sacred Heart. This was the time of the French Revolution when many priests and nuns had abandoned the Church, as did Talleyrand and the father of Baudelaire. These were difficult times, as difficult as those, discussed previously, when several Byzantine provinces fell under the rule of the Moslems, and subsequently of crusaders. And the response was similar.

VIII Now—Present, Under DP, and *Jesus of the People*

1. Iconography

Figure 15 shows *Jesus of the People* ("Jesus 2000") by Janet McKenzie, 1999. The painting, based on a woman as model, depicts a modern dark-skinned Jesus. This Jesus was selected by the *National Catholic Reporter* (an independent news weekly and one of the best known Catholic publications in the USA) as the winner of its contest to update the image of Christ for the new millennium, specifically for the cover of a special millennium issue published on Dec, 24, 1999. *Jesus of the People* was chosen from 1,678 entries submitted by 1,004 artists from 19 countries and 6 continents. A panel of three art experts selected 10 finalists, and Sister Wendy Beckett—well known around the world for her British Broadcasting Corporation art documentaries—picked the winner.

"'My goal was to be as inclusive as possible,' said the artist, Janet McKenzie. 'At first glance, he is a black or African-American Jesus, but looking more deeply you see many people in it.' The painting shows a robed and haloed Jesus. *Against a pale pink background are a yin-yang symbol, intended to represent perfect balance, and a feather symbolizing the American Indian spirituality that McKenzie learned about during a stay in Mexico*" (Crary, 1999, p. 4D; my italics).

"'The pink in the background," McKenzie said, "is both a feminine reference as well as being the color of blood—hinting at both suffering and redemption. Despite wearing a crown of thorns, McKenzie's Jesus does not seem anguished. 'It's a total acceptance of his fate , and that's what the painting is about: acceptance' she said" (Schaeffer & Allen, 1999, pp. 3-4).

"In the view of one of the judges, Sherry Lynn Best, who directs an art gallery at Rockhurst University: 'It's not real masculine. It's not real feminine. It's not really androgenous, either.'... [In the words of McKenzie:] 'This painting is about love. *It's about reminding all of us about the importance of celebrating our differences.*' The judges who reviewed the entries said they were struck by the variety: abstract design, computer graphics, Jesus as a homeless person, Jesus in outer space cradling the Earth, a Jesus with superhero biceps. 'I think Jesus would have liked this contest,' said one of the judges, Pattie Wigand Sporrong of Catholic Theological Union of Chicago. '*It didn't have a lot of boundaries and barriers*'" (Crary, 1999, p. 4D; my italics).

The high division of power of our times, indeed, produced a strikingly *visitor* Jesus.

2. Contemporary Religious Division of Power

Fewer Boundaries:

> The [1962] Vatican Council's 'universal call to holiness' drastically weakened the definitional boundaries that had made religious life a separate and privileged status within the Church. The impact of this delegitimation was swift and profound.... A symbol—and a consequence—of this weakening of the 'wall of separation' which had formerly divided the religious from the secular life, was the modification or discarding of a distinctive religious habit. By 1982, for example, 56 percent of U.S. sisters had made major changes such as shortening the habit's skirt to street-length, abolishing scapulars and wimples, and simplifying the veil. Another 34 percent had ceased wearing any form of habit at all. Other aspects of the boundaries between religious and secular life were also blurred. Communities adopted psychological screening processes for new applicants that were borrowed from the secular world; entrants retained their given name instead of adopting new, 'religious' ones. Some religious orders began to resemble self-help groups. (Wittberg, 1994, pp. 233, 235)[92]

[92] "Membership in the Society of Jesus in the United States peaked in 1965, at 8,393 men. By 2000 it had been cut by more than half, falling to 3,635." The Jesuits are not unique in this. The steep decline in their numbers is equaled or surpassed by those in other Catholic ministries. The number of aspirant priests in all Catholic seminaries

Less Obedience:

> A new and more internalized definition of obedience was formulated, one that explicitly repudiated the old model of subjugation to a superior as 'false and naive.'... The former belief in the divine origin of communal authority was de-emphasized.... By 1982, for example, 75 percent of women's religious congregations in the United States reported having no local superior in some or all of their houses. By 1989, only 11 percent of the sisters believed that 'a truly obedient religious need seek no source other than her rule and the will of her superiors to know what she should do. (ib., pp. 243-44)

A New Understanding, *Jesus Forsaken*

> May 14, 1944: a young Italian woman sits on a hillside above the city of Trent and watches her home destroyed by wartime bombing. Inside her, the first stirring of a vision which would found an international movement dedicated to religious unity, peace and devotion. Today, the *Focolare Movement* begun by this determined woman, Chiara Lubich, has grown to number more than 4 million supporters, and more than 150,000 committed members, among the "consecrated" people living in residential communities. (Gallagher, 1997, back cover)

Fifty-five years later, the University of Malta conferred the Degree of Doctor of Literature (honoris causa) on Chiara Lubich. On that occasion, Prof. Mark Borg, Dean of the Faculty of Psychology said:

> The goal of the Movement thus delineates itself and becomes evident to one and all: striving to realize the testament of Jesus in John 17: "May they all be one." That is, achieving unity of all people—unity in the family, in the world of work, among

---

(mainly diocesan) between 1966 and 1993 dropped by 85 percent. Without fresh recruits to the priesthood, the average age of priests has been climbing sharply. *The falloff in women's orders was even more abrupt. In 1955, there were ten times as many nuns as priests in America. Now there are fewer, and they are even older than the aging priests. Approximately fifty thousand nuns left in the decade between 1966 and 1976*" (Wills, 2002, p. 12).

generations of people, and among people of various religions, cultures and nations.[93] What binds this unity together is love: the kind of selfless love that emanates from the crucified Christ. Thus, *forsaken Jesus* became the motivating force of the group. This is the essence of Chiara Lubich's work. This is the essence of the Focolare Movement. (p. 3)

In her acceptance speech Chiara said:

> Through a circumstance which we believe was foreseen by God, we learned that Jesus' greatest moment of pain was on the cross when He experienced the abandonment of His Father. "My God, my God, why have you forsaken me?" (Mt. 27:46) We were touched by this fact. And under the impulse of youthful enthusiasm and above all by a grace of God *we first focolarine chose Him in His abandonment* as the way to follow in our lives. From that moment on, we discovered Him everywhere: in the pain in our own hearts (which we tried to love because it was an expression of Him), and in that of our neighbors, especially those who were suffering. *He had felt the Father far away*, experiencing the separation which divides people from God and from one another. We saw Him in all the divisions of the world, big and small: in families, between generations, between rich and poor, even in the Church itself between various groups, and among the various Churches, and furthermore, among different religions, and between believers and non-believers... But, significantly, all these divisions did not frighten us. *On the contrary, we were attracted to them because of our love for Him Forsaken.* (pp. 9-10; my italics)

Chiara's theology is definitely non-triumphalist—as non-triumphalist as the *Lament of the Madonna* by Jacopone da Todi—the result of the profound difficulties suffered by the Catholic Church from Socialists, Communists, Fascists, Nazis, and the contemporary modern life.

---

[93] We are immediately reminded of "Jesus 2000," the "Jesus of all people, celebrating our differences."

Unrelated to Rome

Back to the "Jesus 2000" by Janet MacKenzie, and to the theme of the new profound DP within the Catholic Church:

> Michael Farrell, editor of the *National Catholic Reporter* said there may be more to this dark, indigenous Jesus [2000] than meets the eye. "When the church was overwhelmingly a Western institution, we in the West made Jesus in our likeness. But now at last Christianity has spread to the ends of the earth as the founder once prayed it would. Much of the church's energy, and new vocations, have moved from Europe and the United States to the Third World, so perhaps this work of art is a preview of how Christianity will flourish, and what kind of divinity it will look up to, as the next millennium unfolds". (Schaeffer & Allen, 1999, p. 3)

These comments can be strengthened, in terms of the present higher division of power around and inside the Catholic Church, through the following by John Allen Jr.: "*There is great vitality at the church's grassroots today unrelated to decisions made in Rome,*[94] *and perhaps that is the more important story*" (1999, p. 18, my italics to stress this very high DP condition of actual power and of scripts).

Yes, the *Jesus of the People* ("Jesus 2000") is celebrating many differences; he does not have a lot of boundaries and barriers, and he is a perfect reflection of the present condition of high DP.

That "Jesus 2000" clearly says that Christianity has evolved for twenty centuries under all kinds of power situations, DP or UP, which this chapter has tried to highlight.

---

[94] He was referring to the rulings by Cardinal Joseph Alois Ratzinger, head of the Congregation of the Faith—formerly known as the Holy Office and before that as the Holy Inquisition.

Figure 1 *Jesus as the Good Shepherd*, late 3rd century; wall-painting. Catacomb of Priscilla, Cubicolo dei Bottari, Rome.

Figure 2 *Jesus entering Jerusalem*, detail, 4th century; sarcophagus, Museo Lateranense, Vatican.

Figure *3 Jesus as the Good Shepherd*, ca. 430-450; mosaic. Mausoleum of Galla Placidia, Ravenna.

Figure 4 *Jesus resurrecting Lazarus*, early 6th century; mosaic. St. Apollinare Nuovo, Ravenna.

Figure 5 *Jesus enthroned flanked by Angels*, early 6th century; mosaic. St. Apollinare Nuovo, Ravenna.

Figure 6 *Jesus between San Vitale and Bishop Ecclesius*, early-mid 6th century; mosaic. St.Vitale, Ravenna.

Figure 7 *Emperor Justinian with his retinue*, detail, early-mid 6th century; mosaic, St. Vitale, Ravenna.

Figure 8 *Empress Theodora with her retinue*, early-mid 6th century; mosaic, St. Vitale, Ravenna.

Figure 9 *Jesus Pantocrator*, ca. 1100; mosaic in the dome of the Monastery Church, Daphni, near Athens.

Figure 10 *Jesus and the Samaritan Woman*; detail, tempera. Duccio, Thyssen-Bornemiza Collection, Madrid.

Figure *11 Jesus Pantocrator*; fresco. Giotto, 1266-1327. Scrovegni Chapel, Padua.

Figure 12 *The kiss of Judas*; detail, fresco. Giotto, Scrovegni Chapel, Padua.

Figure 13 St. *Francis of Assisi preaching to the birds*; detail, fresco. Giotto, Upper Church, Assisi.

Figure 14 *Jesus paying the tribute money, by instructing St. Peter*; fresco. Masaccio, Florentine, 1401-1428. Brancacci Chapel, S. Maria del Carmine, Florence.

Figure 15 *Jesus of the People* ("Jesus 2000"). Painting by Janet McKensie, Island Pond, VT.

Figure 16 Ruler (Charlemagne?) and Church. *Collection of laws and capitularies,* made in Aquilea in 816-819. St. Paul im Lavantal, Austria: Benediktinerstift St. Paul.

Figure 17 Emperor Charles II the Bald (r. 840/875-877) enthroned; detail. *Codex Aureus of St. Emmeram of Regensburg*. München: Bayerische Staatsbibliothek.

Figure 18 *Presentation of the Vivian Bible to Emperor Charles II the Bald* (r. 840/875-877). Bible from the School of Tours, c. 846. Paris: Bibliothèque Nationale.

Figure 19 Emperor Otto III (r. 983-1002) enthroned. Miniature in *Liuthar Gospels of Otto*; ca. 975, from Reichenau. Aachen: Cathedral Treasury.

Figure 20 Emperor Otto I (r. 936/62-973) offering a model of Magdeburg cathedral to Christ in majesty. Ivory plaque ca. 970. New York: Metropolitan Museum of Art.

Figure 21 Christ in majesty, with St. Mauritius and the Virgin, adored by Emperor Otto II (r. 973-983) and his wife Theophanu and their infant son. Ivory panel, 980/3. Milan: Castello Sforzesco.

Figure 22 Enthroned Christ with the Imperial pair: Conrad II (r. 1024-39) and his wife Gisela. *Codex Aureus*, written (1043-46) for Emperor Henry III. Madrid, El Escorial: Nuevos Museos.

Figure 23 Emperor Henry III (r. 1039-56) presents the manuscript of his Gospels for the Cathedral of Goslar to Saints Simon and Jude, patron saints of the Cathedral. Uppsala: University Library.

Figure 24 St. Peter gives the pallium to Pope Leo III (r. 795-816) and the standard to Emperor Charlemagne (r. 768/800-814). Rome: Mosaic, commissioned by Pope Leo III; in the Triclinium of the Lateran Palace.

Figure 25 A nineteenth-century engraving of a meeting of the Society of Friends (Quakers) in Philadelphia. Washington, D.C.: Library of Congress.

11

Contrasting Imperial Iconographies (Dwarf or Semigod?) Paved the Way for the DP of Canossa

1. Introduction

The early Western imperial iconography differed substantially from the Eastern/Byzantine imperial iconography, a reflection of the fundamental differences between the Western and Eastern branches of Christianity.

Among others, the Roman/Byzantine emperors ruled long before the birth of Christianity, accordingly bishops, patriarchs, and popes were accustomed, since early youth, to honoring and obeying the emperor. In the West instead, the first emperor, Charlemagne, was crowned eight hundred years after the birth of Jesus, and was crowned by a pope. In the West, the emperor—especially before his coronation—was a sinner in need of the Church to attain salvation; and even after his coronation, he never was "the living icon of Christ," he never was *isapostolos* (the equal of the apostles). Yet, already from the time of Charlemagne, the Western emperors played a double game: that of the humble Christian, and that of the mighty emperor blessed directly by God and thereby in charge of everything under the sky. Consequently, there was, in the West, a dual imperial iconography:

- Type A, showing the emperor big, proud, in full command of laity and clergy, blessed by God and his angels; and
- Type B, showing him small, humble, obediently worshiping Christ and honoring his saints.

These two iconographies presented the emperor in his dual function: as *Realpolitiker* in charge and control of the Empire and its Church; and as *Seelenpolitiker*[95] managing himself as a Christian, in need of God's pardon, in search of eternal salvation.

As *Realpolitiker*, the emperor was constantly faced with human reality, including the non-saintliness of many priests, bishops, monks, abbots, and popes. As *Realpolitiker*, the emperor was constantly in search of money, arms, power, glory; all the usual things that kings and emperors seek. As *Seelenpolitiker*, the emperor needed the grace of the sacraments, the protection of the saints, the help of the spiritual relics.

The sharp contrast between the two iconographies speaks clearly the language of the division of power, because the imperial iconography Type B reminds the viewer of eternal life, of Heaven and its keys; and the holder of these keys was not the emperor but the successor of St. Peter: the pope. The emperor may proclaim himself supreme: *rex et sacerdos (king and priest)*, but in the West that claim was contested by the pope. And if the emperor, in iconography Type A, was all mighty over the Church, that could easily indicate simony. Also, if the emperor was *rex et sacerdos*, why did he need to humiliate himself in iconography Type B?

The problem, as already intimated, was that the emperor in the West had not been told by everybody surrounding him, from birth onward, that he was made of superior stuff. On the contrary, all learned people (i.e., the clergy) had stressed his sinful condition and his absolute need of God's pardon and the constant support of the Sacraments administered by the Church. Aging and sick emperors, facing their approaching death, knew better than to play too much the role of the directly-blessed-by-God: deep in themselves they knew they were plain

---

[95] *Realpolitik*, is a German word indicating a foreign policy determined by expediency rather than ethics or world opinion, (i.e., power politics). I plan to use this word to stress the imperial control on every aspects of social life, including many aspects of church life. The opposite word is *Seelenpolitik* (or Soulpolitics) indicating all plans and actions related to the Christian soul.

sinners![96] And yet, already starting with Charlemagne, Western emperors soon discovered that more than one pope had feet of clay, as was the case of Leo III who in April 799 had to flee across the Alps to Paderborn and beg Charlemagne's protection. So, for centuries, the situation remained unclear, as reflected by the two types of iconography under discussion.

On the other hand, while the two imperial iconographies illustrated what might be called the normal dual aspect of the emperor as ruler and plain Christian, their sharp contrast raised doubts on the nature and extent of imperial power: of an earthly power so inferior to that of the pope who holds, as said by Jesus to Saint Peter, the keys of Heaven.

Yet, there were attempts to say that the emperor was *rex et sacerdos* (king and priest), crowned directly by God and therefore superior to the pope, but that novelty for the West was crushed at Canossa in 1077 with the humiliation of Emperor Henry IV by Pope Gregory VII. Henry IV was later able to get even with the Pope, and from the continued war between the two great powers arose the Western DP: the separation between the emperor (State), in charge of people's happiness on this earth, and the pope (Church), in charge of people's happiness in the next life (to use Dante's definition and explanation of the great division of power).

In turn, this contrast between the two imperial iconographies had contributed to the movement which led to Canossa: weakening the emperor and strengthening the Church and its pastor, the pope.

2. Two Types of Pre-Canossa Imperial Iconography

Type A shows the emperor big, proud, *Realpolitiker*, in full command of laity and clergy. Four examples of this type are Figures 16, 17, 18, and 19.

---

96 The contrast with Constantine is evident, who "was a Christian of a very peculiar type, a type that would hardly be recognized as Christian at all today. For the God he believed in was a God of power, who had given him victory, and he would have had little sympathy with the idea that Christianity meant love, or charity, or humility, of which his 'middle-brow' view of religion did not have the slightest comprehension. Furthermore, he was utterly confident that he himself was the man of God, God's servant and representative *who was constantly in touch with him and was told by him what to do-and how*, by doing accordingly, he could avoid divine anger. This made Constantine a difficult man for other mere human beings to deal with, being on a direct line with God, he must always be right" (Grant, 2000, pp. 221-22, my italics). Successive Byzantine emperors were not so idiosyncratic, yet many kept much of Constantine's confidence.

Figure 16 comes from the *Collection of laws and capitularies made in Aquilea in 816-819*, presently in the Benedictine monastery of St. Paul in Lavantal, Austria; and was exhibited, in the Summer of 1999, in Paderborn at the exhibition *799 - Kunst und Kultur der Karolingerzeit. Karl der Grosse und Papst Leo III*. This figure shows a mighty Ruler and a timid, diminutive *Ecclesia* (Church). The *Ecclesia* stands to the left of the ruler, i.e. on his less important side, and seems to be receiving orders, or admonitions, or criticism from the dominating finger-pointing Ruler. At Paderborn, this figure was illustrated by two captions, the first of which said: "*Dargestellt ist ein Herrscher (Karl der Grosse ?) und Ecclesia* [shown is a Ruler (Charlemagne?) and the Church]"; and the second caption: "*Aufgabe eines Herrschers ist es Gesetze zu geben und zu erlassen und Recht zu sprechen und auch der Kirche Gesetze zu geben*" [The task of a ruler (or monarch) is to issue laws and regulations, to provide for justice, and also to give laws to the Church].

Figure 17 from the *Codex Aureus of St. Emmeram of Regensburg*, presently in the Bayerische Staatsbibliothek, Munich—shows a mighty Emperor Charles II the Bald (r. 840/875-877), enthroned, disproportionately large, blessed by the hand of God, and honored by two angels.

Figure 18 from the *Vivian Bible*, presently in the Bibliothèque Nationale, Paris—shows once more Emperor Charles II the Bald: "Within a great arched frame, the dedication miniature of the Vivian Bible presents the king, in the presence of court figures and clergy, receiving his new manuscript. Charles, the largest figure in the composition, [blessed by the hand of God] sits between two princely men and two royal bodyguards at the back of the circle of men. Before him are eleven clerics. Three of them carry the magnificent volume in veiled hands and the other hails the king. At the right, a layman joins these religious figures in saluting the ruler" (Kessler, 1977, p. 125).

Figure 19 from the *Gospel of Otto*, a work of the Reichenau school, presently in the Treasury of the Aachen Cathedral—shows Emperor Otto III (r. 983-1002), the largest figure in the composition, enthroned, and surrounded by the symbols of the four Evangelists. Terra supports his throne. Two kings pay him homage. Below are two princes and two bishops. "At the same time, the Hand of God is reaching down, from above, either to impose or to touch and bless the diadem on the emperor's head. The divine aureola framing the Hand of God intersects wit the imperial aureola, thus allowing the emperor's head to be placed in the spandrel which is formed by the intersecting haloes.... In the central part, right and left of the footstool, there stand, flagstaffs with purple pennants on their shoulders, two male figures in gesture of veneration, if not adoration. That they are of very high rank is suggested by their crowns—feudatory dukes or perhaps rulers of the *regna*, the plurality of which was needed to indicate imperial dignity. At any rate, they are princes or *reguli* dependent on the young kosmokrator who himself is raised toward heaven or into heaven... (*usque ad celum erectus* [standing up to heaven]), all earthly powers inferior to his, and he himself nearest to God" (Kantorowicz, 1957, pp. 62-64).

In the words of Charles Little: "Here the image of the Emperor is completely assimilated with that of Christ so that he is now enthroned within a mandorla and surrounded by the symbols of the four Evangelists, while the hand of God places the crown upon his head" (1977, p. 112).

This Gospel was commissioned by Abbot Liuthar who then dedicated it to the Emperor as follows: "*Mit diesem Buch möge Gott dir, Kaiser Otto, das Herz bekleiden. Erinnere dich, dass du es von Liuthar empfingst*" [With this book, may God protect you, Emperor Otto. Remember that you received it from Liuthar]" (translated from the Latin, and quoted by Lepie & Minkenberg, 1995, p. 60). In plain language, the gift was a *quid pro quo*, from a partisan of Emperor Otto III looking for some benefits, and willing to be very generous in the religious adulation of his master.

Type B shows the Emperor humble and as *Seelenpolitiker*. Here too, I give four examples: Figures 20, 21, 22, and 23.

Figure 20 an ivory plaque, 13x11 cm, carved in Milan between 962 and 973, presently in the Metropolitan Museum of Art, New York City, a gift of George Blumenthal, 1941. The plaque shows Emperor Otto I the Great (r. 936-973) offering a model of the Magdeburg cathedral to Christ in Majesty for his blessing.

This work, part of a set, is "said to have been given by him to his favourite Magdeburg. As a monastery, it was founded in 937 by Otto and his wife Edgith, the daughter of Edward of England, and dedicated to St. Peter, St. Mauritius, and St. Innocent. In 955, shortly before the Battle of the Lechfeld [in which he decisively defeated the Magyars], Otto began the building of a large cathedral and decided to make Magdeburg an archbishopric.... From the start, Otto intended the cathedral to serve as his and his queen's burial place and to be of unrivalled magnificence" (Lasko, 1972, pp. 87-88).[97]

The plaque is one of a group of 40 to 50 similar ones—from a workshop in Milan—which were used to decorate the antependium of an altar, an ambos, or a cathedra. All have a religious theme from the life of Jesus, culminating—so it would seem—in the plaque in question: a *Majestas Domini*, in which Emperor Otto I presents to Christ a model of the Magdeburg cathedral.

According to most authors, the figure to the left of the Emperor is St. Mauritius who is both Otto's personal saint as well as the patron of the Magdeburg Cathedral. The figure to the right is St. Peter holding the keys of the kingdom of heaven. Behind St. Mauritius is the Archangel Michael, and behind St. Peter are two additional nimbed saints. In the center is Christ in majesty, enthroned upon a circle, the sphere of the cosmos, with his feet upon a smaller sphere-circle, the earth.

When I first saw a photograph of the Magdeburg plaque, I was surprised. The mighty Emperor Otto I the Great was represented as an insignificant dwarf! How was that possible?

In search of an answer, I went first to Magdeburg, then to Paderborn, and finally to Aachen, and there spoke with the experts, and from them I realized that even if exceptional and rare, this humble representation of an emperor was not unique.

Clearly, if Otto was "the Great" in the human world, he was not so in the spiritual world. On the contrary, there: "Blessed are the poor in spirit: for theirs is the kingdom of heaven,... Blessed are the peacemakers: for they shall be called the children of God" (Matthew, 5: 3, 9); "Whosoever shall exalt himself shall be abased; and he that shall humble himself shall be exalted (Matthew, 23:12); strongest are the words of the *Magnificat*: "*Dispersit superbos, et exaltavit humiles*" ([the Lord] hath scattered the proud, and exalted the humble) (Luke, 2:52-3). So, if

---

97 The plaque is also discussed by Mayr-Harting (1991), and Fillitz (1993).

Otto's Magdeburg cathedral—and his tomb in it—had to serve their stated purpose of helping him attain eternal salvation, then he had to move from *superbo* to *humile*.[98] Accordingly, his direct, or indirect, instructions for the ivory plaque had to specify the greatest degree of humility, not of pride, because, if in the *Realpolitik* he was a giant, in the *Seelenpolitik* he was (he had better be) a dwarf. If for the "real ending" what counted was Christ's mercy and Christ's acceptance of his cathedral, then he should not present himself in imperial glory, but in Christian humility. It is also better to ask his patron saints—the spiritual giants St. Mauritius and St. Peter—to be there with him, and intercede for him.

Figure 21 an ivory panel, from a workshop in Milan, of 980/3; now in the Castello Sforzesco, Milan—shows Christ in Majesty, with St. Mauritius and the Virgin, adored by Emperor Otto II (r. 973-983) and by his wife Theophanu and their infant son (born in July 980, later Emperor Otto III).

"The panel originally no doubt decorated a book-cover, and the presence of St. Mauritius and the Virgin may mean that such a cover was given by Otto to one of his favourite abbeys, either that of Einsiedeln in Switzerland, which was dedicated to these saints, or perhaps the abbey of St. Mauritius in Milan" (Lasko, 1972, p. 93).

The similarity of this figure with the previous one is striking: again a diminutive, very humble emperor, not only smaller than Christ and the Virgin, but also smaller than St. Mauritius.

The Emperor's position is also very humble: on his knees kissing Christ's feet; a Christ totally disinterested in Otto and his family.

Figure 22 shows Christ enthroned, adored by the Imperial pair Conrad II (r. 1024-39) and his wife Gisela, from the *Codex Aureus*,

---

[98] A small remnant of this humble attitude is described by Rosenstock-Huessy: "By a late ritual in Austria the corpse of the emperor was ordered to be carried to the door of an abbey [the Capuchin crypt in Vienna]. The chamberlain who leads the cortège knocks at the door. A friar opens the window and asks: 'Who knocks?'— 'The Emperor.'— 'I know no man of that name.' The chamberlain knocks again. 'Who is there?' — 'The Emperor Francis Joseph.' — 'We do not know him.' Third knock, and the same question. After reflection, the chamberlain now answers: 'Brother Francis.' Then the door opens to receive a comrade in the army of death, on equal terms with all souls" (1938/1950, pp. 55-56).

written (1043-46) for Emperor Henry III; now in the Nuevos Museos, El Escorial, Madrid. Again we have two dwarfs; not even acknowledged by the majestic Christ.

Figure 23 shows Emperor Henry III[99] (r. 1039-56) presenting the manuscript of his Gospels, for the Cathedral of Goslar, to the Saints Simon and Jude, patrons of the Cathedral of Goslar; now in the University Library, Uppsala. In sharp contrast with Figure 18, it is now the Emperor who humbly presents his Gospels to the two saints.

The Consequences of the Imperial Iconographic DP

To the religious elites of those times, the contrast between the two types (A and B) of imperial iconography was strikingly evident, part of the wider contrast between the *Two Cities* or *Two Powers* of St. Augustine: "The love of self extending to the contempt for God, and the love of God extending to contempt of self" (Burnaby, 1978, p. 367).

That great rulers would show themselves to be big and mighty was expected; it had always been done, as discussed, for instance, by Henry Frankfort in *Kingship and the Gods*: "[In primitive art, and in Ancient Egyptian art in particular] there is a gradation of scale according to the importance of the persons. The king is the dominating figure. This 'hierarchical scaling' has nothing to do with the peculiar nonperspective method of representation of pre-Greek art. It derives from an emotional attitude toward the king.... In one context only does Pharaoh appear on an equal footing with other figures, namely, in company of the gods.... Pharaoh belongs not to the order of mankind but to that of the gods" (1948, p. 7, 10-11). Yet, the Medieval *Seelenpolitik* demanded a Type B iconography, thereby creating contrast, or DP, in the mind of the elites. Then seen from the perspective of the imperial *Realpolitik*, priests, bishops, monks, abbots, and popes were very human—thereby

[99] Henry III died at the age of 39, leaving behind only one surviving son, a six-year-old boy who then became the Henry IV of Canossa. The early death of Henry III, and the long minority of King Henry IV, was one of those fundamental lucky events in the birth of Western civilization. Without the difficulties that Henry IV had with the German princes, pope Gregory VII would not have had his triumph at Canossa, and the Investiture Controversy would have easily been won by the Emperors. The West would not have had its fundamental division of power, and would have gradually evolved in a copy of Byzantium.

deserving to be maneuvered and squeezed for profit, as any other subjects. Seen from the point of view of the religious purists, the emperor, in so doing, was not better than Satan: he was God's archenemy, and the Type B iconography was a sham, either propaganda or a vile attempt to buy Paradise with money extracted from the Church. The religious purists had very good reasons to complain, especially of Otto I the Great who already,

> at the very beginning of his reign, [had] positively indicated that episcopal elections were to be subject to royal domination.... By reserving this right to himself, he would be in a better position to [have] a Church controlled by and employed in the service of the State.... For 200 years, there never had been a time when the Western kings and emperors did not, more or less, exercise an arbitrary control over the candidates for the episcopal dignity.... [However] during Otto's reign this right was so magnified and its application so strained that hardly a candidate could acquire a German bishopric who was not a *persona grata* to the king;... By far the greater abuse of royal authority, and a practice which practically robbed the Church of all power to determine who would be the leaders of souls in the German dioceses, was the direct nomination of the candidates by the king himself.... If it was at all possible, Otto made every effort to appear in person in order to dominate the election. (Gallagher, 1938, pp. 55-70)

In the same vein: "The princes of the Church, themselves the nominees of the emperor, were expected to provide not only cash for the imperial treasury and entertainment for the peripatetic imperial court but also a full complement of men-at-arms whenever the host might take the field" (Lasko, 1972, p. 82). Correspondingly, any truly religious churchman must have been deeply offended by the imperial Type A iconography: by pictures like Figure 16 (Charlemagne?) and Figure 19 (Otto III) which depicted a religion that had nothing to do with the Gospels.

To the truly religious churchmen, a gift like that by Abbot Liuthar to Otto III was blatant simony. Abbot Liuthar was selling a "coronation by God" for an unspecified, but hopefully substantial, remuneration.

Indeed, simony became the primary and immediate cause of the Investiture Controversy. The moving spirit behind Pope Gregory VII[100] was Humbert of Moyenmoutier, Cardinal Bishop of Silva Candida: "The ideas which inspired his activity are to be seen in his treatise *Against the Simonists* (c. 1058), which is at once the earliest, the ablest, and the most extreme statement of the programme of the reformers. To Humbert, simony was not merely a sin; it was the supreme heresy, since it denied the spiritual character of the Church and subordinated the gifts of the Spirit to money and worldly power.... To meet these evils he called for a return to the old canonical principles of free election and emancipation of the Church from the control of the secular power and from the custom of lay investiture" (Dawson, 1964, p. 51).

3. An Inspirational Piece of Papal-Imperial Iconography

The reformers who went to Rome could take inspiration from the mosaic of *Figure 24*: a work in the *triclinium* of the Lateran in Rome—commissioned by Pope Leo III in ca. 796-98 (and reconstructed in the eighteenth century), showing Pope Leo III with St. Peter in the center, and Charlemagne on the other side.[101]

A. Tu es Petrus

In the mosaic—as in the Magdeburg plaque—St. Peter appears much bigger than the Emperor, who is the same size as the Pope. These last two, therefore, should be seen to be of equal importance. However, there are significant differences in the mosaic which point to the supremacy of Leo III over Charlemagne. The Pope stands to the right of St. Peter—the more important position—and his hands are covered by a stole,[102] identical to St. Peter's, and handed down to him by the saint. The military lance and blue standard with red roses of Charlemagne is held in St. Peter's hand, but the lance-standard with roses is not an attribute of St. Peter. More importantly, St. Peter holds the two keys,

---

100 "Even before Gregory [VII] became pope, two causes of friction between the papacy and the royal government in Germany had arisen. In a disputed election to the great see of Milan, Pope Alexander II had supported one candidate, the king [Henry IV] another. Also, Alexander had occasion to rebuke the young king for tolerating simoniacal practices at his court" (Tierney, 1964, p. 53).

101 Both the Pope and the Emperor have the squared aureola of those presumed saints, but still alive.

102 "St. Peter the vicar of Christ handing Leo III the pallium, the symbol of supreme jurisdiction in the Church" (Folz, 1974, p. 115).

signifying his role and power, on his lap. He is the Peter of Matthew, 16:18-19:

> *Tu es Petrus et super hanc petram aedificabo ecclesiam meam, et portae inferi non praevalebunt adversus eam. Et tibi dabo claves regi coelorum. Et quodcumque ligaveris super terram erit ligatum et in coelis; et quodcumque solveris super terram, erit solutum et in coelis*[103].

In the mosaic, therefore, the "*Tu es Petrus*," applies also to Pope Leo III, his successor and alter Petrus: "*Et tibi dabo claves regi coelorum.*"

The equivalent, of this pictorial emphasis on the direct link between St. Peter and the Pope, can be found in the writings of Pope Gregory VII, for instance:

- in his letter of 1075 to Bishop Cunibert of Turin, in which he said: "It was your duty, beloved brother, in view of the reverence you owe to St. Peter, prince of the Apostles, and master of the Church Universal,..."
- in his 1076 excommunication of the bishops of Lombardy: "The bishops of Lombardy who, in contempt of canonical and apostolic authority, have joined in a sworn conspiracy against St. Peter, prince of the Apostles,..."
- in the introduction to his 1076 excommunication of King (later emperor) Henry IV: "O blessed Peter, prince of the Apostles, mercifully incline thine ear, and hear me, thy servant, whom thou hast cherished from infancy and hast delivered until now from the hand of the wicked who have hated and still hate me for my loyalty to thee,..." and
- in the introduction to his 1080 excommunication of Henry IV: "O blessed Peter, chief of the Apostles, and thou, Paul, teacher of the Gentiles, deign, I pray, to incline your ears to me

---

103 "Thou art Peter, and upon this rock I will build my church; and the gates of hell shall not prevail against it. And I will give unto thee the keys of the kingdom of heaven: and whatsoever thou shalt bind on earth shall be bound in heaven: and whatsoever thou shalt loose on earth shall be loosed in heaven."

> and mercifully to hear my prayer...." (Gregory VII, 1073-1085/1932, pp. 73, 90, 149).

This strong papal reliance on St. Peter, and the derived *Tu es Petrus*, were not new: already Pope Leo I (440-61) had said of his office: "'We are in St Peter's place' (*vice Petri fungimur*) — the first pope overtly to claim this vicariate" (Canning, 1996, p. 31).

The *Tu es Petrus*, is also at the base of Gregory VII's *Dictatus Papae* of 1075, of which I report a few of its 37 points:

> 2. That the Roman Pontiff alone can depose or reinstate bishops.
> 10. That his name alone is to be recited in churches.
> 11. That his title is unique in the world.
> 12. *That he may depose Emperors.*
> 13. That he may transfer bishops, if necessary, from one See to another.
> 23. *That the Roman Pontiff, if canonically ordained, is undoubtedly sanctified by the merits of St. Peter* (quoted by Tierney, 1964, pp. 49-50, my italics).[104]

Within this interpretation, it made sense that the Emperor be shown quite small in the Magdeburg ivory plaque: he is only a layman compared to St. Mauritius and to the larger St. Peter (with his keys clearly displayed).

B. Tu es Caesar

We encounter the opposite in figures 16 and 19 where the big "protector" is the Emperor, and the diminutive "protected" is the Church (in Figure 16), and a couple of bishops (in Figure 19). In a letter to Charlemagne in 799, Alcuin put in writing the hierarchical situation presented in these two pictures:

---

[104] Gregory VII was not an original theoretician. He pulled ideas together from several sources and synthesized them into his formidable total program. "He liked to remark that he did not enunciate any new rule, *nihil novi, nihil adinventione nostra statuentes*, that in any manner he did not abandon the way his predecessors had traced for him" (Fournier, 1964, p. 20).

> Up to now, there have been three persons at the summit of the hierarchy in this world. First, the representative of the Apostolic Sublimity, Vicar of the blessed Peter, Prince of the Apostles, whose See he occupies. What has befallen the present occupant of this See [Leo III], your Excellence has taken care to inform me. Next, the holder of the imperial dignity, who exercises secular power in the second Rome [Byzantium]. In what impious manner the ruler of the Empire has been deposed, not by strangers but by his own people and by his fellow citizens, has become known everywhere.[105] *In the third place comes the royal dignity which our Lord Jesus Christ has reserved for you because you rule over the Christian people. This dignity prevails over the other two, eclipses them in wisdom and surpasses them.* It is now upon you alone that all churches of Christ depend, from you alone that they expect well-being; from you, avenger of crimes, guide to those who have lost their way, councelor of the afflicted, supporter of the good. (quoted by Halphen, 1977, p. 88; my italics)

Yet, even Alcuin was not breaking new ground: "In these respects the Carolingians were developing Merovingian ideas. The king was the vicar of God (*vicarius dei*) as Cathwulf reminded Charlemagne in a letter of c. 775:

> Always remember, therefore, my king, with fear and love for God your king, that you are in his place to look and rule over all his members and to give account on judgment day even for yourself. And a bishop is in second place: he is only in Christ's place. Ponder, therefore, within yourself how diligently to establish God's law over the people of God.

Such kingship involved a religious role. At the Diet of Frankfurt of 794, for instance, Paulinus of Aquilea addressed Charlemagne as 'king and priest' (*rex et sacerdos*)" (Canning, 1996, p. 49).

Given the strong unity of power of the Carolingian, Sassonian, and Salian emperors—more and more acting, and being accepted, as *vicarii*

[105] He is referring to the arrest and blinding, in 797, of Emperor Constantine VI by order of his mother, the Empress Irene, widow of Emperor Leo IV.

Dei, with the Pope as their local bishop in Rome—something radical had to happen to change direction, and bring the world out of the universal tendency toward the unity of power, out of the rule of the supreme *rex et sacerdos*,[106] out of rendering to him both what is Caesar's and what is God's. Something radical had to be done to achieve the opposite, (i.e., a high level of division of power, and related civic society), and a major part of this effort was the desacralization of the emperor by showing him as only a diminutive rex. That "something radical" was the *War of Supremacy* of 1075-1313, in particular its first battles of the *Investiture Controversy* of 1075-1122 as discussed in chapters 11 and 23 of volume 1.

The idea of fighting the emperor was far from natural in those times; definitely not to the Christian mind which was constantly reminded of the instructions of St. Paul to obey authority: "Let every soul be subject unto the higher powers. For there is no power but of God; the powers that be are ordained by God. Whosoever therefore resisteth the power, resisteth the ordinance of God: and they that resist shall receive themselves damnation.... Render therefore to all their dues: tribute to whom tribute is due; custom to whom custom; fear to whom fear; honour to whom honour" (Romans, 13:1,7).

Much was needed for things to change, especially to undo that natural tendency to obey the greatest of rulers, the emperor, and one of these—even if not the principal—was the contrast between the two imperial iconographies: *One picture is worth more than ten thousand words*, says an ancient Chinese proverb. That contrast spoke of the Christian notion of this life on earth as a preparation for the eternal one, it reminded the faithful that "It is easier for a camel to go through the eye of a needle than for a rich man to enter into the kingdom of God" (Matthew 19:24); even more, that contrast brought to mind the well known words of the *Magnificat*: "*Dispersit superbos, et exaltavit humiles*" ([the Lord] hath scattered the proud, and exalted the humble) (Luke, 1:52-3). As long as that faith was alive, the proud rulers in the West had to do something to save their soul, such as offering a cathedral

---

106 The unity of power of *rex et sacerdos* is particularly dangerous in isolated countries like ancient Egypt, Peru, and Mexico. It is less dangerous if the religion is polytheistic, or if the country is relatively small and with active contacts with culturally advanced countries. If there are many gods, it is impossible for the *sacerdos* to please every god and his/her faithful. Frequent contacts with surrounding countries honoring other gods will bring doubts about the extent of the power of the *sacerdos*. However, things are dangerous for everybody if the religion is monotheistic and the country very big, as was the case with the Byzantine empire.

to Christ, and showing themselves humble, in at least a few pictures; and this contributed to the DP of Canossa.

Later kings and emperors did not have the same faith and did not offer cathedrals to Christ, but they had learned, in the meantime, not to play the dual role of *rex et sacerdos*. And this helped significantly in maintaining and expanding the DP of Canossa (1077), of Canterbury (1174), of Legnano/ Venice (1176/1177), and of Runnymede (1215).

12

UP Saints versus DP Saints

1. Introduction

A UP society will create UP saints, while a DP society will discover DP saints in its midst. Accordingly, the mild UP of the Middle Ages created—out of nothing—a mild UP saint: Saint Katherine of Alexandria, the imaginary daughter of an imaginary King Costus of Alexandria. Later and stronger UP times upgraded this same St. Katherine to a reigning queen with strong imperial connections, and greatly embellished her sainthood and friendship with Paradise: she was a favorite of Empress Virgin Mary who organized Katherine's mystical marriage to her son Jesus Christ. The rule became that the content of royal blood determined the level of sanctity.

About the same happened when, on orders of Emperor Frederick Barbarossa, an unsaintly person, Emperor Charlemagne, was canonized because this official sainthood would be marvelous propaganda for Barbarossa and his followers.

On the opposite side of the spectrum, the strong DP of communal Lucca made its people discover the virtues of a humble maid, the daughter of simple people, and venerate her as Saint Zita of Lucca.

In many respects, this chapter is a continuation of chapter 10 in this volume, which traced the way in which Jesus was represented at various times, in function of the ruling power (DP or UP). As discussed, DP conditions bring forth an iconography which highlights Jesus' humanity and martyrdom, while UP conditions bring forth images of Jesus as king, emperor, or mighty judge, always richly dressed (this last in stark contrast to the simple clothes of the DP Jesus).

Based on a purely sequential order of cause and effect, I should first discuss the times (UP or DP) behind the creation/discovery of the four saints under discussion (the two variants of St. Katherine, St. Charlemagne, and St. Zita of Lucca), then move to authors who wrote the most famous biographies of those saints (this is very important for the two versions of St. Katherine), and finally discuss the saints themselves as they are described in these biographies. However, this process would be cumbersome and add little to what I said in chapter 10 on the evolution of the iconography of Jesus. What I plan, instead, to do is:

- focus on the narrative of the two main variants of St. Katherine;
- discuss the authors of each variant: the communal *visitor* Jacobus de Voragine of Genoa for the first, and the ducal/royal *insular* William Caxton of London-Bruges for the second;
- highlight the critical thinking of the *visitor* Voragine which would have deserved the praise of Boccaccio, and, instead, how the *insular* Caxton—who spent thirty years in Burgundy, and offered his first book to the Duchess of Burgundy, the wife of the infamous Charles the Bold discussed in chapter 2 of volume 2—eliminated any trace of critical thinking in his translations from Voragine;
- conclude with a discussion of "St. Charlemagne", courtesy of the UP of Emperor Frederick I Barbarossa, and of the discovery of St. Zita in Lucca, a most DP city of Tuscany, very active in the *War of Supremacy* (1075-1313) between popes and emperors, in particular in the *Investiture Controversy* of 1075-1122.

Then–together with showing that the more the times are UP, the more they will invent UP saints; and the more DP, the more they will discover DP saints—I will highlight a number of fascinating points of contact with what was said in previous volumes: on the critical thinking of communal Voragine, so close to that of communal Boccaccio; and, vice versa on the *insular* thinking of Caxton, Charlemagne and Barbarossa as discussed in volumes 1 (chapter 23) and 2 (chapters 1 and 11).

The counterpart of these fascinating connections is that this chapter will prove difficult for those who did not read the prior volumes. Furthermore, the reader faces an intrinsic difficulty in following a discussion of the same saint, Katherine of Alexandria in two versions, which for simplicity I call A and B, whose main difference is the addition for Katherine B of three features: her "imperial" origins, her Parliament, and her mystical marriage with Jesus. These are the sections that Caxton added to Voragine's text while eliminating the critical section that Voragine had written at the end of the life of the Holy Innocents (those killed on orders of King Herod).

After the above introduction, let me report, as preliminary comment, two sections from Weinstein and Bell's *Saints & Society: The Two Worlds of Western Christendom, 1000-1700* on the kinds of saints who were venerated at different times.

> In the eleventh century saints were predominantly power holders—kings, princes, great prelates, and leaders of the monastic orders. They were upper-class males, and they were distributed fairly equally across all parts of Christian Europe. Royalty was believed to possess sacerdotal as well as temporal power, and prelates exercised sway over the bodies as well as the souls of their flock. The prince who exercised power in this world, it was believed, was likely to do so in the next. All power came from God, and few saw any distinction between its temporal and its spiritual manifestations. Sanctity was power, and power was holy[107]....
>
> The great model of twelfth-century saintliness was Bernard of Clairvaux, monastic reformer, preacher of crusade, powerful counselor to kings and popes, and censor of excessive preoccupation with material beauty in churches....The model saint of the thirteenth century was, of course, Francis of Assisi, who in contrast to Bernard sought to make the gospel accessible to every believer. Francis brought his quest for Christlike humility out into the streets, where by embracing Lady Poverty he sought to com-

[107] "The consistent association of temporal power with saintliness, while it contradicted one aspect of the legacy of the gospel, was congruent with another theme of Catholic theology. In Christian thought, power is an attribute of God, and medieval thinkers tended to acknowledge that kings as well as priests exercised a sacramental function" (Weinstein & Bell, 1982, p. 158). "Another example of the developmental process associated with the Germanization of Christianity may be found in the increasing attribution of military virtues and noble origins to saints, known as *Adelsheilige*, whose *Lives* were written in the seventh and eighth centuries" (Russell, 1994, p. 168).

> bine ascetic self-abnegation with a life of Christian charity. Whereas the faithful of earlier generations had tended to associate holiness with visible power, thirteenth-century believers found sanctity among the lowliest, those who not only gave up worldly status to follow the apostolic way, but did it in the public world rather than in the cloister. The preponderance of saints of this period came from the mendicant orders that, in propagandizing their own heroes, made asceticism a public and social virtue. The social attributes of saints underwent a radical transformation. The prelate gave way to the mendicant tertiary as a spiritual hero; the awesome king to his ascetic consort; the feudal magnate to the urban worker. *Whereas saints had formerly come from every part of Europe, they now were overwhelmingly Italian.* (pp. 246-47, my italics to stress the DP contribution of communal Italy)

On the other hand, even in Italy at the same time, there were major differences, and the Genoa of Jacobus de Voragine, the author of the best known life of St. Katherine [A] was not in the same DP league as communal Lucca, the city of St. Zita.

Therefore, in this chapter, I will first contrast the saintly nature of our four saints (or how that saintly nature was imagined) and then relate each type to the kind of power structure (UP or DP) in which the saint or the writer of the saint's story lived. Specifically, I will discuss:

- the early medieval "royal saint" Katherine [A] of Alexandria, as described by a mild DP writer—yet at times sharply critical *visitor* thinker— of the 13th century, the Dominican friar Jacobus de Voragine, first prior of convents at Como, Bologna and Asti, then archbishop of Genoa;
- the late medieval "imperial saint" Katherine [B] of Alexandria, as described by the UP writer William Caxton living in 15th century ducal Burgundy, then in royal England;
- the UP saint Emperor Charlemagne (742-814), "canonized", in 1165 by "pope" Paschal II, on orders of Emperor Barbarossa; and
- the humble (proletarian) saint Zita of Lucca who lived under the strong DP of communal Lucca from 1218 to 1272.

2. A "Royal Saint": St. Katherine of Alexandria [A]

In *The Catholic Encyclopaedia,* Leon Clugnet (1908) summarized St. Katherine's basic life (the early medieval nucleus accepted by de Voragine and Caxton) as follows:

> Katherine of Alexandria, Saint, a virgin and martyr whose feast is celebrated in the Latin Church and in the various oriental churches on 25 November, and who for almost six centuries was the object of a very popular devotion. Of noble birth and learned in the sciences, when only eighteen years old, Katherine presented herself to the Emperor Maximinus who was violently persecuting the Christians, upbraided him for his cruelty and endeavoured to prove how iniquitous was the worship of false gods. Astounded at the young girl's audacity, but incompetent to vie with her in point of learning the tyrant detained her in his palace and summoned numerous scholars whom he commanded to use all their skill in specious reasoning that thereby Katherine might be led to apostatize. But she emerged from the debate victorious. Several of her adversaries, conquered by her eloquence, declared themselves Christians and were at once put to death. Furious at being baffled, Maximinus had Katherine scourged and then imprisoned. Meanwhile the empress, eager to see so extraordinary a young woman, went with Porphyry, the head of the troops, to visit her in the dungeon, when they in turn yielded to Katherine's exhortations, believed, were baptized, and immediately won the martyr's crown. Soon afterwards the saint, who far from forsaking her Faith, effected so many conversions, was condemned to die on the wheel, but, at her touch, this instrument of torture was miraculously destroyed. The emperor, enraged beyond control, then had her beheaded and angels carried her body to Mount Sinai where later a church and monastery were built in her honour. (p. 445)

This description applies to St. Katherine in general, and specifically to St. Katherine [A] of the *unexpanded life*. Indeed, the stories of St. Katherine fall into two main groups: *unexpanded lives*, and *expanded lives*. "The *expanded life* comprises a full *vita* as it adds episodes describing Katherine's genealogy, birth, early life, conversion and mystical marriage [to Christ]" (Lewis, 2000, p. 14). However, the *expanded lives* not only tell more about her life, but—by the nature of what they say—they present a different St. Katherine. She is far more royal/imperial than the one of the *unexpanded lives*. This is in line with the changes discussed in chapter 10 of this volume, which transformed Jesus from the friendly preacher of the Gospel into an awe-inspiring Pantocrator, or a richly dressed emperor sitting on a Byzantine throne.

The best known *unexpanded life* of St. Katherine is that of ca. 1266,[108] by Jacobus de Voragine (in reality "da Varazze"[109]), born in Varazze, a small town on the Ligurian coast, twenty miles west of Genoa.[110] That *life* is part of a set of 152 short biographies of saints, which—with 26 stories on various religious festivities—constitute de Voragine's famous *Legenda Aurea (The Golden Legend)*, the main basis for William Caxton's *The Golden Legend* of 1483.

In de Voragine's *unexpanded life* (of which the above by Leon Clugnet is a good summary), no mention is made of Katherine's grandfather: we are simply told that she was the daughter of an unspecified King Costus who—within the Roman Empire of the IVth century—ruled over Alexandria. This Katherine, with no Roman imperial ancestors, is—in my own words—*St. Katherine-A* with no connection to England, no queenship over Cyprus, and no mystical marriage with Christ.

3. An "Imperial Saint": St. Katherine of Alexandria [B]

While in the *unexpanded lives* Katherine's parents were simply King Costus of Alexandria and his unnamed wife, in the *expanded lives*, Costus has been considerably elevated: he is now the half-brother of the great Emperor Constantine whose mother Helen (later St. Helen) is the

---

108 The *Medieval Sourcebook* (2002, based on Robert Blackmon, 1900) says 1275; Richard Hamer (1998) says ca. 1266.

109 Some authors refer to him as Varagine, Varazze, de Voragine, de Varagine. For simplicity I have converted all quoted names to either Voragine or de Voragine. I did the same with our saint whose name is given as either Katherine or Catherine. Here, I have opted for Katherine.

110 "In 1276, Jacobus was elected provincial of the Dominican province of Lombardy, an office which he fulfilled for nearly twenty years.... In 1288 Pope Nicholas IV commissioned him to free the Genoese from the ban of the Church. In the same year, at the death of the Archbishop Charles Bernard, the Chapter of Genoa elected him archbishop.... Genoa at that time was at the height of her glory. She had conquered Pisa a few years before, and her fleets ruled the Western Mediterranean. Her commerce extended to Constantinople, the Black Sea, and the cities of the Levant" (Ryan & Ripperger, 1941/1969, pp. V-VI).

daughter of King Coel of Great Britain.[111] So, indirectly, yet very powerfully, Katherine is endowed with imperial blood and a connection with England in the *expanded lives*. Katherine's mother is now the daughter of the King of Cyprus; and Katherine herself is described as Queen of Cyprus from the time of the death of her father when she was fourteen. As narrated by Caxton:

> In the meantime tidings came to Rome how that Great Britain, which now is called England, rebelled against the empire. Wherefore, by the advice of the consulate, it was concluded that Constantius, King of Armenia, should go into Britain to subdue them; which addressed him thither, and in short time after he entered into the land, by his prowess and wisdom he appeased the realm and subdued it again to the empire of Rome. And also he was so acceptable to the king of Britain, named Coel, that he married his daughter Helen, which afterward found the holy cross, and in short time he gat on her Constantine, which after was emperor. And then soon after died Constantius. And Constantine, after the death of King Coel, by his mother was crowned king of Britain. And Costus, the first son of Constantius, wedded the king's daughter of Cyprus, which was heir, of whom, as shall be hereafter said, was engendered St. Katherine, which came of the lineage of Constantius. (1483/1900, 7, pp. 2-3)

To expand on the English connection, Caxton tells us that Constantine, first became King of Great Britain at the death of his grandfather Coel, and only later Emperor of Rome. Accordingly, while Katherine herself had no English blood, her half-uncle Constantine's blood was half English, and her great-aunt St. Helen was full blooded English. So, from the ducal Burgundy and royal England of Caxton we get a Saint Katherine [B] "of the noble lineage of the emperors of Rome" (1483/1900, 7, p. 1) who is a ruling Queen of Cyprus (ib., p. 6) from age 14 to 18 (when she was decapitated).

---

[111] It was Geoffrey of Monmouth (d. 1154) who claimed that St. Helen "was of British origin and the daughter of Coel, legendary king of Colchester; this was widely believed in England, but Trier also was claimed as her birthplace" (Farmer, 1997, p. 231). In fact, Helen was born at Drepanum (later Helenopolis) in Bithynia (in NW Asia Minor, now Turkey), possibly an innkeeper's daughter who, indeed married the Roman general Constantius (ib., p. 231).

Given her queenship, we are even told—quite early in the story—of a request by the Parliament of the Kingdom of Cyprus [sic], that she should marry and produce heirs to the throne.

> Bray[112] notes that *the introduction of a parliament into the mystical marriage is a uniquely English addition, and is directly related to the important part which parliaments had in English, as opposed to Continental politics*. Moreover, she notes that the speeches which St. Katherine makes to her parliament 'contain echoes of the phraseology used in...contemporary [parliamentary] records'. In addressing her parliament St. Katherine is deliberately constructed as a fifteenth-century English monarch and this may have been of interest to the actual rulers of later medieval England. *The use of St. Katherine within royal ritual and pageant and her association with royal imagery would have served to make her more visible to the people of England. Displaying devotion to St. Katherine may therefore have served to connect the devotee to the royal family on some level, and also provide a means of expressing loyalty to its members.* By the end of the period devotion to St. Katherine would also have provided a means of displaying pride in and loyalty towards the country as a whole, as part of developing nationalist and patriotic discourses. (Lewis, 2000. p. 79; my italics)

So, while, on one side Katherine [B] is by far more royal than Katherine [A] of de Voragine, on the other that major increase in UP is slightly compensated by the presence and intervention of a parliament. However, in the description of Katherine's encounter with the Empress Virgin Mary, we are immersed in so much royal UP language that the memories of the previous debate with Parliament are eclipsed:

> The noble company of martyrs with the fellowship of virgins, which led the virgin Katherine, fell down flat tofore this royal empress with sovereign reverence, saying in this wise: "Our most sovereign Lady Queen of heaven, Lady of the world, Empress of hell [sic], Mother of Almighty God, King of bliss, to

---

112 Jennifer Relvyn Bray (1982), "The Legend of St. Katherine in Later Middle English Literature," Birbeck College, University of London, 1984.

> whose commandment obey all heavenly creatures and earthly, liketh it you that we here present to you our dear sister, whose name is written in the Book of Life, beseeching your benign grace to receive her as your daughter chosen and humble handmaid, for to accomplish the work which our blessed Lord hath begun in her". (Caxton, p. 12)

Even more astonishing is Katherine [B]'s mystical marriage with Jesus, as specifically demanded by the Empress Virgin Mary:

> "Now mine own daughter be glad and joyful, for ye lack no thing that longeth to the wife of a heavenly spouse, and now I shall bring you to my Lord my son which abideth for you." And so our Lady led her forth unto the quire door whereas she saw our Saviour Jesu Christ with a great multitude of angels,... To whom our blessed Lady benignly said: "Most sovereign honour, joy and glory be to you, King of bliss, my Lord, my God and my son, Lo! I have brought here unto your blessed presence your humble servant and ancille Katherine, which for your love hath refused all earthly things,..." Then our Blessed Lord took up his mother and said: "Mother, that which pleaseth you, pleaseth me, and your desire is mine, for I desire that she be knit to me by marriage among all the virgins of the earth...." And there our Lord espoused her in joining himself to her by spiritual marriage, promising ever to keep her in all her life in this world, and after this life to reign perpetually in his bliss, and in token of this set a ring on her finger, which he commanded her to keep in remembrance of this, and said: "Dread ye not, my dear spouse, I shall not depart from you, but always comfort and strengthen you." (ib., p. 13-15)

William Caxton is right. Indeed, as noted by Lewis (2000), "Katherine alone was worthy to be selected by the Virgin Mary as Christ's bride. Katherine's suitability derives, in part, from *her unique status as a highly educated sovereign queen.* Consequently, her standing as *the* bride of Christ can be seen either as a result of her matchless popularity, or as the explanation for it" (p. 1, my italics).

So, thanks to Katherine's noble blood, and her virginity, everything was possible, including—already in de Voragine's story—so much knowledge and rhetorical power for a complete victory over the fifty orators and masters sent against her by the emperor:

> And then one that was master above all the others said to the emperor: "Know thou, sir emperor, that never was there any that might stand against us but that anon he was overcome. But this maid in whom the spirit of God speaketh hath so converted us that we cannot say anything against Jesu Christ, ne we may not, ne dare not. Wherefore, sir emperor, we acknowledge that but if thou mayest bring forth a more provable sentence of them that we have worshipped hitherto, that all we be converted to Jesu Christ." And when the tyrant heard this thing he was esprised with great woodness, and commanded that they all should be burnt in the midst of the city. And the holy virgin comforted them and made them constant to martyrdom, and informed them diligently in the faith, and because they doubted that they should die without baptism, the virgin said to them: "Doubt ye nothing, for the effusion of your blood shall be reputed to you for baptism, and garnish you with the sign of the cross, and ye shall be crowned in heaven." (ib., pp. 20-21)

Never before had so many virtues been assembled in one person, so many great deeds and miracles accomplished! But how was all this possible: imperial blood, absolute beauty,[113] astonishing intelligence and

---

[113] On the subject of her beauty both de Voragine and Caxton are in agreement, given her royal/imperial blood. For instance, according to Caxton, Katherine's uncle had said to her: "in bodily shape and beauty there is none like you" (1483/1900, 7, p. 6). Our two authors were, indeed, in perfect harmony with the general UP thinking of their times: "Charged with a divine mission, the royal personage was supposed to conform in appearance, gesture and voice to the image of his office, an image recognized by all Christians" (Braunstein, 1988, p. 559).

"[Medieval] hagiographers and chroniclers associated noble standing primarily with birth and assumed that the well-born possessed inherently superior physical and moral qualities that destined them to position of leadership. The belief in such a 'noble culture' persisted through the Middle Ages" (Evergates, 1987, p. 147). Even much later, as discussed by Maravall for the Spanish *Siglo de Oro* (1550-1650), the noble-born was endowed with a whole package of virtues: "*la riqueza, la cortesía como educación social de fondo, la sabiduría, el amor y el honor, y toda una constelación de virtudes—la generosidad, el valor, etc. ... la pretendida correlación social riqueza-posición-linaje-valor-virtud* [wealth, courtesy as basic social education, wisdom, love and honor, and a whole constellation of virtues—generosity, valor, etc. ... the pretended social correlation of wealth-status-lineage-valor-virtue" (1972/1990, pp. 25, 65).

knowledge, virginity, marriage with Jesus,[114] and miracle after miracle during her life and after her death?

The answer is simple: Saint Katherine (both A and B) never existed. She is a figment of the medieval imagination. Being free from any link to reality, the hagiographers had no restrictions on the amount of virtues and capabilities that the heroine could possess. She could have royal blood in the unexpanded versions of her *life*, and imperial blood in the expanded version. That wonderful blood could produce all those virtues, and elicit from God the highest love, appreciation, blessings, and miracles.

> The cult built on this legend, which strongly appealed to the imagination of artists, flourished exceedingly throughout Europe in the Middle Ages under the influence of the Crusaders and later of the *Golden Legend*. Her intercession was valued because she was considered to be (*a*) the bride of Christ, (*b*) the successful advocate who triumphed over the philosophers, (*c*) the protectress of the dying. Hence she was the patron of young girls, of students (and hence the clergy), especially philosophers and apologists, of nurses (because milk instead of blood flowed from her severed head), and of craftsmen whose work was based on the wheel, such as wheelwrights, spinners, and millers. In England her cult was as widespread as anywhere in the West. Sixty-two churches were dedicated to her and 170 medieval bells still bear her name.... (Farmer, 1997, pp. 91-92)

Later in history—as happened with several other beloved great saints of antiquity—scholars realized that the story of St. Katherine, especially in the Caxton version, did not make much sense. Already *The Catholic Encyclopedia* of 1908—while still devoting about 830 words to St. Katherine, and still stressing that she had truly existed—was forced to admit that much was unclear: "So far the Acts of St. Katherine.... Unfortunately we have not these acts in their original form, but transformed and distorted by fantastic and diffuse descriptions which are entirely due to the imagination of the narrators who cared less to

---

[114] Marriage with the adult Jesus (not with the child Jesus as shown in almost all paintings dealing with this subject).

state authentic facts than to charm their readers by recitals of the marvelous" (p. 445).

The *New Catholic Encyclopedia* of 1976, could not even be so generous: its article is only of about 165 words, and, at the very beginning, includes a caveat on her very existence: "Data of her life derive from two works without historical value." Further on, we read that "The earliest evidence of her cult, apparently introduced by Eastern monks who had fled from iconoclasm, appears in a painting of the early 8th century in Rome. After the 10th century her cult became very popular, especially in Italy" explaining therefore the text that Jacobus de Voragine devoted to her.

4. The UP Environment which gave Origin to the *Expanded* Versions of St. Katherine's Life

"Caxton's *Golden Legend* was partly based on the *Légende Dorée*, partly on a Latin text of the *Legenda Aurea*, and partly on an 'englishe boke' which was probably a manuscripts of the *Gilte Legende*" (Lewis, 2000, p. 234n). In turn "The *Gilte Legende* containing the prose life of St Katherine constitutes the first complete translation of *Legenda Aurea* into English" (ib., p. 10).

> Nevertheless, if [Caxton] had used all the material found in the three texts, his edition would have reached gargantuan proportions. It was necessary to do some pruning. But many of the omissions may not reflect a desire to prevent the edition from growing too large; they were made because in some places, Caxton felt out of sympathy with his sources. One of the characteristics of the Latin *Legenda Aurea* is a certain reserve on the part of the author towards the veracity of the stories. *It is this expression of a critical spirit which modern readers of Voragine's work most admire. Although these expressions of doubt are generally kept by Vignai* [the author of the French translation, the main text used by Caxton], *they are frequently omitted by Caxton who obviously felt that they were unsuitable. Thus Voragine's comment that many of the bones said to be of the Holy Innocents are too big to belong to children under two is omitted. The exclusion of this type of passage 'not only impairs the quality of his version and misrepre-*

> *sents its original but also detracts from his editorial stature'.*[115] *Probably Caxton was an uncritical admirer of the Church who would tolerate no doubt or backsliding, for whom accuracy was less important than truth, though he may have omitted such passages because he assumed that his clientele would not approve of them.* (Blake, 1969, pp. 119-20; my italics)

Specifically, the DP *visitor* text by Voragine on the bones of the Holy Innocents said:

> Herod then returned from Rome emboldened by the confirmation of his authority, and forthwith ordered the killing of all the male children in Bethlehem who were two years old and under, figuring from the time he had learned from the Magi. But the phrase 'two years old and under' can be taken two ways.... Chrysostom says that the star appeared to the Magi for a year previous to the Savior's birth, and that after Herod had heard the Magi, he was away for another year on his trip to Rome. He thought that Jesus was born when the Magi saw the star and would now be two years old, so he killed all the males two years old and down to the age of five but not under two. *There seems to be some likelihood to this interpretation because some of the Holy Innocents' bones have been preserved and are so large that they could not have come from two-year-olds. Yet it might be thought that at that time men grew far larger than they do now.* (1483/1993, *1*, pp. 57-58; my italics)

This last suggestion that "at that time men grew far larger than they do now" can be interpreted in many ways, but coming from a man of the maritime republic of Genoa—with its extended trade with many distant countries where people had all about the same size—may simply be a warning against exaggerated claims. In essence, he may be saying "yes, but ..." De Voragine is no Martin Luther, yet he is sufficiently a *visitor communal man* not to accept large bones as holy relics of the Holy Innocents. Similarly, he is not the man who would have added that mystical marriage to Christ as demanded by Empress Virgin Mary.

---

115 Jeremy (1946, p. 215).

What is important in this whole story is a different evolution between, on one side, *visitor-communal* Italy, and, on the other, *insular-ducal* Burgundy (see Huizinga's *The Waning of the Middle Ages: A study of the Forms of Life, Thought and Art in France and the Netherlands in the XIVth and XVth Centuries* (as discussed in chapter 1 of volume 2). The reader may remember what I wrote there on *Saint Ciappelletto of Burgundy* whom we can now relate to the *editor/translator/printer William Caxton also of Burgundy*.[116]

With some irreverence, I am tempted to compare the Boccaccian canonization of Ser Ciappelletto, by the Burgundians, to the expanded life of St. Katherine by the "Burgundian" Caxton, especially when related to Caxton's exclusion of Voragine's critical passage on the bones of the Holy Innocents which, as reported, "'not only impairs the quality of his version and misrepresents its original but also detracts from his editorial stature'" (Blake, 1969, pp. 119).

Indicative of the strong UP under which Caxton operated for nearly thirty years before editing Voragine and expanding on the life of St. Katherine, is an engraving showing Caxton kneeling, in front of a young noble lady, and humbly offering her two books. The caption of that figure (e.g., in Painter's *William Caxton: A biography* (1977); opposite the title page) says: "Caxton presents his *Recuyel of the Histories of Troy* to Margaret Duchess of Burgundy".

> Caxton's association with Margaret had very probably begun at the time of her marriage to Charles in 1468, and continued ever since. He had served on diplomatic missions throughout the 1460s with the same high Yorkist dignitaries, including her brother-in-law Rivers, Wenlock, Vaughan, Hatcliff, Russell, who had negotiated her match and attended her wedding, and

---

[116] Born c. 1422, he was apprenticed, in 1438, to Robert Large, a rich mercer, who the following year became lord mayor of London. Large died in 1441, and Caxton moved to Bruges, the center of the European wool trade. During the next 30 years he became an increasingly prosperous and influential member of the English trading community in Flanders and Holland. In 1463 he took up the duties of Governor of the English Nation of Merchants and Adventurers in the Low Countries: "the statutory body to which the English government granted wide executive powers whereby they effectively controlled all merchants both at home and abroad" (Childe, 1976, p. 81). Some time in 1470 he ceased to be governor and entered the service of the duchess of Burgundy, possibly as her financial adviser. Toward the end of 1476 he returned to England and established his press at Westminster.

> he was head of the English colony who welcomed her to Bruges. No resident Englishman was more fitted by experience, position, contacts and loyalties to act as her political and commercial advisor in her strange new home of Burgundy, where he had lived for more than twenty years.... Then again, Margaret was a patroness of authors and scribes, a devoted reader of chivalric romance and pious literature, like Caxton himself, and no doubt delighted to welcome this uniquely literate merchant to her court.... Caxton's first customers ('diverse gentlemen and my friends') were no doubt found in Bruges and the Low Countries among the English retinue at Margaret's court, the English colony of merchants, and visiting English envoys. (Painter, 1977, pp. 44-45, 63)

At that time, practically nowhere in Western Europe was there more UP than in the Dukedom of Burgundy, especially under its last ruler, Duke Charles the Bold, the husband of Margaret of York, the sister of King Edward IV.[117]

Then, from 1476 onward, Caxton resided back in England, where "under the patronage of Edward IV, he brought in and set up the first printing press at Westminster" (Bowle, 1971/2000, p. 231).

There, when it was his turn to choose between a "royal unexpanded life" of St. Katherine, like the one by Voragine's, and an "imperial expanded life" like Capgrave's, he naturally opted for the latter, viewing, like Capgrave, "his writings as an extension of his responsibilities toward King Edward IV" (Winstead, 1999, p. 5). Caxton's straightforward preference for the imperial expanded life of St. Katherine, was the result of his adoption of ducal/royal UP scripts which brought him the patronage he sought:

> Caxton did not enter upon his new adventure of printing books without good and able patronage. Edward IV, as we have seen, paid him a sum of money for certain services performed; and Caxton printed "Tully" and "Godfrey" under the king's "protection". Edward's sister Margaret, Duchess of Burgundy, was his friend and supporter, and perchance may have paid a visit to her

[117] I refer the reader to what I wrote in volume 2, in the chapter entitled "The Contrasting Results of the Unity of Power of the Dukes of Burgundy, and the Division of Power of the Tuscan Communes.

> old servant at the "Red-pale," when she visited England in 1480. Margaret, Countess of Richmond, mother of King Henry VII, also favoured his designs. Earl Rivers, brother to the queen, was a fast friend, with whom Caxton seems to have enjoyed a considerable degree of intimacy, and the Earl of Warwick likewise must have had some knowledge of him as Caxton dedicated to him the "Chess-Book". The "Order of Chivalry" was dedicated to Richard III. Henry VII personally desired Caxton to translate and print the "Fayts of Arms," and the "Eneydos" was specially presented to Arthur, Prince of Wales. Master William Daubeney, King Henry VI's treasurer, was his "good and synguler friend." William, Earl of Arundel, took great interest in his progress, and allowed him the "yearly fee" of a buck in summer and a doe in winter. (Blades, 1971, pp. 80-81)

Painter (1977) had similar observations: "Caxton's propaganda was calculated to please many great persons, and perhaps each of these rewarded him.... Caxton's outright, twice repeated dedication of his book to the King is unique, although he had already produced three books (*Jason, Mirror, Tully*) 'under the shadow of his noble protection'" (pp. 115-16). Specific to *The Golden Legend*, and in Caxton's own words, and final comments by Painter (1977):

> "Forasmuch as this said work was great and overchargeable to me to accomplish," says Caxton, "I feared me in the beginning of the translation to have continued it, by cause of the long time of the translation and also in the emprinting of the same, and in a manner half desperate to have accomplished it was in purpose to have left it after that I had begun to translate it and to have laid it apart, had it not been at the instance of the puissant, noble and virtuous earl my Lord William Earl of Arundel, which desired me to proceed and continue the said work and promised me to take a reasonable quantity of them when they were achieved and accomplished, and sent to me a worshipful gentleman a servant of his named John Stanney, which solicited me in my Lord's name that I should in no wise leave it but accomplish it, promising that my said Lord should during my life give and grant to me a yearly fee, that is to wit a buck in summer and a doe in winter, with which fee I hold me well content. Then at contemplation and reverence of my said Lord I have endeav-

> oured me to make an end and finish this said translation, and also have emprinted it in the most best wise that I have could or might, and present this said book to his good and noble lordship as chief causer of the achieving of it, praying him to take it in gree of me William Caxton his poor servant, and that it like him to remember my fee."...So, with *Order of Chivalry* and *Golden Legend*, Caxton charmed the support of the monarch and the topmost earl, just as he had in the previous reign with Edward and Rivers; and in due course, with the Earl of Oxford and Henry VII, he would bring off the hat trick. (ib., pp. 143-45)

With this in mind, we can understand why his expanded life of St. Katherine is so UP, so different from the unexpanded life by Voragine. Not only did Caxton eliminate every part from Voragine's text which could indicate doubt or backsliding, but he added all kinds of episodes of which his aristocratic clientele would approve.

We do not know from where Caxton took the expanded portions of St. Katherine's life: if he took them from Capgrave, he compressed the text considerably. For instance, while Capgrave devoted 1498 lines of rhyme-royal, for ca. 11,400 words to "The Marriage Parliament" section, this same part of the story is told by Caxton in only about 1500 words. But in the end, it is not important from where Caxton took his text. In England, at that time, he was not the only one who felt the need to say more on saint Katherine: more of what would please a world indoctrinated with the admiration of the royalty.

We are faced here with that sad phenomenon discussed in the above mentioned chapter on Burgundy. In the two hundred years which elapsed from Voragine's text of ca. 1266 and Boccaccio's *Decameron* of ca. 1350, that wonderfully critical thinking engendered by the DP of the people of the Italian communes not only was unable to extend its influence to Burgundy and England, but went into a nose dive in Italy itself under the UP of the Renaissance *signori*.[118] At the same time,

---

[118] In Florence the rule of the *signori* had already begun before the advent of the Medici, whose principate started in 1434: "Cosimo traditionally has been accused of destroying the Florentine liberties; but these ancient liberties, more of an illusion than a reality, had already ceased to exist in the Florence of the Albizi. Cosimo only had to perpetuate the formula of those he was evicting, in other words, to maintain the appearance of a constitutional régime. But in order not to be taken by surprise like the Albizi, he perfected the system" (Jourcin, 1978, p. 816).

Burgundy and England, and most other countries in Europe, moved further toward establishing *absolute monarchies* (the rule of Charles the Bold over Burgundy was already *absolute*).

5. A Very Royal Saint: "Saint" Charlemagne

The transformation of Charlemagne into a saint is as prodigious as creating St. Katherine out of nothing or very little. Charlemagne had no claim whatsoever to sainthood with the exception of his imperial status which, under normal conditions, is an impediment: "It is easier for a camel to go through the eye of a needle, than for a rich man to enter the kingdom of God" (Matthew 19:24), and emperors are nearly always very rich; and "Blessed are the meek.... Blessed are the merciful.... " (Matthew 5: 5, 7), and emperors are rarely meek or merciful, and Charlemagne certainly was not.

Had Emperor Frederick Barbarossa not lost the battle of Legnano against the Lombard communes which were backed by Pope Alexander III (as discussed in volume 1), the Catholic church would now be saddled with a most unsaintly person: Saint Charlemagne.

> Charles caught pneumonia and died at Aachen on Jan. 28, 814. His corpse was borne in a sarcophagus to the church that he had had built. It remained there undisturbed until its formal translation in 1165 on the occasion of Charles's canonization at the request of the emperor Frederick I Barbarossa by the antipope Paschal III. The decree of an antipope did not of course bind the church; nevertheless, the popular cult of Charlemagne was fairly widely observed in Germany and France during the middle ages; so that he is now regarded as having been informally beatified. His feast is kept at Aachen on Jan. 28. (Devisse, 1970, p. 291)

> The idea to have Charlemagne canonized might have originated with Rainald [von Dassel, Frederick's chancellor]. It was designed to provide divine sanction for the greatest of Frederick's predecessors and by honouring Charlemagne some of the glory was to be reflected upon Frederick.... If the primary object of the canonization was now to establish the great predecessor of Frederick as a saint and thus invoke divine sanction for whatever policies Frederick was about to pursue, Rainald and Frederick had also a more narrowly political and immediate ob-

> ject in view: they wanted to prove that Frederick was superior to the king of France.... Louis might claim direct descent from Charlemagne and he might have the holy oil and his kingship might be considered especially sacred. But Frederick now had Charlemagne as his personal saint, and with the canonization Aix-la-Chapelle, the place where the kings of Germany were crowned, was elevated at the expense of St. Denis. (Munz, 1969, pp. 242-43)

However, there was nothing saintly in Charlemagne except the political interests of Emperor Frederick Barbarossa. Charlemagne had been extremely cruel against his enemies. As reported in volume 2:

> Charles's Saxon wars [772-85] were the most protracted and most bitterly fought of the numerous campaigns of his reign.... The conflict with the Saxons in this period involved features such as forcible mass-conversion and large scale deportation, together with the massacre of prisoners, that were not normal components of Frankish warfare. (Collins, 1998, p. 43)

> By 782, in response to an increased formalizing and application of militant Christianity, Witikind returned again from his refuge in pagan Denmark and summoned the northern tribes to revolt. Once again Charles applied superior force, and many of the rebels immediately submitted. But churches had been burnt, priests killed, and proud Saxons had ritually washed off their enforced baptism. Charles resolved upon a severe punishment. The most active rebel leaders and warriors were brought as captive to Charles on his command; 4,500 men were bound and taken by their countrymen to a camp at Verden on the river Aller. Charles ordered the beheading of the entire company, bound and helpless as they were. (Stewart, 1988, p. 31)

"Charles's own followers, perhaps even his own executioners, were horror-stricken" (Cabaniss, 1972, p. 39). Pierre Riché spoke of infamy (1993, p. 104); Russell Chamberlin noted that "the Saxons saw the King of the Franks now not as a civilizing force but one of destruction and terror" (1986, p. 121).

Also, Charlemagne's caesaropapism was completely unsaintly: "Charles certainly regarded the Frankish Church as his domain, in which he was largely his own master, without reference to the pope, whom he did not consult in the establishment of at least one new bishopric" (Eyck, 1999, p. 44).

Also, as reported in volume. 2: "The second Council of Nicaea was, in accordance with Charles's wishes, denied the title of Seventh Ecumenical Council of the Church, and its doctrines were condemned. Pope Hadrian's only alternatives were to swallow the offense to his authority or to break with Charles, and he was in no position to make a break. The papal legates cast their vote also for the rejection of Irene's council, and Charles's highhanded revision of the dogma was put on record with the tacit consent of Hadrian if hardly with his good will" (Winston, 1956/1969, p. 225).

The same is true for the letter that Charlemagne wrote in 796 to the new Pope Leo III, in which: "the ruler did the work while the pope turned the prayer wheel" (Arnold, 1977, p. 81). And before Pope Leo, "although Hadrian, a proud pontiff, protested at times, after a few years he was reduced to the role of docile auxiliary to the policy conducted by the Frankish king" (Eyck, 1999, p. 47).

On sexual matters, Charlemagne was as unsaintly as Alexander VI Borgia:

> There was, of course, a series of mistresses. Himiltrud, mother of his firstborn, Pepin the Hunchback, may have been a lawful wife, but if not, in addition to her, whose name Einhard does not give in his biography, there were three, probably four, other concubines whose names he does mention, and still another whose name he said he had forgotten. Gersuinda, a Saxon, gave birth to Adaltrud; Regina, to Drogo, Bishop of Metz in 823, who died in 855, and Hugo, later abbot of St. Quentin, who died in 844; and Adallind, to Theodoric, a monk who died about 819. There may have been Madelgard, whose daughter Rothild was abbess of Faremoutiers in 840 and who died in 852. The unknown concubine whose daughter was Rothaid may have borne the name Sigrada. It is impossible to establish dates for these mistresses. Charles thus had eighteen children by four (known) wives and six mistresses. Only two wives, the daughter of King Desiderius and Liutgard, left no issue. Of the children, eight were sons: four (known) legitimate and four illegitimate; and ten were daugh-

> ters, seven legitimate and three illegitimate. (Cabaniss, 1972, pp. 90-91)

Obviously, there is no Saint Charlemagne mentioned in such standard books as *The Oxford Dictionary of Saints,*[119] nor in the book by Jacobus de Voragine and William Caxton.

6. An Admirable DP Saint: St. Zita of Lucca (1218-72)

As narrated by Benigni (1912):

> At the age of twelve she entered the service of the Fatinelli family of Lucca. Her piety and the exactitude with which she discharged her domestic duties, in which she regarded herself as serving God rather than man, even supplying the deficiencies of her fellow servants, far from gaining for her their love and esteem and that of her employers rather brought upon her every manner of ill-treatment of both the former and, through their accusations, of the latter. The incessant ill-usage, however, was powerless to deprive her of her inward peace, her love of those who wronged her, and her respect for her employers. By this meek and humble self-restraint she at last succeeded in overcoming the malice of her fellow-servants and her employers, so much so that she was placed in charge of all the affairs of the house.... After her death numerous miracles were wrought at her intercession, so that she came to be venerated as a saint in the neighborhood of Lucca and the poets Fazio degli Uberti (Dittamonde, *III*, 6) and Dante (Inferno, *XXI*, 38 ["Here's an alderman of St. Zita's coming down]) both designate the city of Lucca simply as "Santa Zita". (p. 762)

David Farmer adds interesting information, especially also on her cult as far away as England:

> Many miracle stories were told of her, including the attribution to angels of the baking of her loaves while she was rapt in

[119] However Charlemagne is listed in the German *Reclam Lexikon der Heiligen und der biblischen Gestalten* (Reclam Dictionary of Saints and Biblical figures).

> ecstasy. Soon after her death a popular cult grew up round her tomb.... Her popular cult had already spread to other countries in the later Middle Ages, testified by chapels in her honour as far afield as Palermo and Ely. In England she was known as Sitha and was invoked by housewives and domestic servants, especially when they lost their keys or were in danger from rivers or crossing bridges. She occurs in mural paintings. (1997, p. 526)

A. The Stimulating Division of Power which Existed at Lucca

Few cities were so actively involved in the *War of Supremacy* (1075-1313) between popes and emperors—in the century and a half before the birth of St. Zita, and during her life—as Lucca in Tuscany (40 miles west of Florence). In those years, Lucca carefully shifted its allegiance, from emperors to popes and vice versa. But it sided with the emperors more often, and from them it gained major political and economic advantages, together with major ideological humiliations. In the words of Tammaro de Marinis:

> Characteristic is the position of Lucca in the conflict between Church and Empire. The imperial party, aiming at making Lucca independent of the Marquisate of Tuscany, opposed strongly Popes Alexander II [r. 1061-1073) and Gregory VII [r. 1073-1085]. Lucca, at those times, as Pisa, was a city faithful to the Empire, and this even after Canossa (1077), thereby obtaining conspicuous gains (1081 [i.e., in the middle of the war between Gregory VII and Emperor Henry IV]). Only later, it returned within the papal circle, and Pope Urban II, on his return from Clairmont stopped there to bless the crusader knights (1096).... Also Barbarossa (soldiers from Lucca fought for the Emperor against the freedom of Milan) was generous toward Lucca.... Lucca even surpassed Pisa in its fidelity to the Emperor by siding decidedly for the imperial anti-popes. (1934, p. 561)

This last point is important for the development of the *visitor personality* (see volume 1) because—with Legnano (1176), and Venice (1177) when Barbarossa recognized Pope Alexander III as the only and true Supreme Pontiff—all those imperial popes that Lucca had accepted one after the other (Victor IV, Paschalis III who canonized

Charlemagne, Calixtus III, Innocent IV) automatically became false popes, and all their decrees invalid.

Few cities were so well acquainted with the human limitations of emperors and popes, more expert in trading their allegiance for specific advantages, as Lucca was. Accordingly, few cities were richer in *visitor thinking*, and therefore prepared to rediscover sainthood in simple people like the housemaid Zita. "Blessed are the meek.... Blessed are the merciful.... " had said Jesus (Matthew 5: 5, 7), and Zita was both meek and merciful: a true DP saint like St. Francis of Assisi, the son of a merchant, and, like St. Katherine of Siena, the daughter of a dyer.

As was to be expected, neither Voragine nor Caxton included Zita of Lucca (1218-72) among their saints, while including Queen Elizabeth of Hungary (1207-31). For both, Zita's origins may have been overly simple, thereby negating the relation between noble blood and saintly virtues.

B. Pirenne's Indirect Contribution to the Understanding of Italy and its St. Zita

The eminent Belgian historian Henri Pirenne noted, in his study of the European merchant class, how the European nobility had nothing but disdain for commerce and banking: "Save in Italy, where aristocratic families did not hesitate to augment their fortunes by having an interest in commercial operations in the capacity of moneylender, the prejudice that it was degrading to engage in business remained deep-rooted in the heart of the feudal caste up to the time of the French Revolution" (1925/1969, p. 123).

These comments on the Italian aristocrats being willing to be moneylenders, reminded me of Dante's background: the moneylending done by his father, and the aristocratic status of his family. Specifically, "his father Alighiero may have been a notary and was certainly a moneylender, doing well enough out of his transactions to buy property in the city and in the *contado*, the income from which supported Dante without the necessity of his practicing a profession up to the time of his exile" (Anderson, 1980, p. 53); and: "The family certainly was a noble and ancient one, though it did not belong to the 'Grandi'.... Besides, the rapidity of Dante's political career, the circumstance that he could marry a lady from the first house of the town, that of the Donati, moreover that the young King of Naples was one of his personal friends—all these facets seem to prove his noble birth" (Federn, 1902/1970, pp. 196-97).

Why were the Italian aristocrats, and so many other Italians, different? In the words of William Anderson: "The [Italian] communes arose in response to one crisis in papal and imperial relations [the *Investiture Controversy*]; they were to find their maturity in another. Frederick I Barbarossa, ..." (1980, p. 30). Then, an additional explanation may be found in my interpretation (in chapter 1 of volume 2) of the sang-froid of Boccaccio's Ser Ciappelletto in making, shortly before his death, a fake confession. I stressed that the *Decameron* had been written in a country and times saturated with partisan propaganda that sharpened the general thinking: the propaganda of emperor against pope and pope against emperor, in a country divided between Guelphs and Ghibellines.

Not that in Italy, cardinals, bishops, priests and friars were much more corrupt than elsewhere, but people there were able to look at their clerics with critical eyes, with the eyes of *visitor people* (see chapter 11 of volume 1). In other words, those Italian aristocratic money-lenders probably had the same doubts as Ser Ciappelletto about the ability of priests and friars to be God's agents: to give valid confessions and absolutions.

The majority of the Italians certainly did not have as radical a position as Ser Ciappelletto, nor so relaxed an attitude toward money-lending as the aristocrats discussed by Pirenne; yet they used their *visitor critical thinking* in their wish to return to the virtues of the Gospel by honoring such saints as Francis of Assisi (1181-1226), Clare of Assisi (1191-1273), and Zita of Lucca (1218-72), and by keeping away from the aristocratic propaganda of the stories of St. Katherine [A], St. Katherine [B], and St. Charlemagne.

7. Conclusion

Once more we seem to encounter the distortions brought by the Unity of Power: from totally inventing a saint who will proclaim the intrinsic virtues of royal-imperial blood, to establishing, for political propaganda, the sainthood of an unsaintly person. The UP in Burgundy and England led to the "upgrading" of St. Katherine from A to B. In Germany, the long shadow of the UP of Emperor Barbarossa continued: even today St. Charlemagne (*Der Heilige Karl der Grosse*) is venerated there,[120] leading many faithful—ignorant of the historical

---

120 "Veneration permitted, but not recognized" by the Catholic Church (Keller, 1968, p. 307).

Charlemagne—to associate imperial power with sainthood, and national glory with the will of God.

In the other direction, the strong DP of Lucca meant a return to the virtues of the Gospel, and the discovery of a true saint in a house servant!

13

Why Did Venice Have No Great Writers When it Had Great Painters, Architects, and Musicians?

1. Introduction

Venice is an extraordinary place, praised for centuries by such connoisseurs as Goethe and Ruskin, for its architects, painters, and composers. Everybody comes back enchanted. However, little of this praise goes to its writers. On this point, the silence is nearly universal. Indeed, none of the great Italian writers were Venetian; not Dante, Petrarch, Boccaccio, Machiavelli, Ariosto or Tasso. Even the stars of second and third magnitude of Italy's *golden centuries* (14th, 15th and 16th), like Guicciardini, Castiglione, Pulci, Boiardo, Vasari, were not Venetian.[121]

Others have noticed this too. Even the Venice-friendly William McNeill, in his *Venice: the Hinge of Europe, 1081-1797*, was forced

---

[121] Dante (born in 1265), Petrarch (1302), Boccaccio (1313), Machiavelli (1469), were all Florentines; Ariosto (1474) came from Mantua-Ferrara, and Tasso (1544) from Sorrento-Naples; Pulci (1432) and Guicciardini (1483) were again Florentines; Castiglione (1478) from Mantua in Lombardy; Boiardo (1441), from Scandiano-Ferrara; Vasari (1511) from Tuscan Arezzo, 40 miles south-east of Florence. A partial exception could be Paolo Sarpi (b. 1552), the author of *History of the Council of Trent* first published in London in 1619. Later there will be such writers as Goldoni (b. 1707), and autobiographers like Casanova (b. 1725), but they already belong to a different world: Goldoni wrote his *Mémoires* in French and dedicated them to King Louis XVI, and Casanova also wrote his *Histoire de ma vie* (History of my life) in French.

to admit that "until after 1481, when Venice's imperial power was already entering on a downward path, the city remained a follower, not a pace-setter in matters cultural" (1974, p. 92).

Similarly, the Venice-friendly Vincent Cronin—in his *The Flowering of the Renaissance*—could only assemble four names in praise of the Venetian classical revival: "Aldus Manutius [b. 1450], who typifies Venetian freedom of the press; Sperone Speroni [b. 1500], a characteristic Venetian humanist in that he thought for himself; the poetess Gaspara Stampa [b. 1525], who expressed without inhibition the sensuous, passionate element in the Venetian character; and lastly, Pietro Aretino [b. 1492], an outspoken champion of political and intellectual freedom" (1969, p. 180). However, the first is neither a writer nor Venetian (he was born of Roman parents, and educated in Rome and Ferrara), the next two are little known even in Italy, and the last, again, is no Venetian given that he was born in Arezzo (Aretino means "of Arezzo" from the Latin Arretium), moved first to Perugia in his adolescent years, then to Rome, and finally to Venice at the relatively advanced age of 35.

On the other hand, as soon as we move to painting and architecture, Venice comes into its own: Giovanni Bellini (1430), Carpaccio (1465), Giorgione (1478), Titian (1490), Palladio (1508), Tintoretto (1518), Veronese (1528), were all Venetian.[122] Later came a string of great musical composers: Andrea (1510) and Giovanni (1556) Gabrielli, Vivaldi (1665), Alessandro (1684) and Benedetto (1686) Marcello, Tartini (1692), Galuppi (1706).

Usually, this combination of few great thinkers (poets, novelists, essayist, philosophers, historians) and many great painters, sculptors, architects and composers is the result of many centuries of unity of power. In the case of Venice it was the unity of power of its oligarchy and of its cultural Byzantine roots.

2. The Strong Unity of Power of the Venetian Oligarchy

As if living in pre-Canossa times:[123]

---

[122] On the other hand, there is no *either/or*, and great painting, sculpture, and architecture can go along with great writing as proved by the following eminent Florentine (or Tuscan) artists: Giotto (1266), Donatello (1382), Fra Angelico (1395), Ghiberti (1378), Masaccio (1401), Piero della Francesca (1415), Pollaiuolo (1431), Botticelli (1445), Leonardo da Vinci (1452), Michelangelo (1475).

[123] My italics hereafter.

- *The doge had the right to appoint the Patriarch of Venice and the Canons of St. Mark* and had, among other titles, that of "Most Serene Prince." (Montanelli & Gervaso, 1967, p. 164).

- The strenuous demands Venetian magistrates made on their citizens' loyalty left scant room for other ties. In particular, *the Venetian clergy were strictly subordinated to secular authorities*. The pope figured more often as a foreign power than as an ecclesiastical superior; and the religious enthusiasm associated wit the Dominican and Franciscan friars lapped up against the well-consolidated patterns of Venetian political and economic behavior without the one much affecting the other, so far at least as public event are concerned. A Venetian Savonarola never arose; religion remained always subject to civic and secular control. "*Veneziani, poi Christiani!* [Venetians first, then Christians!]" accurately sums up the city's relationship to Latin Christendom, insofar as Latinity found its Christian expression in obedience to the pope. (McNeill, 1974, p. 92-93)

- [Some] historians still speak of a lay and thoroughly politicized Venice, of a "cesarismo veneziano" dedicated to protecting the overriding interests of the state against monastic and ecclesiastical interests;... The exercise of jurisdiction over ecclesiastics did not derive, in Sarpi's eyes,[124] from an abstract principle of political sovereignty, but from a living tradition which was at once political and religious: "The ground whereby the Republic judges ecclesiastics lies in the fact that it received this power from God at its birth, and that it has exercised it continuously and without interruption whenever it seemed to be to the advantage of the common good, and in the records that survive we see that at all times all kinds of ecclesiastics have been subject to the magistrates on account of any crimes it seemed proper to charge them with. And if popes have passed laws subsequently, the city has not renounced the power given it by God..." (Prodi, 1973, pp. 410-11)

---

[124] Paolo Sarpi (1552-1623) patriot, scholar, and state theologian during Venice's struggle with Pope Paul V.

Fascinating! Sarpi's argument is nearly identical to that used by Emperor Henry IV against Pope Gregory VII, during the *Investiture Controversy*, which said, in essence, that the (German) king and emperor had always been *rex et sacerdos* by the pious ordination of God, and that kings and emperors had always designated and invested bishops. Gregory's opposite answer was equally clear: "The Lord hath not said 'I am Tradition,'[125] but 'I am the Truth'" (quoted by Tellenbach, 1936/1991, p. 164).

Even stronger than the position of Emperor Henry IV, is the one taken by Venice as "direct religious follower of Byzantium":

> Sarpi's frequent references to the model of the imperial church of Constantine, Theodosius and Justinian are well known.... Venice still regarded itself as the true heir of the [Roman/Byzantine] imperial Church: the doge, *princeps in republica* and *princeps in ecclesia*, a latter-day Constantine, is responsible for the unity and prosperity of both church and state in an equilibrium elsewhere lost in both east and west. (Prodi, 1973, p. 413; italics in the original)

There is, indeed, no stronger unity of power than that which combines in one person or hierarchy the command of the state and of the church! At this point we can refer to what the eminent Belgian historian Henri Pirenne said on the Venetian Byzantine roots:

> At the date the city [Venice] was founded, all Italy still belonged to the Byzantine Empire. Thanks to her insular situation, the conquerors who successively overran the peninsula—first the Lombards, then Charlemagne, and finally, still later, the German emperors—were not successful in their attempts to gain possession. She remained, therefore, under the sovereignty of Constantinople, thus forming at the upper end of the Adriatic and at the foot of the Alps an isolated outpost of the Byzantine civilization. While Western Europe was detaching herself from the east, she continued to be part of it. And this circumstance is of capital importance. The consequence was that Venice did not cease to gravitate in the orbit of Constantinople. Across the

---

[125] Indeed, "Lay control had centuries of tradition behind it" (Brooke, 1964, p. 27).

> waters, she was subject to the attraction of that great city and herself grew great under its influence. (1925/1969, p. 83-84)

The consequence was, also, that Venice's literary output was not much higher than that of Byzantium which was little indeed, as discussed in chapter 14 of volume 1 "The Low Creativity and High Unity of Power of the Byzantine Civilization". Just as in Byzantium, but not elsewhere in Christendom: "Art was at the service of the state in Venice, with an intensity not paralleled elsewhere in Italy.... In order to celebrate themselves, influential men had to glorify the state by showing themselves in service of it" (Wills, 2001, p. 21). Art at the service of the state is not too bad when done in painting and sculpture, and music, but not in literature because it quickly turns into propaganda.

3. Down-to-Earth *Fox* Ideas

> The most general element in the political ideal to which Venetian writers exposed their audience was a ubiquitous secularism.... expressed rather in an antipathy to speculative systems that impose an artificial coherence on all values and experience and thereby claim a right to supervise, among other matters, the political order. They were enemies not of religion but of metaphysics, and of the notion that the conduct of human affairs should be determined by some comprehensive vision of the nature of things[126].... The Venetian approach to human affairs was earthbound and empirical. Its refusal to force the data of human experience into large systems was notably exhibited in the preference of Venetian writers for exposition through dialogues in which various points of view may find expression without explicit resolution. (Bouwsma, 1990, p. 267)

4. Complicating Matters: The Secret Police

> [The members of the all powerful Council of Ten] could not leave Venice except on emergencies, and three times a week

---

[126] One seems to read from Isaiah Berlin's definition of the *fox thinkers* "who pursue many ends, ... moving on many levels, seizing upon the essence of a variety of experiences and objects for what they are in themselves, without, consciously or unconsciously, seeking to fit them into, or exclude them from, any one unchanging, all embracing,... unitary inner vision" (1953/1967, pp. 1-2).

they held audience for spies and informers. In other words, they were a real secret police, and early version of the NKVD. (Montanelli & Gervaso, 1967, p. 166)

There was no appeal against their decision. The only constitutional hold on them was the fact that they were elected by the Great Council, annually and then every six months, and for a limited period which started at ten days and then by stages became permanent. Really it created a police state; and some people date Venice's decline from this point. It was started [in 1325] to stamp out the last traces of a conspiracy in 1310. Everyone outside the nobility now knew that their role was to watch and listen. And it says volumes for the genius of the noble class that the ordinary people and the middle class were happy to do this for another half millennium....

The dark Venice that has frightened historians ever since came into being in the fourteenth century; it was government as conceived by the military, a régime supported by spies and secret murders. The three "inquisitors of state", as they came to be called, grew out of the Ten, as its secret committee.... The Three could punish an offender secretly or publicly as they thought expedient for the state. The stifling *piombi* (cells under the prison roof) and the sodden *pozzi* (at canal level) were at their service. They could draw any amount of money from the treasury of the Ten without explanation. Even the identity of the Inquisitors was not revealed....

The fact was that the governing families could never rely on themselves to keep order. That was why they elected despots, and needed the despotism while hating and fearing it. That was the reason first of all for the Ten and then for the brisker version of the Ten in the Three.... But no inflexible police state could have lasted even the five hundred years Venice did last after the Closing of the Council, let alone a thousand. Her secret was that her nobles shared not only the fruits of despotism but the discipline as well. (Rowdon, 1970, pp. 16-19)

[The Council of Ten] took the initiative in internal police. It was on the look-out to suppress not only any possibility of armed insurrection, but also of any noble acting as if above the law, and any attempt to organize a faction or party even if merely by

> soliciting and swapping votes. They permitted no organized opposition.... Both the Inquisitors and Council of Ten had strict rules of procedure to follow in passing judgment and inflicting punishment; but their proceedings were secret. The accused had no chance to face accusers, to hire lawyers, or to know fully the case against him. Judgments could not be appealed, and they were sometimes executed with unseemly speed. Some glaring cases of injustice shook confidence in the procedure. Antonio Foscarini, an ambassador accused of selling state secrets, was speedily tried and executed. A few months later it was discovered that he had been framed, his accusers had been lying. The Council of Ten publicly admitted its mistake, restoring honor but not life to the accused. In some other cases, there were shocking contrasts between the severity of sentences imposed on minor nobles and on members of the most powerful families. (Lane, 1973, pp. 117, 404)

5. In the End

I have no doubt that the major cause of Venice's lack of great writers was the first of the three discussed above, that of the UP of state and church, of the Venetian oligarchy. A major DP, instead, as it existed in Florence, would have limited the power of the aristocrats and much reduced the power of the police. Equally important, an intellectual DP church-state would have modified the Venetian earthbound and empirical approach to human affairs. As the reader knows, I have nothing against the Berlinian *foxes,*[127] but a world composed of only small *foxes*—with no *hedgehogs,* big or small—cannot have great literature, as it also happened in Edo during the Tokugawa centuries.

Also, the DP provides conflicting ideas and data, which can then be compared, accepted, refused, or overcome. Only the DP provides the valid elements for a network of dialectical creativity. The UP, instead, provides one single truth, the one that the *supremo(s)* likes.

As discussed in chapter 1 of volume 1, the initial sparks of creativity come from the clash of personal scripts with those of society in a major *war of the scripts*. However, if the scripts of society are so powerful, universally accepted, never questioned, and never debated,

---

[127] On the Berlinian *foxes* and *hedgehogs*, please see chapters 8 and 9 of volume 1, and chapters 5 and 7 of volume 3.

then the individual *war of the scripts*, if any, will be limited to down-to-earth matters (e.g., some technical creativity), or, in the higher spheres of creativity, to the visual and auditory domains (painting, sculpture, architecture, music). As seen above: "Art was at the service of the state in Venice, with an intensity not paralleled elsewhere in Italy" (Wills, 2001, p. 21).

To this same category of "service of the state" belongs the creativity of the previously mentioned Paolo Sarpi who wrote—in defense of the Venetian tradition of independence from Rome—as the theological adviser to the Republic against the Pope. Not surprising, "the *History of the Council of Trent*, like most of Sarpi's writings, is a partisan work, written with one eye on the Protestants whom Sarpi saw as Venice's potential allies against Rome and Spain" (Cronin, 1978, p. 252).

The case of Venice is of special importance because its unity of power lay mid-way between that of Byzantium and Florence, and its creativity was also mid-way. It is worth remembering, that we are comparing three Christian societies at about the same time (Middle Ages to Renaissance), and that Venice was not as isolated, linguistically and geographically, as Byzantium was. As seen above, two of the four "Renaissance Venetians" praised by Cronin came from other parts of Italy: one from Rome, the other from Arezzo. Later, the Tuscan Galileo (born in Pisa in 1564) would teach for 18 years in Venetian Padua, performing there the bulk of his most outstanding work.

Having stressed the importance of the division of power between state and religion, let me briefly expand on McNeill's statement that "a Venetian Savonarola never arose" by adding that Savonarola was able to give so much trouble to Lorenzo de Medici because Lorenzo did not have the UP of the Venetian Council of Ten: "Savonarola preached boldly against the tyrannical abuses of the government. Too late Lorenzo tried to dam the dangerous eloquence *with threats and flattery,...*" (Ridolfi, 1978, p. 298; my italics to stress the limits of Lorenzo's power). In reality, the Venetian Council of Ten never had to dam the eloquence of any "Savonarola" with flattery, threats, *piombi*, or *pozzi*, because, as in Byzantium, it took care not only that friars and priests would never dare to voice the slightest criticism against the system, but also—and more efficiently—that they would enthusiastically back it as Paolo Sarpi did.

14

The DP Origin of Western Medicine: On the History of the Scientific Use of Dissection

I. Introduction

Roy Porter could not have given a more appropriate title to his 1997 book of history of medicine than *The Greatest Benefit to Mankind*. Medicine, indeed, greatly helps everybody, from birth to old age, provided that it is scientific medicine, i.e., based, first of all, on a sound understanding of the human body, its anatomy and physiology.

In reality, the study of the human body through dissection was much more, it was a revolutionary change in the way man understood himself as commanded by the Delphic "Know thyself". Only with this brand new knowledge, and derived self-confidence, could man follow Protagoras's statement that "Man is the measure of all things." It is fair to say that without the DP practice of dissection we would not have the Renaissance as we know it: the rediscovery of the classic authors could not bring more than what the Greeks and Roman knew and dared to know. Indeed, the most impressive drawing of the Renaissance is Leonardo da Vinci's dissectional presentation of a fetus inside the womb: "Faulty as it is in some respects, in others—notably the position of the fetus and umbilical cord—it is so accurate and so expertly drawn that it can still be used as an example in medical textbooks today" (Wallace, 1967, p. 103).

The beginning of this scientific understanding of the human body is relatively new in the history of mankind, and can be dated with some precision to the early years of the fourteenth century when Mondino de' Luzzi made his first public dissections of human cadavers at Bologna, and then described them in his famous *Anotomia Mundini* of 1316-17. Before this date, and elsewhere in the World (with the exception of some work done in Alexandria around 300 B.C.), there were such strong religious/ethical prohibitions against dissection of human cadavers that the beginning of the medical science was blocked.

In chapter 13 of volume 1 of my *GAM/DP Theory*, I briefly discussed the contrast between what had been done in the West and what had not been done elsewhere,[128] and how the *visitor mentality* which began at Canossa 1077 (see chapter 11 of volume 1) and the existence of a greater division of power (DP) in Italy at the time of Mondino de' Luzzi, were essential for the beginning of the medical use of dissection. Still, more needs to be said on this important subject, especially on *forensic (partial) dissections* and on *university public/ book-driven (complete) dissections* (e.g., those of Mondino de' Luzzi) which preceded truly *scientific dissections* (e.g., those of Vesalius). We also need to examine the exceptional conditions in Alexandria at the time of Ptolemy I which favored dissections by Herophilus, and then discuss the rapid surge of the unity of power (UP) under Ptolemy II and the consequent decline of the Alexandrian School of Medicine.

Before proceeding, we need to appreciate how revolutionary those first dissections at Bologna were because, from our present point of view, those dissections may not seem so heroic; they were so un-heroic, in fact, that a number of modern historians have been critical of the slow pace with which things progressed, and astonished that the dissections by Mondino de' Luzzi were not performed more scientifically and did not produce major anatomical discoveries. However, at a minimum, these historians forgot to ask why scientific dissection was not practiced in the other civilizations, nor did they address the strong repulsion against tampering with human cadavers which was also an integral part of the Greco-Roman world and of the first twelve centuries of Christianity:

---

[128] For instance, "it must be remembered that Muslims were strictly prohibited from dissecting either human bodies or living animals. Thus experiment was practically impossible in medicine, so that none of Galen's anatomical and physiological errors could be corrected" (Meyerhof, 1931, p. 344).

> [In mainland Greece] knowledge of human internal anatomy had been seriously hampered by superstitious beliefs which still invested the human corpse. Religious scruples, veneration of the dead and dread of the corpse itself had all combined to bring about a deeply entrenched and powerful taboo against human dissection. (Longrigg, 1993, p. 184)
>
> As long as men have still been capable of humane sensitivity, it has never been possible to dissect without a feeling of inner revulsion. No one, aware that the body into which he is cutting is the body of a human being, can do so without emotional resistance, even though his reason may refute all arguments and show him that his feelings are unreasonable.... The Hellenistic physicians had at least had the help of the kings, who were themselves schooled in philosophy. The Roman Empire permitted no dissections. Only in Egypt, as Galen says, not in the European provinces, could anatomy still be practiced. The latinized world at any rate allowed no dissecting, and it is not by chance that a Latin writer tells that even dissection of animals was sufficient grounds for convicting him of the practice of magic. (Apuleius, *Apologia*, ch. 25 ff.) (Edelstein, 1932/1967, pp. 279, 280, 283)

There had, indeed, been a short-lived exception around 300 B.C. when Herophilus and Erasistratus had been able to practice dissections on human cadavers in Alexandria, with the help of Ptolemy I. However their work was neither repeated nor expanded: "In Alexandria human pathological necropsies were probably obtained without difficulty; in Rome the current prejudice against them was prohibitive, and it persisted even into the fifteenth century" (Allbutt, 1921/1970, p. 161).

The situation was not much different during the early Middle Ages, thereby explaining the admiration by Lino Sighinolfi (1930) for the Bolognese School of Medicine which, toward the end of the thirteenth century had been able to "overcome the obstacles kept for long centuries by the prejudices on the inviolability of the cadavers and the blind faith in the authority of Galen, through the direct study of anatomy" (p. 9).

It is, therefore, against this very restrictive background that we must place our understanding of the first steps in the Western practice of the dissections of cadavers.

Dissections in the West evolved in four stages, each preparing the way for the next:

- *Forensic/post mortem dissections*, to establish the cause of death in uncertain cases
- *(Simple) University-public-bookdriven dissections,* to teach anatomy (Mondino de' Luzzi)
- *(Formal) University-public-bookdriven dissections,* to teach anatomy (post Mondino)
- *Scientific dissections*, to learn more about the human body (Vesalius)

P*ublic dissections*, practiced first at the University of Bologna in 1315, were indeed exceptional, the result of the DP between emperors and popes, which had begun at Canossa in 1077 and had been greatly expanded, after the battle of Legnano of 1176, by the freedoms granted by the *Magna Carta of the Communal Freedoms* of 1183.[129] After Bologna, the practice of public dissection spread throughout Europe: Montpellier, 1377; Lerida, 1391; then Prague, 1460; Paris, 1478; Tübingen, 1485 (see footnote 70, at p. 192 of volume 1).

II. DP and *Visitor Thinking* (1261-1316)

The fist date of the above title section corresponds to the foundation of the *Milizia della Beata Vergine Maria*[130] at Bologna, the second to the completion of Mondino de' Luzzi's *Anothomia Mundini* "the first book of anatomy of the Middle Ages which is based on the dissection of human cadavers" (Wickersheimer, 1926, p. 65).

---

[129] As noted by Montanari (1973), "[Bologna] after a feudal period became a free commune when the emperor recognized its rights (1183)." The emperor, here, is Frederick I Barbarossa, and the date of 1183 refers to the *Peace of Constance with the Lombard Communes* (the Magna Carta of the Communal Freedoms) signed by the Emperor after his defeat in 1176, at Legnano, by the soldiers of the Lombard League (see chapter 23 of volume 1). Bologna is one of the seventeen cities specifically mentioned in the treaty together with Bergamo, Brescia, Faenza, Lodi, Mantova, Milan, Modena, Novara, Padua, Parma, Piacenza, Reggio, Treviso, Verona, Vercelli, Vicenza. (*Pax facta*, 1183/1976, p. 28). These "towns were marked by an unparalleled freedom. They had developed as autonomous worlds and according to their own propensities" (Braudel, 1979, pp. 509-510).

[130] The Militia of the Blessed Virgin Mary; also called the *Ordine Cavalleresco della B. V. Gloriosa* (The Knightly Order of the Blessed Glorious Virgin) or *Cavalieri di Santa Maria* (The Knights of Saint Mary).

1. *Forensic Dissections*

Our story begins in Bologna. The religious order of the *Milizia della Beata Vergine Maria* was founded by Loderingo degli Andalò, Catalano dei Malavolti and other nobles to quell riots of any kind, especially fights between "papal" Guelphs and "imperial" Ghibellines[131]. Its members were divided in two groups: clerical brothers who lived in convents, and lay/married brothers who lived in their homes. The brothers could bear arms like the crusaders, and use them "*pro sedantis tumultibus civitatum*" [to calm riots in the city]. Because of this the brothers were also called *milites* (soldiers).

Loderingo came from a Ghibelline family, Catalano from a Guelph family, a fact that helped their *Milizia* in imposing its will on both parties. In 1261, the *Milizia* was approved by Pope Urban IV, and first in 1265, then in 1267, brother Catalano was elected *reggitore* (governor) of Bologna. Loderingo and Catalano, having been so successful in Bologna, were invited to come to Florence in 1266 to prevent fights between Guelphs and Ghibellines after the death of Manfredi, the son of Emperor Frederick II, at the battle of Benevento (Bosco & Reggio, 1979, p. 344).

As part of their efforts to maintain order and peace, Loderingo and Catalano issued rules in 1265 in Bologna—backed by "the full authority of the Commune and the People"—which ordered that every case of doubtful death be studied by at least two physicians who would then officially report the results of their investigations. *The consequence of this ruling (and related financial compensations or fines) was a number of forensic autopsies as that performed on February 15, 1302 by two doctors* (Bartolomeo da Varignana, the leader, and Giacomo Rolandino) and three surgeons (Giovanni da Brescia, Pace degli Angeli, and Tommaso Grinci) (Sighinolfi, 1930, pp. 1-22; my italics).

A certain Azzolino degli Onesti, after having been visited by a doctor Barufaldino, had suddenly died, and poisoning was suspected; therefore the request of a post-mortem examination. The doctors and surgeons performed a careful dissection, and concluded that Azzolino had not been poisoned but had died of an excess of blood in the vena cava.[132] The five experts went very far in their examination by also

---

131 See section 5, "Ideological Guelphs, Ideological Ghibellines" chapter 13 of volume 3.

132 "*Quo dictum Azolinum ex veneno aliquo mortum non fuisse, sed potius et certius ex multitudine sanguinis aggregati circa venam magnam* ... [The said Azzolino did not die of poison, but certainly from too much blood which had accumulated around the vena cava" (quoted by Sighinolfi, 1930, p. 6).

looking carefully at the viscera, as said in their report "*viscerus eius anatomice circumspectis.*"

"Their report is in existence, and terminates with the words: 'we have assured ourselves of the condition by the evidence of our senses and by the anatomization of the parts.' The statement is given without any remark that the proceeding was unusual. This Bartolomeo da Varignana was a pupil of Thaddeus [of Florence] and was professor of Medicine at Bologna" (Singer 1925/1957, p. 73).

2. From *Forensic/Post-Mortem Dissections* to *(Simple) University-Public-Bookdriven Dissections*

> It is highly probable that—in those same times and thanks to the political endeavors of those same Loderingo and Catalano—the Faculties of Medicine and of the Arts were officially accepted as members of the *Studio*, along the older Faculties of Law of the *Ultramontanes* and *Cismontanes*, given that there is no *Rector in Fisica*, among the rectors of the Studio before 1268. (Sighinolfi, 1930, p. 6)

This, in turn, says that the post-mortem cooperation between doctors and lawyers had progressed to the point of demanding that dissection be taught publicly as part of a university curriculum: "As time went on, post-mortem examination passed into anatomical study" (Singer 1925/1957, p. 71).

It is worth stressing, that in the whole history of this development there is not one word about a prince, king or emperor, about a bishop, cardinal or pope, but only about the common people of a free commune, taking order in their own hands and expanding the scope of their university.

3. Mondino de' Luzzi

> *Janvier et mars 1315, ces deux dates sont à retenir: l'histoire des dissections humaines en Occident n'en connaît pas de plus anciennes.*[133] (Wickersheimer, 1926, p. 68)

---

[133] "January and March 1315, these two dates must be remembered; the history of human dissections in the West does not know of older ones." These are the dates mentioned by Mondino when he anatomized two women.

> The *Anothomia* of Mondino, written in 1316, is the first modern work on the subject. Those who preceded him incorporated their anatomical work in larger treatises on Surgery, and do not refer directly to their own anatomical experiences. With Mondino this is changed. His work is entirely devoted to Anatomy, and is essentially a practical manual of the subject[134]... Though dissecting on his own account, Mondino is yet relying almost entirely on Arabian authorities.[135] He is really dissecting to memorize their works, much as a student nowadays dissects to memorize his textbook, not to enlarge knowledge nor to make discoveries. The scientific spirit has hardly awakened in Mondino. Nevertheless, he has made a great step forward. But lastly, and above all, we would emphasize the fact that Mondino dissected *in person.* (Singer, 1925/1957, pp. 75-76; italics in the original)

Similarly, in the words of Sarton (1947):

> As contrasted with many of the anatomists who followed him, until Vesalius' time, Mondino did his own dissecting; but this was partly due to the fact that anatomical teaching was not yet deemed sufficiently important and the professor had to shift for himself without the assistants *ostensor* and demonstrator, who were associated with him a little later. (p. 842)

4. The *Visitor Mentality* at the Base of those Revolutionary Bolognese Dissections

In chapter 13 of volume 1, I briefly illustrated the *visitor mentality* through a short poem by Cecco Angiolieri who lived at about the same time as Mondino. Here, I will quote from a novella by Franco Sacchetti, born ca. 1335, not long after Mondino (b. ca. 1270).

---

[134] "The *Anothomia* of Mondino is not arranged like a modern anatomical textbook, which deals with the various systems consecutively. It more nearly resembles a modern manual of dissection, in which the organs are described in the order in which they present themselves" (ib., p. 76).

[135] Sadly, by this time these same Arab authorities were losing ground at home: "While the orthodoxy of early Islam tolerated the sciences, we may say that, from the time of the famous religious teacher al-Ghazali (d. 1111) onwards, this tolerance gave place to persecution of these studies 'because they lead to loss of belief in the origin of the world and in the creator'" (Meyerhof, 1931, p. 337).

The story deals with a near contemporary of Sacchetti: Bernabò Visconti (1323-1385), Duke of Milan, famous for his passion for hunting dogs which he gave to others to rear. One of the persons responsible for his dogs is the abbot of the story who seems to have been negligent in caring for two mastiffs. When Bernabò spoke of punishment, the abbot asked for mercy. Bernabò answered that he would spare him if he could explain to him four impossible things. The abbot was desperate, until the miller of his abbacy offered to answer Bernabò in his place, if he could be properly disguised as abbot. This is the dialogue between the miller (abbot) and the Duke:

> "Yea, my lord. Ye asked of me how far it is from here unto the heavens. All things considered, the distance from here to up above is thirty-six millions and eight hundred and fifty-four thousand and seventy-two miles and a half, and twenty-two paces."
>
> Said the lord: "Thou hast measured it very precisely; how canst thou prove this?" The miller answered: "Send ye and have it measured, and if it be not thus, then hang me by the neck. Secondly, ye asked how much water there is in the sea. This has been very difficult for me to determine, because water is a thing which never stayeth still and more is ever pouring in; nevertheless, I have found that in the sea there are twenty-five thousand and nine hundred and eighty-two millions of hogsheads, and seven barrels and twelve gallons and two cups." Said the lord: "How dost thou know this?"
>
> He replied: "I have measured it as well as I knew how; if ye do not believe it, cause barrels to be found and measure it yourself. Thirdly, ye asked of me what things are done in hell. In hell they cut to pieces, quarter, rend and hang, neither more nor less than ye do here." "What proof givest thou of all this?"
>
> The miller made answer: "I held discourse formerly with one who had been there, and it was from this man that Dante the Florentine learned that which he wrote of things appertaining unto hell; but he is dead. If ye do not believe it, send one there to see."...
>
> On hearing this the lord discerned but too well that this could not be the abbot, and looking at him very fixedly, and perceiving him to be a man of much greater learning than was the abbot, he said: "Thou art not the abbot!... It is well; and since he hath made thee abbot and thou art worth more than he, in God's name

I will confirm thee in his place; and I will that from henceforth thou shalt be the abbot and he shall be the miller, and thou shalt have the revenue of the monastery and he shall have that of the mill." And this thing he made to continue all the days of his life, and the abbot became miller and the miller became abbot. (ca. 1378/1908, pp. 10-11)[136]

Several points deserve to be highlighted:

- The notion that very simple persons could think clearly and talk courageously to a mighty lord. This may not have been actually true, but it is indicative of the sense of independence and pride that the people of the free communes had at those times.[137]
- The courage of Sacchetti—and others in those free Communes—in decrying the evils of the great lords: "Thirdly, ye asked of me what things are done in hell. In hell they cut to pieces, quarter, rend and hang, neither more nor less than ye do here."[138]
- A passion for measurement and quantification (as described by Crosby (1997) in his book "*The Measure of Reality: Quantification and Western Society, 1250-1600*", discussed in chapter 10 of volume 3).[139] In turn, such a love for measurement and quantification helped to open the door for

---

[136] I skipped the fourth question and its answer which do not add much to our discourse.

[137] See on this chapter 11 of volume 1 "The Long-Term Effect of Power on Creativity"; and chapter 1 of volume2 "The Contrasting Results of the Unity of Power of the Dukes of Burgundy and the Division of Power of the Tuscan Communes", in particular the discussion of the courage before death of Ser Ciappelletto of Prato. See also chapter 24 of volume 1, "Dante, Petrarch, Boccaccio, Sacchetti, *Visitors*", where I discuss two other *visitor* novellas by Sacchetti.

[138] Equally critical of a great man in power is novella CXXV entitled: "Charlemagne believeth he hath converted a Hebrew to the true faith; but the Hebrew, being at table with him, reproveth him for not observing the Christian faith as he should, whereby the ruler remaineth much disturbed" (ib., p. 105).

[139] "Western civilization's acquaintance with quantification certainly dates at least as far back as the Neolithic (my herd has twelve goats and yours only seven), but millennia passed before it became a passion.... Then, between 1250 and 1350, there came, not so much in theory as in actual application, a marked shift" (Crosby, 1997, p. 18). And indeed, in the novella, both Bernabò and the miller are passionate quantifiers.

dissection by looking at the body as a physical entity (e.g., its length, weight, density) and not only at its celestial or infernal ontological connections.

- A love for providing scientific evidence, even if in fun: "I have measured it as well as I knew how; if ye do not believe it, cause barrels to be found and measure it yourself"; "I held discourse formerly with one who had been there, and it was from this man that Dante the Florentine learned that which he wrote of things appertaining unto hell." How could Bernabò dare to argue with Dante?
- A wonderful revolution in the appraisal of people, which then sadly disappeared under the *signori* of the Renaissance to return only with the Enlightenment (see Da Ponte/Mozart's *Figaro*): the miller—trained in counting bags of grain and flour—is more capable than the abbot, in consequence: "and thou shalt have the revenue of the monastery and he shall have that of the mill."
- The abbot is described not as a religious leader but as the administrator of some land controlled by the Duke, thereby also responsible for the care of the Duke's mastiffs. These worldly interests of the Church are further lampooned by Sacchetti in the novella XI which end with these words: "And perchance, if Alberto had been a rich man, the Inquisitor would have caused him to be so troubled that he would have redeemed himself with his money from being burnt or crucified" (ib., p. 19). If this kind of critical writing had been able to continue (possible only with a continuation of the *visitor communal civilization*) there would have been no burning at the stake of Giordano Bruno in 1600, nor the humiliating recantation of Galileo in 1633.

5. Did the Recovery of Roman Law Play a Role?

According to Roger French, in his *Dissection and Vivisection in the European Renaissance* of 1999, a share of the merit for the new medieval thinking, which among others rediscovered the use of dissection, goes to "the recovery of Roman Law" (1999, p. 13).

However that same recovery of the Roman Law gave much authority to Emperor Frederick Barbarossa against the freedom of the Lombard communes. In 1158, at the diet of Roncaglia, four masters of

Roman law of the Bologna school (Bulgaro, Martino, Jacopo, and Ugo) sided with the Emperor in the matter of the *regalie* and stated that to him only—and not to the communes—belonged "all sovereign rights—the ports, the rives, the taxes, the appointment of magistrates" (Procacci, 1971, p. 18). At Roncaglia, "The Emperor and his advisers could not fail to be affected by Roman concepts so favourable to monarchical power, and traces of the outlook of the Bologna jurists are to be found in the legislation issued at Roncaglia in November 1158" (Hampe, 1909/1973, p. 173). Similarly, in the words of Cantor (1993), "By laying claim to the legal prerogatives of the Roman emperor, Barbarossa justified political absolutism and the enhancement of his authority in Germany; in addition, he could use the evidence of the Justinian code to assert his sovereignty over the Italian states" (p. 311).

Accordingly, if the Lombard League had not defeated Barbarossa in 1176 at the battle of Legnano, and torn apart the impositions which had been scholarly approved by those four Bolognese law professors, there would have been no independent communes, no dissections, and no Mondino de' Luzzi. Europe would have been ruled by the unity of power of the emperor: "the living law" (ib., p. 309), a "living law" not interested in democratic impious procedures.

There was a major forensic contribution to the making of *visitor communal thinking*, and through it to the development of scientific dissections, not from the recovery of Roman Law, but from the development of Communal Law as elaborated, for instance, by Bartolo da Sassoferrato (1313-1357) famous for his *DP teaching* that the authority of a city-state (*civitas*) on affairs conducted within its jurisdiction is parallel to that of the emperor. In Bartolo's celebrated phrase the *civitas* is its own emperor (*civitas sibi princeps*), and does not recognize a superior. Equally important is Bartolo's treatise *De Guelphis* of circa 1340 (discussed in chapter 13 of volume 3) in which he states the revolutionary doctrine that the fight against tyranny must supersede any Guelph or Ghibelline allegiance: "Suppose that in one place there is a tyrant, who, together with his following, is called a Guelph. A good citizen will be opposed to this man as a tyrant, and in that place this citizen is a 'Ghibelline', but, in another city not dependent upon the former, where there is a Ghibelline tyrant, certainly that same good citizen will be opposed to that tyrant and there he will be a 'Guelph'" (1340/1964, p. 277). Within the communal world, it is the good citizen who rules, not the emperor, nor the pope: and dissections will be made because they are needed by good citizens to foster peace and medical progress.

6. Could Mondino de' Luzzi and his Direct Followers have been more Scientific?

According to Weisstein (2003, p. 1):

> Mondino de' Luzzi (1275-1326) was the Italian physician who marked the revival of medical practice in the West following the Dark Ages.... Although Mondino de' Luzzi is historically important as one of the first physicians of note following the Dark Ages, his medical procedures were, in fact, a step backwards. He taught his students while seated on an elevated chair, and employed a barber surgeon to perform the actual dissections. He believed in dissecting from the inside out, since internal organs rot the most quickly. In the process, he inevitably destroyed parts of the body. Furthermore, Mondino de' Luzzi blindly accepted Galen's anatomy, even when a simple dissection would have conclusively proven him to be at odds with actual observations. He wrote a compendium of anatomy, which was basically a guide for understanding Galen. This represented a regression from scientific procedures, and stands out in sharp distinction to Grosseteste's and Roger Bacon's extensive experimentation and questioning of established authorities which were being undertaken in approximately the same period. Unfortunately for medicine, as well as science at large, Mondino de' Luzzi's methods became standard practice in medical schools until they were eventually replaced by the sound observational and experimental practices of Vesalius.

With all due respect, the above includes three errors, the first, but small, concerns the way Mondino operated: as noted by Singer (see above): "Mondino dissected in person".

The second error consists in considering the procedure in which the professor taught his students while seated on an elevated chair and employing a barber surgeon to perform the actual dissection as a step backwards. For the sake of clarity I report the caption that Singer wrote for Fig. 35.: "Dissection scene from the Italian *Fasciculo di Medicina* printed at Venice in 1493. The picture forms the first page of an Italian translation of the *Anothomia* of Mondino. The professor in his robes is reading his lecture from his chair. A menial *demonstrator* dissects, as directed by the wand of the *ostensor*. Students in academic dress stand around and look on, but do not themselves dissect" (p. 76). This elaborate post-Mondino procedure was not a step backwards but the only one which could lift the gory dissection to a scholarly/university status. This gave dissection the needed academic aura and panache. Dissection, before, was too close to butchery, too impious, to be accepted as a respectable academic endeavor.[140] Only thanks to these various ennoblements could dissection be accepted by other universities, and become common practice. I may even go one step further and say that if Herophilus had been able to include in his work some of the fancy public procedures so criticized by Weisstein, dissections would have been practiced for a longer time in Alexandria, and more books would have been written for posterity.[141]

Mentalities change slowly, and it is not surprising that it took more than two centuries for people to become accustomed to dissections commanded by a professor from his high chair, before Vesalius could progress by again dissecting in person and pay attention to details.

The third error is to demand from Mondino the kind of scientific mentality and experimental procedures of Grosseteste and Roger Bacon who never practiced medicine, and never did dissections or anything close to them.

> Grosseteste [c. 1175-1253] introduced into the world of European Christendom Latin translations of Greek and Arabic philosophical and scientific writings.... [His works] include a commentary on Aristotle's *Posterior Analytics* and *Physics*, many independent treatises on scientific subjects, and several scriptural commentaries. (*Encyclopaedia Britannica*, 2002, *5*, p. 512);

---

[140] Porter saw it correctly when he wrote that: "Public dissection was spectacle, instruction and edification all in one. The corpse would be that of an executed criminal, presupposing municipal cooperation. It was sometimes staged in a church, usually in winter, since cold slowed putrefaction. Mondino's order of dissection of the three main bodily cavities — first the lower abdomen, then the thorax and the skull — was designed with decay in mind. In illustrations of dissections, a physician resplendent in academic robes sits on a throne, intoning from a Galenic anatomical text, while a surgeon slits the cadaver with his knife, and a teaching assistant points out notable features. Whether or not dissections were actually conducted in this way, what is conveyed is the ritual of the performance: religious, civic, *and university authorities agreed that the occasion must be accorded due gravity*. Book-driven anatomy—a demonstration of what was already known, within the explanatory framework of learned medicine—served many purposes, providing guidance to the student, who would not have been able to see much for himself." (1997, pp. 132-33; my italics).

[141] "The works in which [Herophilus's] anatomical views were expressed have all been lost, the most important quotations from them being in Galen, Rufus, Soranus and Celsus" (Potter, 1976, p. 45).

> Bacon [c. 1220-1292] studied mathematics, astronomy, optics, alchemy, and languages. He was the first European to describe in detail the process of making gunpowder, and he proposed flying machines and motorized ships and carriages.... His studies on the nature of light and on the rainbow are especially noteworthy. (ib., *1*, p. 777-78)[142]

There is, indeed, a huge difference between studies of astronomy, optics, or alchemy, and dissections lasting four days in a row on human bodies, as practiced by Mondino de' Luzzi: "In the absence of preservatives, dissection had to be performed hurriedly, and this especially with the abdominal viscera. Work was sometimes continued through the night, and the whole process completed in four days, a day each being given to the belly, the thorax, the head, and the extremities in that order" (Singer, 1925/1957, p. 75). Also, as reported before: "No one, aware that the body into which he is cutting is the body of a human being, can do so without emotional resistance" (Edelstein, 1932/1967, p. 280). None of these emotional restrictions apply to studies of astronomy, optics,or alchemy.

There were also strong religious fears associated with dissecting human cadavers. As noted by Gordon:

> The first anatomical teacher to struggle with the authority of the Church which prohibited dissection was Mondino de' Luzzi, professor of anatomy at the University of Bologna. He aroused the curiosity of the medical profession by his well ordered demonstrations of the different parts of the body. In January 1315, he dissected and demonstrated the parts of two female bodies, and, in the course of the following year, he accomplished the same task on the person of a single female cadaver. Acts such as this were done in defiance of the authority of the Church, for a Bull issued in 1301 by Pope Boniface read: "Persons cutting up bodies of the dead and barbarously cooking them in order that their bones be separated from their flesh are by the very act excommunicated." In 1319, criminal prosecution was initiated

---

[142] In fairness there is a text by Roger Bacon that can be considered medical: his *De Erroribus Medicorum Secundum Fratrem Rogerum Bacon de ordine Minorum*, but as the title indicates it is mainly a list of things to beware of as "Physicians foolishly trust ignorant apothecaries who overcharge, substitute and dispense stale drugs ... " (quoted by Gordon, 1959, p. 272).

> against a professor and students of Bologna for body snatching. (1959, p. 422)

Can the same be said of studies of astronomy and optics? Even today, few people would feel comfortable dissecting a cat or dog. One way to do it is in an aseptic university laboratory, surrounded by modern equipment and guided by a professor or one of his assistants: the contemporary equivalent of the medieval professor instructing from his high chair.

7. In Praise of DP

At this point, let me stress that my hero is not Mondino de' Luzzi, but the DP and derived *visitor mentality* at the times of the Italian communes:

> *The credit of being the first country in Europe to revive the study of anatomy goes to Italy. While other European nations were either profoundly ignorant or grossly indifferent of this science so basic to medicine, the Italians made great advances in the science of anatomy, in spite of the great difficulties in opposing the papal edicts that prohibited dissection.* (Gordon, 1959, p. 420; my italics)

> From the thirteenth until the sixteenth century the history of Anatomy remains largely in the custody of the Bologna School. At Bologna dissection received official recognition in the University Statutes of 1405 and the same event took place at Padua in 1429.[143] (Singer, 1925/1957, p. 87)

Neither under imperial nor papal UP would there have been dissections: neither for forensic nor for scientific purposes, at a minimum because neither emperor nor pope could risk the reciprocal accusation of impiety. As to the administration of the law, both emperor and pope

---

[143] Padua was the first among the Italian cities to revolt against Emperor Frederick I Barbarossa in 1164. Padua is where Vesalius went in 1537 and remained until 1546 working at his the *De humani corporis fabrica*: "Indeed, Harvey's discovery of circulation might have been impossible if it had not been for Vesalius' correct anatomical description of the circulatory apparatus" (Gordon, 1959, p. 629).

always felt so superior, so sure of themselves (the "law incarnate"), that they did not need a scientific judgement based on dissection.

III. Andreas Vesalius and the Beginning of *Scientific Dissection*

> Born Andreas van Wesele in Brussels, where his father was pharmacist to Emperor Charles V, Vesalius learned Latin and Greek and enrolled in the Paris Faculty of Medicine, studying under the conservative humanist Sylvius, then Galen's great champion.... Vesalius learnt his dissecting skills from Guinther von Andernach, and when in 1536 war forced him to flee Paris, he returned to Louvain where he introduced dissection. He showed his anatomical zeal by robbing a wayside gibbet, smuggling the bones back home and reconstructing the skeleton. In 1537 [at the age of 23] he moved to Padua, where he made his anatomical name.... Thereafter Vesalius grew more critical. Familiarity with human anatomy drove him to the unsettling conclusion that Galen had dissected only animals, and forced him to see that animal anatomy was no substitute for human. He now began to challenge the master on points of detail: for instance, the lower jaw comprised a single bone not two, as Galen, relying on animals, had stated. Evidently, human anatomy had to be learned from dead bodies not dead languages.
>
> In 1539 [still at Padua] he acquired a larger supply of cadavers of executed criminals and worked on his great masterpiece, the *De humani corporis fabrica.* (Porter, 1997, p. 179)

All these travels, and working in many different countries at a young age, allowed Vesalius to detach himself from the old medical scripts and become open to the new, in a creativogenic DP process (new ideas fighting against the old) which I discussed in detail in chapter 6 of volume 1 "Creativity via a Detour Abroad".

> Vesalius soon recognized that the system of teaching anatomy was wrong: "The detestable procedure now in vogue is that one man has to carry out the dissection of the human body while another gives a description of the parts. The lecturers are perched up aloft in a pulpit like jackdaws and with a notable air of disdain. They drone out information about facts they never approached at first hand but which they merely commit to memory

> from the books of others or of which they have descriptions before their eyes; the dissectors are so ignorant that they are unable to explain the dissections to the onlookers and botch what ought to be exhibited in accordance with the instruction of the physician who never applies his hand to the dissection and steers the ship of the manual, as the saying goes. Thus, everything is wrongly taught, days are wasted in absurd questions and in confusion; less is offered to the student than a butcher in his stall could teach the doctor."
>
> At last he was persecuted by the authorities who checked his future scientific activities. In a fit of indignation, he left Padua and accepted the post of court physician to Emperor Charles....Vesalius was disgusted with the controversy over his work and particularly hurt by the hostility of his teachers. In 1544 when only thirty years of age he decided to resign his professorship. He burned his medical books and some of his unpublished manuscripts. (Gordon, 1959, p. 628-29)

More than two hundred years after Mondino, it was still difficult to do something new in the field of dissection without running into trouble. Could therefore anybody have done what Vesalius did at the time of Mondino, or even one hundred years after? Probably not. In those times, people needed plenty of time to adjust to new ideas, and it should not seem strange that it took nearly two hundred years of public/book-directed dissections before moving to scientific dissections made *in person* as Vesalius did.

Yet, let me repeat, the real revolutionary step was the one between forensic and public dissections. The next one, of the professor looking carefully at what he found while dissecting in person was no doubt very important, however it was a shorter step in comparison with the earlier one. Sadly, we lack a report of the conversations which led to the first public dissection. One way or the other, it was a superb act of communal creation which integrated the interests and contributions of many people.

IV. Herophilus and Erasistratus, and the Alexandrian School of Medicine

In the third century BC Greek rational medicine was transplanted into Egypt and at Ptolemaic Alexandria it achieved its greatest success. Expatriate Greek doctors displayed, on the one

> hand, the same rational attitudes towards medicine that we have previously encountered at Cos, Athens and elsewhere in classical Greece; but *being uprooted from their native environments and thus no longer constrained by traditional customs and attitudes, they also reflected in their pioneering approach to medicine the new freedoms they encountered in the new and cosmopolitan city*. In consequence, new levels of sophistication in anatomy and physiology were attained which were not surpassed until the Renaissance. (Longrigg, 1993, p. 177; my italics)

> It was at Alexandria that Anatomy first became a recognized discipline. The two earliest and greatest Alexandrian anatomical teachers of whom we hear are Herophilus and Erasistratus, who were both Asiatics and who both flourished in the first half of the third century B.C. They inaugurated what we may call the "Alexandrian period" of Anatomy. The works of both these men are lost, but we gain a good idea of them from passages gleaned from Galen. Herophilus of Chalcedon (about 300 B.C.) was, so Galen assures us, *the first to dissect both human and animal bodies*. (Singer, 1925/1957, p. 28-29; italics in the original)

> *Anatomical research at Alexandria flagged after the first generation, but the city long remained a great teaching centre, and minor advances were made. The stagnation in medical matters at Alexandria is in sharp contrast to the continued activity there in Mathematics, Astronomy, Mechanics, and Geography*. (ib., p. 33; my italics)

We have here something that we can relate to my criticism of Weisstein: anatomical research through dissection belongs to a different psychological, moral and religious category than mathematics, astronomy, mechanics and geography. These last can still flourish while anatomy is neglected or even persecuted.

The question then becomes: What was in Alexandria at the time of Herophilus and Erasistratus, and gone shortly after? The probable answer rests in the difference between Ptolemy I and his son Ptolemy II. Ptolemy I (b. ca. 366 B.C.) was, at first, only a small Macedonian noble, then a general of Alexander the Great. At the death of Alexander, in 323 B.C., he became ruler of Egypt, and finally king of Egypt in 305 B.C.

> [Ptolemy] found it necessary from the outset to pursue a conciliatory policy toward the Egyptians, since Egyptians had to be recruited for his army, which initially numbered only 4,000 men. Ptolemy won over the Egyptians through the establishment in Memphis of the Sarapis cult, which fused the Egyptian and Greek religions; through restoration of the temples of the pharaohs, which had been destroyed by the Persians; and through gifts to the ancient Egyptian gods and patronage of the Egyptian nobility and priesthood. Finally, he founded the Museum (Mouseion), a common workplace for scholars and artists, and established the famous library at Alexandria. Besides being a patron of the arts and sciences, he was a writer himself.... (Werner, 1978, p. 181)

Ptolemy seems also to have taken a strong interest in what was studied at the Alexandrian School of Medicine, so much so that "Celsus, a sober and competent writer, has a horrible story that Herophilus vivisected criminals given him by Ptolemy I (anaesthetics were unknown), and the same is implied of Erasistratus" (Tarn, 1927/1961, p. 306).[144]

There was therefore around 300 B.C., a rare combination of a new ruler, (not born in the purple, who had participated in many battles, who had major problems in balancing his power, and who had strong intellectual interests),[145] and a brand new capital (Alexandria) which had been founded by Alexander the Great in 332-331 B.C. These conditions, while not a division of power per se, can, at a minimum, be considered a *creativogenic detour abroad*: strong cultural elements from Macedonia and the rest of Greece, a brand new capital and harbor, and the steady influence of old Egypt. In such a diverse and lively environment many new endeavors could be undertaken.

With Ptolemy II, things changed: He was born in the purple in 308 B.C., and already in 285 B.C., at the early age of twenty-three, he became king, first together with his father, then, at the death of his

---

[144] Singer—mainly from the study of the books by Galen—denies this charge of vivisection (1925/1957, p. 35).

[145] As was the case with Napoleon, especially in his pre-imperial years.

father, the sole ruler from 283 B.C.. It was not promising for progress that Ptolemy II

> was the first king of the Macedonian dynasty of the Ptolemies in Egypt to set up the Hellenistic divine cult of the living ruler.... He displayed a vivid interest in Greek as well as in Egyptian religion, paid visits to the sanctuaries in the countryside, and spent large sums erecting temples. Anxious to secure a solid position for, and religious elevation of, his dynasty, the King insisted upon divine honours not only for his parents but also for his sister and wife Arsinoe II and himself as the *theoi adelphoi* ("brother gods"). He thus became one of the most ardent promoters of the Hellenistic ruler cult. (Heinen, 1978, p. 182)

Such a "god" may become a famous patron of the arts, but less of science, and even less of anatomical studies through dissection of dead or even living humans.

Ptolemy's divinity was in part Egyptian: "Arsinoe was made a 'co-templar god' *(synnaos theos)* in all the temples of Egyptian deities, and certain royal revenues were creamed off to finance the cult. Despite the royal initiative that lay behind these cults, there was Egyptian precedent for ruler-worship; so the royal cult of the Ptolemies, as practiced and documented in native Egyptian temples, developed somewhat autonomously, adapting prior pharaonic practice to new requirements" (Shipley, 2000, p. 160). However, dissection was not favored by the native Egyptians given "the fact that in Egypt the religious taboos surrounding the dead and the disposal of the body were so strong that it is most unlikely that corpses or even human organs would have been readily from this source" (Longrigg, 1993, pp. 186-7).

So much royal/divine power, predictably, brought strong corruption. The ancient historian Phylarchos wrote: "[Ptolemy II] was so deceived in judgement and so corrupted by his boundless luxury that he believed he would live for all time and said that he alone had found immortality" (quoted by Shipley, 2000, p. 263). A similar judgement was passed by Bevan : "Ptolemy the son was a very different character from Ptolemy the father. The softening of fibre which became more pronounced in several of the later kings already showed itself in the son of the tough old Macedonian marshal" (1927/1968, p. 56).

In essence, there was nothing in the world of Ptolemy II that would favor a continuation of the anatomical studies of Herophilus and Erasistratus. Lacking the broad DP that Western Europe enjoyed from 1077, the discoveries and methods of Herophilus and Erasistratus not only were not expanded, but were soon in good part forgotten. Later,

> with the advent of the Empire and of Imperial ideas, the tone of Science changes. There is a waning in enthusiasm for all save "useful" knowledge. The so-called "practical" outlook has been and is a great obstacle to the advance of Medicine. The plaintive cry for what is "useful" as against "theoretical knowledge" echoes down the ages. We hear it still. But when and where that cry swells into a chorus, then and there Science dies... Dissection of the human body was still occasionally practiced at Alexandria towards the end of the first century B.C., but it had ceased by the middle of the second century A.D. There is evidence that it was not then practiced elsewhere. Considering the indifference to human life which the Romans exhibited, considering their brutality to slaves and the opportunities afforded by gladiatorial combats, considering the value of anatomy for surgical practice and the demand for surgical skill involved in the organization of medical service throughout the Empire, it is truly remarkable that the anatomical knowledge of antiquity was allowed to lapse. Its passing is one of the innumerable instances that illustrate the danger of entrusting things of the mind to the tender mercies of the "practical man".[146] Anatomy did not revive till the rise of the medieval universities. (Singer, 1925/1957, pp. 37-38)

---

[146] However, Singer himself tells us three pages later of Pliny the Elder (A.D. 23-79) who had said that "in his time the examination of the human viscera was looked at as impious" (p. 40); not that such an examination was not useful. We must remember that already Augustus (63 B.C.-A.D. 14) "In spite of his own vulnerable morals, wished to be seen as a guardian of sound private, sexual, morality, which he sought to revive by an extraordinary range of enactments" (Grant, 1975/2000, p. 75). "Augustus rebuilt temples, restored neglected ceremonies and priesthoods and, in general, attempted to revivify the old state religion with its patriotic associations. In the moral sphere he tried to restore the sanctity of marriage and to stimulate the birth rate by two laws,..." (Jones, 1973, p. 761). This is not the ruler who will provide the kind of "impious patronage" that Ptolemy I gave to Herophilus and Erasistratus. In this matter, the Roman emperors had less power than Loderingo and Catalano had in Bologna.

V. Conclusions

While it is true, as Singer said, that anatomy did not revive till the rise of the medieval universities, it is essential that we also fully acknowledge the contributions of the communal DP starting with its demands from the medical profession: "In Italy, autopsies for the purpose of elucidating the cause of death or settling doubts regarding the manner in which death came about and especially for the discovery of poison, were made with increasing frequency in an age when mysterious homicide was common" (Gordon, 1959, pp. 420-21).

Forensic autopsies were performed in communal Bologna because the Guelph-Ghibelline team of Loderingo degli Andalò and Catalano dei Malavolti, together with other nobles, had founded the *Milizia della Beata Vergine Maria* to quell riots of any kind, especially fights between Guelphs and Ghibellines, and later had issued rules for checking the cause of all sudden deaths. A democracy demands, far more than autocratic governments, that justice be administered fairly: a kind of justice which will only result from DP, when neither the state (emperor, king, prince) nor the church (pope, cardinal, bishop) has full power, but when these two powers oppose each other giving the cities the possibility to rule themselves:

- The very engine of communal life was impelled by the urge to elaborate collective norms" (Becker, 1981, p. 166),

- One presupposition was constant intervention by the commune in every aspect of social and economic activity. This was connected with the confident belief that problems were soluble. Discussion in council could find solutions which could then be enforced by legislation. If experience showed that this was not the correct solution then another would have to be tested.... A number of communes made agreements with doctors who came on to the city's pay-roll as a sort of 'medical officer of health'. Thus in 1214 Hugh of Lucca reached terms with Bologna whereby he accepted citizenship and agreed to buy a house and live there for at least six months each year in return for an annual salary of 600 *lire* ... It was hoped that one of his sons might succeed him and should this not eventualise he was to pay Bologna 200 *lire* on resigning his office. (Waley, 1969/1988, pp. 59, 63)

This is the *DP world*, and derived *visitor mentality*, which is at the origin of Western medicine through first forensic dissections, then university, public/book-driven ones, and finally through scientific dissections.

15

The DP Contribution to the Abolition of Slavery in the 18th and 19th Centuries

1. Introduction

In chapter 11 of volume 1—on the effect of centuries of division of power, and the derived *visitor* ethnopsychology—I wrote that "there is no need to describe the continuous high level of creativity in the West in science, philosophy, and the arts. What, instead, should be stressed is its high creativity, especially in the last centuries, in the most important fields of law, morality, humanitarian help, and peace. First there was the abolition of slavery and torture, then reduction in racial, ethnic, gender prejudice; then the foundation, in 1864, by Henry Dunant of the Red Cross,..." (p. 156).

Let me expand this praise by showing how DP shaped mentalities and institutions directly involved in the abolition of slavery, as was the case with the Quakers.

2. DP and the Quakers' Contribution to the Abolition of Slavery

> Despite their small numbers, the Quakers provided an essential nucleus of leadership and financial backing for the fledgling abolitionists, *and the long tradition of Quaker female activism and separate women's meetings gave women a noticeable role in the effort.* (Games, 1997, p. 531; my italics)

"From any historical perspective, this was a stupendous transformation.... *The progress of emancipation from the 1780s to the 1880s is one of the most extraordinary events in history*" (Davis, 1984, pp. xvi; my italics).[147] The Quakers were major participants in it as early as 1657, when George Fox, considered the founder of Quakerism (*The Society of Friends*), advised the Friends who owned slaves to convert them as brothers.

Then, on February 18, 1688, the famous *Germantown Protest* occurred when a group of Pennsylvania Quakers declared openly at their monthly meeting that slavery was contrary to Christian principles, and signed an antislavery resolution to that effect. "This document is viewed as the first public condemnation of the institution and practice of slavery in the Western Hemisphere" (Rodriguez, 1999, p. 204).

Next, the four signatories of the Protest petitioned the Philadelphia Yearly Meeting (hereafter PYM) of the Quakers to take measures against slavery and the slave trade, but the PYM refused to consider their demands because it believed that Pennsylvania and New Jersey Friends could not outlaw slave trading as long as Quakers living elsewhere were involved in the trade.

Five years later, it was the turn of the Quaker George Keith who presented to the PYM a paper entitled *An Exhortation and Caution to Friends Concerning Buying or Keeping of Negroes*. It was Keith's desire that those Quakers who owned slaves should free them as soon as possible.

Forty-five years later, in 1738, at the PYM, to underscore the violence inherent in perpetual bondage, Benjamin Lay plunged a sword in a hollowed-out book disguised as a Bible and filled with pokeberry juice, sprinkling "blood" on Friends nearby.

> In 1774 the Yearly Meeting ruled that Friends who bought or sold slaves should be disowned and those who refused to free

[147] In the words of Tocqueville: "We have seen something absolutely without precedent in history—servitude abolished, not by the desperate effort of the slave, but by the enlightened will of the master, not gradually, slowly, through successive transformation which by means of serfdom led insensibly toward liberty;... *What Christianity itself did only over a number of centuries has been accomplished in a few years*. If you pore over the histories of all people, I doubt that you will find anything more extraordinary or more beautiful" ("On the Emancipation of Slaves" of 1843; quoted by Davis, 1984, p. 110; my italics)

> their enslaved black should be excluded from leadership. Two years later, PYM prohibited slaveholding outright, directing local meetings to disown members who refused to manumit their slaves.... With the battle against slavery within the meeting largely won, antislavery friends sought to eradicate the institution in the larger society.... At the same time Quakers furnished much of the initiative for abolition in Great Britain. (Soderlund,1998, p. 753-54)

3. But Why the Quakers?

> Among the numerous religious groups which appeared in Europe after the Protestant Revolt were the Quakers, or Friends. Rejecting many of the religious forms practiced by both Protestant and Catholics, they guided their actions by an 'inner light' or direct revelation from God. *Extremely democratic in their forms of worship and their theory of life*, they refused to take oaths and even to doff their hats in the presence of magistrates. (Faulkner & Kepner, 1934, p. 35; my italics)

The central concept of Quakerism was that potentially anyone could experience God directly and inwardly, regardless of gender, nationality, or social status. Friends rejected the Calvinist concept of predestination, and instead believed that all people who sought to lead blameless lives could receive the Light and be saved.

The Friends have no formal creed, rites, liturgy, or priesthood, and reject violence in human relations, including war. Friends have always given women equality in worship. They also have rejected a formal or salaried clergy. Meetings to worship God and await his word (always open to anyone who wishes to come) are essential to Quaker faith and practice. It is in the pregnant silence of the meetings of the true waiters and worshipers that the Spirit speaks.

Figure 25, from the Library of Congress, is a nineteenth century engraving of a meeting of the Society of Friends: "Quakers were renowned both for the place of silent meditation in their worship and for their acceptance of women preaching and testifying. Here a woman rises up in a meeting to relate an opening of light from God" (Goustad & Schmidt, 2002, p. 87). Long rows of benches, on one side the women, on the other the men; no altar, no pulpit, no special place for clergy or minister: it would be difficult to conceive something more democratic, more DP.

In stark contrast,

> the colonial church [building], favored especially by evangelical Protestants—Congregational,[148] Lutheran, Baptist, and Presbyterian—is often an actual example or copy of the traditional New England church/meetinghouse featured on countless New England calendars.... It is simple and rectangular, with a steeple and a shingled roof, and is constructed of wood and painted white. It has many windows and is usually almost without interior decoration except for an occasional curly brass or pewter sconce or chandelier. There may be a balcony running around three sides of the building, supported by columns. *The most striking interior feature is the large, impressive pulpit, which may have a canopy or sounding board and is usually elevated three to eight feet above floor level, silently indicating the dominance of the preacher, and of his sermons and prayers.*
>
> *At first most of these churches contained parallel ranks of box pews whose sides were sometimes so high that most of the occupants could not look at their neighbors: all they could see was the preacher.* From a Puritan theological point of view, this was not a disadvantage: worldly curiosity was a distraction from the divine. As Peter W. William puts it in *Houses of God*, "the preached Word [was] the chief point of intersection between the sacred and secular realms." Martin Luther had also stated that written and spoken language was the exclusive representation of God's power, and that the purpose of worship was to praise God and understand his Work. (Lurie, 2003,, p. 30; my italics)

In the prior instance, there was the DP of worshippers sitting, in silence, in front of others like them: their brothers and sisters in Christ. Now, we have the UP of a professional preacher: all worshippers paying attention to him, trusting his words, accepting his choice of subject for the sermons, following his leadership.

In addition, the Quakers had invented an ideal machinery for moving petitions from the simplest local meeting up to the Philadelphia

[148] Congregationalism was the dominant form of Christianity of New England: basically the Puritanism of the Pilgrim Separatists of Plymouth.

Yearly Meeting, as the four brethren of Germantown did in 1688 when asking that all Friends give up slavery:

> The Society of Friends had a hierarchical structure of meetings. At the lowest level were the local *meetings for worship* that met at least twice weekly and were held in a member's home until a meeting house was built. At these gatherings, Friends and anyone else who desired to attend waited in silence for God to inspire someone—usually a seasoned minister but on occasion a young person or other less-experienced speaker—to preach or pray. Men and women attended meetings for worship together, but sat on opposite sides of the room.
>
> The next higher meeting was the monthly meeting for business, which was composed of Friends from several local meetings for worship in the same township or group of adjacent townships. All adult Friends who were not under discipline could take part....[149]
>
> The *hierarchy of Quaker business meetings* went from preparative to monthly, then to quarterly, and finally to the Yearly Meeting. The monthly meetings sent representatives every three months to the quarter, which in turn sent delegates to the annual meeting. In the Delaware Valley, there were six quarterly meetings until 1758 when Chester Quarterly Meeting was split in half.... For the most part questions of policy reached Philadelphia Yearly Meeting when a monthly meeting encountered a problem that it could not solve by itself or when a member raised an issue that the meeting believed should be discussed by the entire Society. (Soderlund, 1985, pp. p. 189-192)

That, in itself, is a wonderful DP system, capable of bringing, within a year, the ideas of a small local group, to the attention of the most important members at their yearly great meeting—something non existant in other confessions.

---

[149] "*Weighty Friends,* the Society's most respected members who included *ministers, elders,* and *overseers,* performed most of the meeting's work. The overseers were charged specifically with monitoring the behavior of the congregation; but committees assigned to treat with miscreants usually included ministers and elders as well. The chief duty of ministers, of course, was to preach. *They were unpaid except for expenses when they traveled.* Elders were appointed to ensure that both ministers and more ordinary members were guided by the Light" (Soderlund, 1985, p. 190; my italics).

As early as 1688, those four Quakers of Germantown, Pennsylvania had written their protest against slavery for discussion at the next higher meeting: "This is from our meeting at Germantown held ye 18th of the 2d month, 1688, to be delivered to the monthly meeting at Richard Worrell's" (quoted by Rodriguez, 1999, p. 414). Those four Quakers were also thinking about the opinion of their Christian brothers in Europe: "Then is Pennsylvania to have a good report, instead, it hath now a bad one, for this sake [of slavery], in other countries; Especially whereas the Europeans are desirous to know in what manner the Quakers do rule in their province" (ib., p. 414).

> Despite its inhumanity and size, the slave system aroused little protest until the 18th century. Rational thinkers of the Enlightenment then began to criticize it for its violation of the rights of man, and pietistic or evangelical religious groups condemned it for its unchristian and brutal qualities. *In Britain and America, the Quakers, who began their criticism in 1671, were the first significant opponents of slavery, and the dynamics of the antislavery movement were largely religious throughout....* The first phase [to outlaw the slave trade] involved British and American efforts to prohibit the importation of African slaves into the British colonies and the United States. *English Quakers began actively to campaign for such a prohibition in 1783. In 1787 the Abolition Society, consisting mostly of Quakers, was formed. Under William Wilberforce, who led the movement in Parliament, and Thomas Clarkson, who devoted many years to the tireless collection of evidence concerning the evils of the trade, the antislavery forces waged an unremitting contest against powerful opposition. The trade to the British colonies was finally abolished in 1807.* Meanwhile, in the United States, the Constitutional convention in 1787 had considered placing in the constitution a prohibition of the trade, but, in order to conciliate Southern interests that opposed immediate action, the convention agreed to a provision that congress might prohibit the trade after 20 years. Accordingly, in 1807 the United States also prohibited it. (*Encyclopaedia Britannica,* 1978, *16*, p. 862)

Clearly, the Quaker's DP brand of Christianity contributed much to the abolition of slavery, while less-DP-brands of Christianity contributed less or nothing. One brand of Christianity that contributed nothing was the Byzantine, probably the most UP brand, as discussed in chapter 14 of volume 1 "The Low Creativity and High Unity of Power of the Byzantine Civilization." In Christian Byzantium, slavery continued with few changes from the Roman times (Hellie, 1998, pp. 136-137).

4. In Praise of the High British DP

As praiseworthy as the Quakers's DP was, we cannot properly understand it without relating it to the centuries-long DP in Britain. However, to avoid repeating things already said in my previous volumes, let me start by referring to two authors—Temperly and Williams— with a different perspective than mine.

Howard Temperley, in his (interim[150]) defense of Eric Williams's *Capitalism and Slavery* of 1944, wrote that: "Much of the plausibility of Williams's account, indeed, derives from the fact that Britain, the first nation to industrialize, also took the lead in the campaigns to abolish the slave trade and slavery. This is a development which the intellectual diffusionist account virtually ignores" (1981, p. 23). However, Britain—England in particular—was significantly different from the rest of Europe for more reasons than being the first to industrialize. Specifically, England was different:

- because of Magna Carta first, then of its Parliament;
- for its harsh handling of its kings, e.g., publicly flogging Henry II in 1174, beheading Charles I in 1649;
- for its *DP reading* of the Bible in English (as discussed in chapter 11of volume 3).

Accordingly, in Voltaire's own words:

> The English are the only people upon earth who have been able to prescribe limits to the power of Kings by resisting them;

[150] Interim, because later followed by Temperley's own and different interpretation: "What I am suggesting, in other words, is that antislavery can be seen as a secularized or semi-secularized form of Christian evangelism, the secular components of which reflected the ideological set of the dominant metropolitan cultures. Broadly viewed, it was part of that process, observable in relation to other issues too, which led to evangelical beliefs becoming more secular and secular beliefs becoming more evangelical" (1981, p. 32).

> and who, by a series of struggles, have at last establish'd that wise Government, where the Prince is all powerful to do good, and at the same time is restrain'd from committing evil; where the Nobles are great without insolence, tho' there are no Vassals; and where the People share in the government without confusion. (1733/1999, p. 34)

Also, Voltaire, in his *Lettres anglaises* of 1733, "Saluted England as a 'nation of philosophers' and the cradle of liberty, tolerance and sense" (Porter, 2001, p. 6).

Similarly in the words of Montesquieu: "I am here in a country which hardly resembles the rest of Europe. This nation is passionately fond of liberty ... every individual is independent[151] (quoted by Porter, 2001, p. 24).

> Prévost[152] was really the first French writer to become fascinated with that free, wise, philosophical and in other respects quite ideal England which was the Salentum of the eighteenth century. Everything connected with the country delighted him—its air of liberty to begin with. "What a lesson to see a lord or two, a baronet, a shoemaker, a tailor, a wine-merchant and a few others of the same stamp," all seated together round the same table in a coffee-house and chatting familiarly, pipe in mouth, on matters of public interest! Verily, "the coffee-houses are, as it were, the seat of English liberty." It is true that the common people are somewhat coarse. But it is also true that "there is no country where one finds such integrity, such humanity, and such sound notions of honour, prudence and happiness as among the English. Love of the public wealth, a taste for practical science, and a horror of dependence and of flattery, are virtues which are almost innate in these fortunate people; they descend from father to son like an inheritance." The English, in short, are "one of the first nations in the universe." (Texte, 1899, p. 48)

---

[151] Every individual was independent, because of Britain's *visitor ethnopsychology* thanks to centuries of DP (see chapters 11 and 23 of volume 1; chapter 12 of volume 2; chapter 11 of volume 3).

[152] Abbé Prévost (1697-1763): French novelist whose fame rests primarily on one work: *Manon Lescaut* (1732); full title: *Histoire du Chevalier des Grieux et de Manon Lescaut*.

In all this, there is no trace of "the fact that Britain [was] the first nation to industrialize."

Later, in 1843, when Tocqueville insisted "that the emancipation of the slaves was a parliamentary reform, the act of the nation, and not of its rulers", when he praised "the philanthropic and especially Christian conscience [*sentiment*]"[153] that had produced the great event of British emancipation, he was definitely not talking of industrialization, but of a conscience which by now we know to relate to the *DP visitor ethnopsychology*.

5. Some Important Creativogenic DP from a Reduction of Power

> But while Quakerism furnished a complex set of conditions which made an antislavery movement possible, it was essential that there be some triggering force to set off a cumulative reaction.... The change was sudden, when it came.... In 1760 New England Quakers made the importation of slaves an offense subject to discipline. This flurry of legislation culminated in 1761, when London Yearly Meeting announced that slave dealers merited disownment. It is clear that this sudden stiffening of official policy was the direct response to a crisis brought on by the Seven Years' War [of 1756-63].[154]
>
> There were a number of reasons why the war strained the Quaker conscience. The British and colonial governments, recognizing the grim seriousness of the conflict, were understandably impatient over the Quakers' refusal to fight or pay taxes. Both in England and America, Friends endured the unpleasant experience of being branded as traitors and fanatics....A strengthening of in-group discipline was accompanied by despair, by yearnings for withdrawal from evil, and by millennial prophecies. The crisis was most acute in Pennsylvania, where the pres-

---

[153] "On the Emancipation of Slaves", quoted by Davis, 1984, p. 110, from S. Dresher (Ed.), *Tocqueville and Beaumont on Social Reform* (1968, New York: Harper & Row).

[154] A conflict between the major European powers—France, Austria, and Russia on one side and Great Britain and Prussia (of Frederick the Great) on the other. The Franco-British aspect of the war can be understood as an extension of the French and Indian War, a colonial struggle that broke out in North America and in India before 1756. As a result of the conflict, Great Britain became the undisputed leader in overseas colonization, and Prussia emerged as a powerful force in Europe.

> ence of Friends in the colonial government forced a decision on the compatibility of politics and morality. For the Quaker members of the assembly there could be no middle ground between compromise and withdrawal. To remain in office and to obstruct the war effort would be disastrous for the Quakers' reputation, and might conceivably prevent a victory which even pacifists could consider essential. Subjected to heavy pressure from English Friends, who feared retaliatory measures from their own government, the Pennsylvanians resigned from office and surrendered their original idea of a Holy Experiment. This decision for withdrawal and self-purification led to a significant linkage of ideas: Quakers who continued to hold public office were subject to the same disciplinary rules as slaveholders....
>
> It was, then, in a time of severe crisis that the Society of Friends took the first significant steps of commitment to antislavery principles. In a period of intense soul-searching, of desire for self-purification, and of concern over their image in the eyes of others, a decision to refrain from dealing in slaves was a means of reasserting the perfectionist content of their faith. (Davis, 1966, pp. 330-332)

This, by Davis, describes what I would call a fascinating case of "creativogenic DP from a reduction of power in a DP environment." This major loss of power by the Quakers (loss of prestige, humiliations, resignations from important political positions, surrender of their dream of a Holy Experiment in America) was creativogenic in that it fostered a more substantial anti-slavery attitude: in being, as Davis said, a "triggering force to set off a cumulative reaction."

It is a fascinating case study because it parallels what happened to the Scots at the time of the Scottish Enlightenment. I am referring, here, to the creativogenic DP from a reduction of power which I discussed in chapter 18 of volume 1, where I wrote that "the most formative misfortunes at the origin of the Scottish Enlightenment were those that barred the way to a purely Scottish development and forced a number of fine Scottish minds to take an Anglo-Scottish path." I then listed a string of military, political, and economic misfortunes, and linguistic and literary conflicts. None of these events were crushing, yet jointly they forced the Scots to those new forms of thinking which originated the high creativity of the Scottish Enlightenment.[155] On the other hand, not every loss of power is a creativogenic form of division of power. Losses of power are not creativogenic, for instance, when they call for a return to past forms of unity of power, in something that Toynbee labelled "archaism". Scotland was saved from this faux pas when King James IV of Scotland became also—at the death of Queen Elizabeth in 1603—King James I of England, i.e., King James I of Great Britain. From then on, the English were no longer enemies but "brothers", and London a place where Scots could go and prosper (see section 7 of chapter 18 of volume 1).

In the case of the Quakers, they were too much a part of the Anglo-American business and political world to be tempted by an archaic solution of the Amish type; and, more importantly, their high DP fostered a forward reaction in the form of the strong moral anti-slavery attitude and rules described by Davis.

---

[155] "[The Scottish Enlightenment was] one of the great achievements of European culture. In philosophy, law, economics, politics, linguistics and the physical sciences, Scots were key players in changing the way the world was viewed. And this explosion of activity still reverberates. It was the age of David Hume, Thomas Reid, and Adam Smith, of Adam Ferguson, James Hutton, and Sir John Sinclair" (Broadie, 1997, back cover page).

16

An Ethnopsychological Explanation of Argentina's Difficulties

This chapter is divided into the following sections:

I. Introduction
II. *Mindbinding* by the Inquisition
   1. A First Set of Key judgments
   2. Sarmiento's Judgment
   3. Montesquieu's and Voltaire's Judgments
   4. In Sicily Under the Spaniards
   5. The *Anaconda Effect*
   6. Véliz's and Salvador de Madariaga's Whitewashing
III. The *Vertical Component* of the Hispanic Character
   1. As seen by Madariaga
   2. As seen by José Antonio Primo de Rivera
   3. As seen by López Ibor
   4. As seen by Bunkley
   5. *Vertical/Essential* People are Intransigent
   6. The Inquisition, and the Emphasis on Dying Well
   7. In Answer to Sarmiento: The Great *Vertical Men* of Unamuno and Salvador de Madariaga
   8. The *Vertical Component* as Medieval Thinking
   9. Is this Uniquely Spanish?
IV. Key Terms: *Ser entero, hombre total, hombre integral, método de la pasión,arbitrariedad, complejo de*

I. Introduction

Few countries seemed to have a more assured future than Argentine. Yet things went so wrong so quickly:

> For many decades many Europeans believed that Argentina offered an opportunity equal to, if not greater than, North America. The pampas *estancieros* enjoyed the reputation that Texas or Arab oil magnates have today, and the expression *riche comme un Argentin* remained a commonplace among the French until the 1930s....[However,] for at least the past two decades economists have classified Argentina in the under-developed or "third" world, and by the 1960s Argentina was becoming a by-word for political instability, inflation, and labor unrest. During the 1970s a sudden procession of horror stories emanated from Argentina—unbridled popular riots, guerrilla warfare, assassinations, abductions, imprisonment of dissidents, institutionalized torture, and eventually mass murder. For a time Argentina elicited a single association: los *desaparecidos,* the thousands of students, workers, writers, lawyers, architects, and journalists, men and women alike, who had "disappeared", simply vanished without trace.... *The central, compelling question about Argentina is simply, What went wrong? Why has Argentina failed to realize its promise?* (Rock, 1985, pp. xxi-xxiii; my italics)

There have been many answers to that central question, but few that have gone back far enough in history as demanded by the GAM/DP Theory of Personality and Creativity. There is no valid explanation of what went right or wrong with a nation, without a clear understanding of its distant past. For instance, in any discussion of the origins of the West, there is no way that we can say something meaningful without dealing first with what happened at Canossa in 1077, Legnano/Venice 1176/1177, and Runnymede 1215. It is not that later positive DP events are not important, but that their importance was prepared by the DP ideas and agreements of those key events.

In the case of Argentina, it is less a question of key events and agreements, than, at first, of a UP institution, the Inquisition, that many scholars have decided to forget or minimize, and which in my mind must be faced squarely. Because of this, Part II of this chapter is devoted to the evil perpetrated by the UP of the Inquisition in Latin America, Spain, Sicily, and the Papal States. Accordingly, my discussion will move from

Argentina to Spain and back to Argentina, stressing how Latin America, but especially Argentina, suffered much more of royal/inquisitorial UP than Spain itself, because its distance from Europe prevented the many liberalizing contacts that Spain had with France, England, Italy, and Flanders, including the contacts of war, as for instance the Peninsular War of 1807-1814 which so strongly affected both Spain and Portugal.

Part III, on the *vertical component* of the Hispanic character is equally Spanish at first, yet various key aspects of that *vertical component* have sadly become stronger in Argentina than in Spain itself.

Before leaving the preliminaries of this introduction, something more must be said first on what went wrong, and then on the aims and methods of this chapter.

2. What Went Wrong?

- On March 24, 1976, a junta led by Gen. Jorge Rafael Videla seized power from President Isabel Peron. During the next seven years, at least 8,900 people disappeared, according to a 1985 government report, although human rights groups place the figure at around 30,000.[156]
- A microcosm of international politics, Argentina has experienced domestically the hostilities that citizens of other states experience mainly in the area of foreign relations. More than any other nation, since World War II it has been a battleground of ideas competing for supremacy—liberalism, democracy, socialism, communism, fascism, and the social teachings of the Catholic church—ideologies contributing to a climate of social and political chaos. In Argentina their encounter reached a pitch of violence unsurpassed in the hemisphere. Argentina can boast of the two largest, best-organized, and best-financed urban guerrilla formations on record—the Ejército Revolucionario del Pueblo (ERP), and the Montoneros. The response to their revolutionary war was also record-breaking—the military's dirty war, a revival of the methods of the

---

[156] Sadly, this behavior was not new. Already under Juan Manuel Rosas, who ruled Argentina in 1829-1852, "one source estimates that by 1843 he had been responsible, through executions and civil war casualties (not including the badly wounded who would die shortly afterwards, or the deaths inflicted by Rosas' enemies) for over 22,000 deaths" (Earle, 1971, p. 47).

Spanish Inquisition, and the most systematic form of state terrorism in the New World. (Hodges, 1991, p. xi)

- Borges remarked, some thirty years ago, that the Argentine is not a citizen but an inhabitant; that he lacks an idea of the nation where he resides, but views it as a territory which, owing to its wealth, can be exploited rapidly. To my mind, this is a noteworthy comment on the Argentine problem....

1) If Borges were to comment on his own definition, he would say that the Argentines' error stems from their inadequate understanding of ancient Germanic literature. Borges would claim that it's impossible to create a citizen unless he has read the books of the Veda or at least the Egyptian Mummy's Prayer....
2) Right-wing sectors would accept Borges's remark as an absolute truth, claiming that it was the flood of immigrants that was responsible for the destruction of the roots implanted by the Spanish monarchy—those Hispanic roots of the noble Bourbons and of Franco—immigrants who came merely to get rich, to"make" America. It was they who prevented consolidation of a notion of citizenry.
3) Liberal sectors would accept Borges's remark as an absolute truth, claiming that it was the incapacity of the Argentine ruling class to understand the immigrant phenomenon—with all its contributions to creativity, culture, the republican and democratic spirit, its impulse toward civic activity, the struggle for human rights and man's equality—and, specifically, the battle of the ruling aristocratic groups against unlimited immigrant access to every level of Argentine life, above all political life, that has prevented national consolidation, and hence the creation of an Argentine citizenry.
4) Leftist sectors would accept Borges's remark as an absolute truth, claiming that a man can only genuinely feel that he's a citizen within a Socialist nation, and therefore no bourgeois nation can expect its inhabitants to share any possible common interests.
5) Fascist sectors would accept Borges's remark as an absolute truth, claiming that citizens exist only when a central power organizes them as such, and that it is precisely that central power, which is required to give explanations to no one, that is lacking in Argentina. (Timerman, 1981, pp. 16-17)

In essence, what Timerman has described is not a country under either unity of power or division of power—as discussed in volume 1—but a *divided nation*.

Nicolas Shumway (1991) said all this in a compact way:

> At best, Argentina's divisions lead to a lethargic impasse in which no one suffers too much; at worst, the rivalry, suspicion, and hatred of one group for another, each with different notions of history, identity and destiny, lead to bloodletting like the civil wars of the 1800s and even the "dirty war" of the late 1970s which saw the "disappearance" of thousands of Argentines. (p. 229)

3. Aims and Methods

Expanding on what I said in Section 1, many scholars have attempted a diagnosis of Argentina's predicament, and have developed a number of valid, but—in my opinion—incomplete interpretations: incomplete because they left a number of *UP hidalgo* ethnopsychological features out of the picture (e.g., vertical thinking, shortage of *sentido social*, intransigence), and because they have paid too little attention to the *mindbinding* which was caused by the *Inquisition's Anaconda Effect.*[157]

The reader may remember what I reported in volume 2, about the old Chinese practice of *footbinding*, specifically what John Fairbank said:

> This excruciating practice, which darkened the lives of probably more than half of the women in China in recent centuries, is today generally ignored. *No doubt we China specialists love our China and hate to make it seem sordid. Only this, I think, can account for the neglect of a social custom* carried out by women ostensibly to please the men but with incalculable psychic consequences. Once your feet were bound they were with you night and day, a source of pain and weakness impossible to eliminate and constantly proof of your inferiority. (1990, p. 17)

The same can be said for the Spain and Latin America specialists and for the Catholic scholars who love their Spain, Latin America, and Catholic Church, and hate to make them seem sordid. Also, and under-

---

[157] See Point 5 of Section II.

standably, most historians who favor economic interpretations have a tendency to belittle cultural and psychological factors, especially those of distant origin.

One interesting consequence is that those authors who minimize the evils of the Inquisition (e.g., Véliz, 1980; Shumway, 1991) also have a tendency to blame the founders of the Argentine independence—the Men of '37 in particular[158]—for the origins of Argentina's difficulties. These men are accused of lack of love for their Spanish roots, of their passion for French and English authors, and—in consequence—of not having sought the highest and most unique goals for Argentina.

My approach is different. Given that I am presenting evidence validating the GAM/DP theory, and am not a Spain or Latin America specialist, I will concentrate on those elements which have been neglected to date. Also, as in my other volumes, I will present the consequences of my analysis, and propose, to the Argentines, a number of ways to improve their ethnopsychology, so that they may be better able to work at building a united nation rather than dividing it; so that they may live in a prosperous, just, and fair country.

II. *Mindbinding* by the Inquisition

1. A First Set of Key Judgments

> Many causes contributed to the decay of the Spanish colonies, but among them not the least was the impossibility of settled and orderly administration occasioned by the multiplicity of rival jurisdictions, inherited from the medieval conceptions of the relations of Church and State.... *Of these the most disturbing element was the Inquisition, armed with the irresistible weapon of excommunication, by which it could paralyse its antagonists at will, and the arbitrary power of arrest, which inspired general terror....* The Inquisition, [said the Council of Indies in a report to Carlos II in 1696], was founding a supreme

[158] "The Generation of '37 was a group of impassioned young men, mostly in their twenties and early thirties, who in 1837 organized a Literary Society as part of an ongoing critique of their country that would eventually produce some of Argentina's most durable guiding fictions. Despite the century and a half that separates us from their first writings, the Generation of '37 arguably remains Argentina's most distinguished group of intellectuals. The Men of '37 assigned themselves two major intellectual tasks: to identify without idealization the problems confronting their country, and to devise a program that would make Argentina a modern nation" (Shumway, 1991, p. 112).

> monarchy, superior to all others in the State. *It was regarded with universal hatred in all the regions of the Indies and with servile fear by all, from the lowest to the greatest.* (Lea, 1908, pp. 511-12, my italics)
>
> *The measures taken for the education of the people went side by side with the stupid and brutal machinery for crushing and destroying all freedom of thought with the aid of torture and murder.* In 1569 Philip II decreed the establishment of the "Holy Inquisition" in Peru; with twelve familiars at Lima, and one in every town inhabited by Spaniards. There was one redeeming feature in the decree. All Indians were exempted from the jurisdiction of the inquisition. They were to be considered as catechumens. The familiars were to practice on their own countrymen only; who were exposed to all the rigors of this atrocious tribunal, which shrouded its proceedings in impenetrable mystery, and forced innocents to acknowledge guilt by excruciating tortures. Hanging, burning, flogging, the galleys, and confiscation were inflicted for the smallest difference of religious belief, for suspicion of erroneous doctrine, and even for the hallucination of a weak mind. The first "*auto de fé*" of the inquisition at Lima took place in November, 1573, when a poor old French Lutheran who had led the life of a hermit on a *huaca* in the valley of the Rimac, was burned to death. In 1578 another "*auto de fé*" was celebrated with great pomp, in presence of the gloomy tyrant Toledo and the judges of the Audience. There was a procession of sixteen victims with ropes around their necks, including six priests, a lawyer, and a merchant; the sentences being two hundred lashes on some, burning on others, confiscation on all. The next "*auto de fé*" was in 1581, when there were twenty victims; and so the ghastly work continued during the centuries of Spanish domination, creating a sensation of terror through the land, spreading misery and sorrow broadcast, benumbing thought, and gradually, but very surely, exciting hatred and repulsion. (Markham, 1968, pp. 171-72)

J. T. Medina, in his *El tribunal del Santo Oficio de la Inquisición en las Provincias del Plata* [The tribunal of the Holy Office of the Inquisition in the Provinces of the Río de la Plata, i.e., Argentina] highlighted the fearsome secrecy in which the Inquisition shrouded all its operations:

An impenetrable veil covered all its actions, it was impossible to know what was happening in its courtrooms or in its dungeons.[159] Having established the most absolute seal on its procedures, the accused and the trials were completely withdrawn from anybody's knowledge (at the exception of its own highest functionaries), and harsh punishment was meted on those who infringed this secrecy. Nobody could express an opinion on its decisions, and with the exception of its *autos da fé* to which the people were invited to assist to see the luckless who had been condemned to abjuration or to death, nothing transpired to the outside world. Without express permission by the Inquisitors, nobody was allowed to report, specifically to print a description of those ceremonies. (ca. 1890/1945, pp. 7-8)

Of all the obnoxious features of the Inquisition, however, perhaps the most obnoxious was its natural tendency to generate a climate of mistrust and mutual suspicion peculiarly propitious for the informer and the spy. There were some 20,000

---

[159] That same impenetrable veil was still at work in 1858 at Bologna, in the Papal States, when the Inquisition kidnapped the six-year-old Edgardo Mortara, the son of a rich Jewish family, on the ground that a maid of the Mortara had secretly baptized the child, five years before.

"They implored [the Inquisitor] Father Feletti to tell them why he had ordered the police to take Edgardo. Responding in measured tones, and hoping to calm them, the Inquisitor explained that Edgardo had been secretly baptized, although by whom, or how he came to know about it, he would not say. Once word of the baptism had reached the proper authorities, they had given him the instructions that he was now carrying out: the boy was a Catholic and could not be raised in a Jewish household. Padovani protested bitterly. It was an act of great cruelty, he said, to order a child taken from his parents without ever giving them a chance to defend themselves. Father Feletti simply responded that it was not in his power to deviate from the orders he had received. The men begged him to reveal his grounds for thinking that the child had been baptized, for no one in the family knew anything about it. The Inquisitor replied that he could give no such explanation, the matter being confidential, but that they should rest assured that everything had been done properly. It would be best for all concerned, he added, if the members of the family would simply resign themselves to what was to come. 'Far from acting lightly in this matter,' he told them, 'I have acted in good conscience, for everything has been done punctiliously according to the sacred Canons.'" (Kertzer, 1997, p. 6).

"The case of Edgar Mortara unleashed a storm of protest, but to no avail. The boy was brought to Rome placed first in a Catholic boarding school, then in a seminary from which he emerged as Father Pio Edgardo Mortara. The name Pio is in honor of Pope Pius IX who took a "paternal" interest in his Catholic development and who had "asserted his own paternal right, superior to that of a merely natural (or, in this case a Jew, unnatural) father" (Wills, 1998, p. 20).

familiars scattered through Spain, ever on the alert for manifestations of unorthodoxy; and their activities were supplemented by the unpleasant device known as the Edict of the Faith, by which inquisitors would visit a district at regular intervals and would have a list of heretical and obnoxious practices read to the assembled population. The reading would be followed by an exhortation to the hearers to denounce any such practices as had come to their knowledge, with severe penalties being threatened to those who kept silent. Since victims of the Inquisition were never informed of the identity of their accusers, the Edict of Faith presented an ideal opportunity for the settlement of private scores, and encouraged informing and delation as a matter of course. (Elliott, 1964, p. 211)

No, no cabe negar el daño tremendo producido por el temor a la Inquisición en la devoción de los hispanos por el saber. Impidió con rigor al libre vuelo de las meditaciones filosóficas o de las especulaciones que de algún modo pudieron topar con las vidriosas cuestiones de la fe. Contribuyó indirectamente a sangrar el caudaloso potencial de curiosidad intelectual que el humanismo y los descubrimientos habían creado en los peninsulares. Y a frenar el impetuoso despliegue de tal potencial, promisorio y hasta fecundo mientras duró la relativa libertad espiritual de la época carolina o vivieron los hombres en ella formados.[160] (Sánchez-Albornoz, 1956/1991, pp. 1275-1276)

Ahora bien, la política constante de Fernando el Católico fue la de apoyar a fondo a la Inquisición contra sus súbditos y contra el papa.... La Inquisición fue, para el Rey Católico, la mejor arma contra los fueros, es decir, el agente más eficaz del absolutismo....Sin embargo, la Inquisición no pudo asegurar la unidad espiritual de España, lo que hoy es evidente. En cambio, por su presencia y su tenacidad, por

[160] (No, it is not possible to deny the terrible damage produced by the fear of the Inquisition in the devotion to learning by the Spaniards. It rigorously hindered the free flight of any philosophical thoughts which could in some way lead to touchy religious questions. It contributed indirectly to bleeding the rich potential of intellectual curiosity which humanism and geographical discoveries had created in the Spaniards, and to brake the vigorous unfolding of this potential, so promising and even fertile as long as the relative intellectual freedom of the epoch of Charles V lasted, or as long as those who had been formed in that epoch lived.)

el miedo que inspiraba, hizo de España por mucho tiempo el reino del conformismo. Quiero decir del conformismo político, del conformismo intelectual....Secó las fuentes vivas de la investigación y de la especulación teórica, por lo que la teología sufrió tanto como otras actividades del espíritu hasta el punto de que España, ardiente hogar de los teólogos en el siglo XVI, no los produjo apenas en el XVII. ***Creó una deconfianza ante el libro cuyos nefastos efectos comprobaron los hombres de las Luces en el siglo XVIII y que ha durado hasta una época muy próxima a nosotros. Sustituyó la reflexión, la meditación religiosa, por la afirmación.***[161] (Bennassar, 1981, pp. 323, 339-40; my highlighting)

La Inquisición española fue implacable con la cultura foránea. Su agresividad ha sido muy bien descrita por Sarrailh y Deforneaux....La defensa contra el pensamiento extranjero, especialmente francés, fue intensa.[162] (García Cárcel, 1992, pp. 151, 153)

2. Sarmiento's Judgment[163]

Sarmiento devoted a long chapter of his book *Conflicto y armonias de las razas en America* of 1883 to a discussion of the Inquisition, first in Spain then in Latin America. Already from the first page we read:

---

[161] (The constant policy of King Ferdinand the Catholic was to strongly back the Inquisition against his subjects and against the pope... The Inquisition was, for King Ferdinand, the best weapon against the *fueros* [municipal franchises], i.e., the most efficient tool of absolutism.... However, the Inquisition was unable to achieve the spiritual unity of Spain, something which is evident today. On the other hand, with its presence and tenacity, with the fear that it inspired, it made Spain the realm of conformity for a long time. I mean, of political conformity, intellectual conformity.... It dried up the live sources of the investigation and of the theoretical speculation, in consequence theology suffered as much as other intellectual domains, to the point that Spain, which had been the home of great theologians in the sixteenth century, practically did not produce any in the seventeenth century. **The Inquisition created a mistrust towards the book, whose hateful effects were made evident by the philosophers of the Enlightenment in the eighteenth century, and which lasted until recent times. Reflection and religious meditation replaced affirmation**.)

[162] (The Spanish Inquisition was implacable towards foreign cultures. Its aggressiveness has been well described by Sarrailh and Deforneaux.... The fight against foreign thoughts, especially French, was intense.)

[163] Sarmiento (1811-1888) "Educator, statesman, and writer who rose from rural schoolmaster to become Argentina's first civilian president (1868-74). As president, he

[The Grand Inquisitor] *Torquemada es la figura o la síntesis más terrible de una aberración del espiritu humano.... [Torquemada] quemó todo lo que encontró con forma de inteligente,...¿Y sabéis lo qué es España? La España es la Inquisición!...La Inquisición negó durante tres siglos de horrible, implacable práctica, el derecho del acusado de conocer la acusación, y testigos para recusar los inhábiles y parciales; el derecho de defensa pública, con recusación de juez, sin comisiones speciales....Descartes puso por fundamento de la filosofia, no dando per probada la existencia de nada: "Pienso; luego existo".*

Un español o un americano del siglo XVI debió decir con más verdad: Existo; ¡luego no pienso![164] (1953, pp. 102-04, 115-16, 125).

For Sarmiento,

"the basic Latin-American thinking is still Spanish and suffers from the ills of the Spanish system. The chapter on "The

---

laid the foundations for later national progress by fostering public education, stimulating the growth of commerce and agriculture, and encouraging the development of rapid transportation and communication. As a writer, he is best remembered for his sociological-biographical study *Facundo, o la civilización y la barbarie* (1845; *Life in the Argentine Republic in the Days of the Tyrants; or Civilization and Barbarism,* 1868), the first serious study of the culture of the gauchos of the Argentinean pampas, which initiated an indigenous tradition of gaucho literature that flourished in Argentina for almost a century" (*Encyclopaedia Britannica*, 1978, *8*, p. 904).

Unamuno had the greatest praise for Sarmiento: "*Hai en la obra unamuniana admiración y devoción por Sarmiento, nombre que verbenea en centenares de sus paginas* [There are, in the works of Unamuno, admiration for and devotion to Sarmiento, whose name abounds, in hundreds of his pages]" (Chaves, 1964, p. 91).

164 "Torquemada is the most terrible figure or synthesis of an aberration of the human spirit.... Torquemada burnt everything he met that was intelligent,...and do you know, what is Spain? Spain is the Inquisition!...The Inquisition denied—for three centuries of horrible, implacable practice—the right of the accused to know of what he was accused and the witnesses in order so that he could refute the unqualified and the biased; the Inquisition denied the right to a public defense, refutation of the judge, and avoidance of special committees....Descartes placed at the base of philosophy—due to the absence of proof for the existence of anything—'I think, therefore I am.' A Spaniard or [Latin] American of the XVIth century could not but say in full truth 'I am, therefore I do not think!'"

Inquisition as a Spanish Institution" dwells on this point, as it sets forth Sarmiento's view on the Spanish Inquisition and its consequences not only for the Spanish population, but for its Latin-American successors as well. To show the inquisition's devastating effects upon Spain, Domingo Faustino Sarmiento quotes from an article by Galton, in which it is claimed that 32,000 persons were burned alive, 17,000 thrown in prison, and 290,000 subjected to tortures. Sarmiento examines North American contentions that the Spanish rule enforced unnecessary servitude upon the Indians and he raises the question whether Latin Americans too have not done the same. He concludes: 'One of the most powerful accusations, which as American publicists, we have always made against Spain has been that we ourselves became so much like it....'" (Crowley, 1972, p. 58).

3. Montesquieu's and Voltaire's Judgments

What the philosophes said about the Inquisition is important in understanding the thinking of the *founding fathers* of the Argentine Republic; what they read, for instance in Montesquieu and Voltaire, made them ashamed of being related to Spain, to a country so backward and so evil: "Montesquieu's political criticism of the Inquisition pointed toward an enlightened and secular analysis of religious movements and institutions that perceived the Inquisition as the implacable and irrational enemy of political liberty and social productivity, not just in Spain and Portugal—but, as Montesquieu emphasized—in 'the Inquisition of Europe'" (Peters, 1989, p. 177)

In Voltaire's own words, from his *Philosophical Dictionary* of 1765:

> In order to form some correct idea of the jurisprudence of the Inquisition, and the forms of its proceedings, unknown to civil tribunals, let us take a cursory view of the "Directory of Inquisitors," which Nicolas Eymeric, grand inquisitor of the kingdom of Aragon about the middle of the fourteenth century, composed in Latin, and addressed to his brother inquisitors, in virtue of the authority of his office....Eymeric says, in the fifty-eighth page, "Commiseration for the children of the criminal, who by the severity used towards him are reduced to beggary, should never be permitted to mitigate that severity, since both by divine and human laws children are punished for the faults of their

> fathers....The information and depositions (*procés-verbal)* may be read over to the accused, completely suppressing the names of the accusers; and then it is for him to conjecture who the persons are that have brought against him any particular charges, to challenge them as incompetent witnesses, or to weaken their testimony by contrary evidence. This is the method generally used. The accused must not be permitted to imagine that challenges of witnesses will be easily allowed in cases of heresy, for it is of no consequence whether witnesses are respectable or infamous, accomplices in the prisoner's offence, excommunicated, heretical, or in any manner whatever guilty, or perjured, etc. This has been so ruled in favor of the faith."...Eymeric continues: "Care should be taken never to state in the form of absolution, that the prisoner is innocent, but merely that there was not sufficient evidence against him; a precaution necessary to prevent the prisoner, absolved in one case, from pleading that absolution in defence against any future charge that may be brought against him." (1927, pp. 228-232)

4. In Sicily, Also Under the Spaniards

In his *Philosophical Dictionary*, Voltaire not only quoted at length from the *Directory of Inquisitors* by Eymeric, but also from the *Origin and Progress of the Office of the Holy Inquisition*, printed by the royal press in Madrid in 1589 by order of its author, Louis de Paramo, Inquisitor in the kingdom of Sicily. This, by Voltaire on Louis de Paramo, brought to my mind what Denis Mack Smith (1988) had written on the Spanish Inquisition in Sicily, in his chapter "Spanish Administration 1500-1650":

> The Inquisition was an additional means by which the Church was harnessed to the State and became another buttress in the elaborate edifice of checks and balances. The Inquisitors were directly responsible to the King who appointed them, and were sometimes used by him to counteract the Viceroy and to obtain private information about other royal officials. Until towards the end of the sixteenth century, the Viceroy and the judges could even be made by an Inquisitor to do public penance if ever there was a serious conflict of jurisdiction, and several times a secret report to Madrid by the Holy Office helped to secure a viceroy's

> recall....[165] From the King's point of view, however, this authoritarian organization with its secret and even terroristic methods had some advantages, since it helped to maintain religious and political orthodoxy. For him it was an ancillary spy system, and its separate police force and prisons were a means of inculcating fear and submissiveness. In a country where he could afford only a small military establishment but where fear was an essential ingredient of political power, this was a cheap instrument by which he could recruit Sicilians themselves to keep other Sicilians in subjection. In such a good cause the Inquisitors were allowed to use torture, and their instructions encouraged them not to stop the torture "until the truth should be told": if their victim died in the process, it was held his fault and the judgment of God....About one person a year on average was handed over to the secular authorities for execution, and the choreography was always carefully devised to make each occasion entertaining and instructive for as many spectators as possible. (pp. 164-65)

5. The *Anaconda Effect*

Perry Link in his 2002 article "China: The Anaconda in the Chandelier" has described a strong repressive psychological effect which, with few changes, could be called "Spain and Latin America: The Inquisitional Anaconda in the Chandelier":

> [The cases of Gao Zhan and Li Shaomin remain a puzzle] but the "vagueness" of the charges is hardly new. Such vagueness is purposeful and has been a fundamental tool in Chinese Communist censorship for decades. It has the following four advantages:
>
> - A vague accusation frightens more people. If I, like Gao Zhan, am a Chinese scholar working in the US, and I don't know why she was arrested, then the reason could be virtually anything; therefore it could be what I am doing; therefore I pull back. (Result: many people begin to censor themselves.) If, on the other hand, I could know exactly why Gao Zhan was nabbed, then I could feel fairly confident that my own work was all right—or, if not, how to make it all right (Result: few

[165] What Mack Smith said about the power of the Inquisition over the Spanish viceroys in Sicily, perfectly matches the remark made by Sarmiento (1883/1953, p. 102) on the power of the Inquisition on the Spanish viceroys in Latin America.

people would pull back.) Clarity serves the purpose of the censoring state only when it wants to curb a very specific kind of behavior; when it wants to intimidate a large group vagueness works much better.

- A vague accusation pressures an individual to curtail a wider range of activity. If I don't know exactly why I was "wrong", I am induced to pay more attention to the state's strictures in every respect. This device has been used in literary and social campaigns in China since the 1950s....
- A vague accusation is useful in maximizing what can be learned during forced confessions. When Li Shaomin was arrested, he asked his captors the reason and they answered, "You yourself know the reason." It was up to Li to "earn lenience" by "showing sincerity" through "confession". This word game is standard. The police routinely say that they already possess an exhaustive amount of information on your crimes and that the purpose of interrogating you is not to get information but to measure your sincerity by observing your confession. In fact, though, this is often a lie. Usually the point is precisely to extract new information, which can then be used either against you or against someone else. Clarity about the accusation would obviously destroy this tactic. (p. 67)
- A vague accusation allows arbitrary targeting. Leaders who exercise arbitrary power like to disguise the real reasons for their actions. (p. 67) In sum, the Chinese government's censorial authority in recent times has resembled not so much a man-eating tiger or fire-snorting dragon as a giant anaconda coiled in an overhead chandelier. Normally the great snake doesn't move. It doesn't have to. It feels no need to be clear about its prohibitions. Its constant silent message is "You yourself decide," after which, more often than not, everyone in its shadow makes his or her large and small adjustments—all quite "naturally". (p. 69)

Seventeen years before Perry Link, Julian Marías had said similar things on the Spanish Inquisition in *España intelegible*:

> I believe that the most profound damage done by the Inquisition to the Spanish cultural life was not that it persecuted or reprimanded the great creators. There were some—only some—who were molested or persecuted, and that was all. What the Inquisition did was to *dissuade* people to concern themselves

with those questions which drew the Inquisition special attention, which could become the object of worrisome scrutiny, which were suspicious to its eyes. Rarely did it need to exercise effective violence: its presence was enough, its unwanted vigilance, which immediately and before that fear developed, was able to eliminate at the root the ease, the spontaneity demanded by certain forms of creativity: precisely those that by their nature are not agonistic nor polemic, those that do not go against anybody, but consist in the unruffled, peaceful, and from time to time ludic, research of the truth.The Spanish intellectual, in the orientation of his studies, nearly automatically avoided those subjects which could bring him inconveniences or troubles. The heretic could, however, be led by his passion to face dangers. The pure thinker searched for fields in which he hoped not to encounter unwanted attention, and spontaneously avoided those which could bring him inconveniences and annoyances....[Because of this prudence, much was not done]. This *retraction* was the true handicap of Spain. Because what was lost—or did not arise in new forms—was the *theoretical spirit*, and when it was finally possible to exercise these disciplines, one had before to create the needed intellectual instrument and even more the lively vocations. (1985, pp. 241-42; italics in the original)

Also J. H. Elliott had clearly noticed the *Anaconda Effect*:

[Then, in 1568-69, Philip II, fearful of Lutheran Influences, prohibited the importation of foreign books and forbade Spanish students to go abroad. These measures] administered a drastic shock to Spanish intellectual life. By cutting off the supply of foreign books, and imposing further restrictions on theological and devotional writings, they inevitably undermined the confidence of Spanish men of letters, and added one more set of barriers all over Europe to impede the free circulation of ideas.... One of the most marked intellectual repercussions...was to be found in the realm of science, more dependent than the arts on a collective effort and continuing tradition. In the early seventeenth century the continuity had barely been established, and society and the State had lost interest; and the Castilian science, as a result, was either extinguished or went underground, to be pursued in secrecy by a few dedicated spirits in a mental climate totally uncongenial to their efforts. (1964, pp. 219, 381).

6. Véliz's and Salvador de Madariaga's Whitewashing

According to Véliz, on the other hand, it is wrong to blame the Inquisition for the present difficulties of Latin America because the Inquisition did, relatively speaking, little evil:

> [The Inquisition] was established in the middle of the sixteenth century and abolished early in the nineteenth century. It had therefore a duration of just over two and a half centuries. In this period the total number of victims who died as a result of the activities of the Inquisition in the Indies, including those who were driven to commit suicide or went insane under torture, did not exceed ninety. This is ninety too many. But ninety victims—all of them Spaniards, foreigners, or Protestants, with only one exception—means one execution every three years in the whole of the Indies. Comparisons are odious in this respect, but it should be permissible to note that very few political conflicts of the past three decades, most of which have not been particularly edifying, have claimed less victims than the Inquisition in three hundred years in the Indies. (1980, pp. 214-15)

Similarly, for Salvador de Madariaga:

> In the whole of the Indies, for the whole three centuries of the Spanish rule, the number of fatal victims of the Inquisition, including those who committed suicide or went mad under the physical or moral torture which that antiquated and benighted system entailed, stands nearer to sixty than to one hundred. Let us put it at thirty a century on average. Thirty victims a century for a whole continent and for a population as large as that of England and Spain put together; when in England alone, and only under the Tudors, the victims of religious persecution on both sides exceed five hundred. From our modern point of view the Inquisition cannot be defended in itself. But to single it out from the practices and ideas of the day as "the" institution of persecution and cruelty is unhistorical...When it organized its *autos* the pageant was always carefully regulated, and everyone in the city was eager to take part in it. "At the appointed hour many honourable citizens congregated in the hope that each would be entrusted with a penitent whom he would accompany in the *auto*, thus showing in every way they could the affection

> with which they desired to serve the Holy Office...." Where is the terror and the gloom? What in actual fact was the chief crime of the Inquisition, was that it failed to maintain the standards which it professed to serve: the purity of the faith and the decency of behaviour. (1949, pp. 148, 164-65)

So, according to Véliz's and Madariaga's criteria, what happened to Galileo was nothing: the Inquisition did not burn him on the stake, did not torture him, did not even jail him. It only told him to speak the truth: "otherwise recourse will be had to torture [*alias devenietur ad torturam*" (Santillana, 1955, p. 303). The next day, on June 22, 1633, Galileo abjured:

> I, Galileo, son of the late Vincenzo Galilei, Florentine, aged seventy years, arraigned personally before this tribunal and kneeling before you, Most Eminent and Reverend Lord Cardinals Inquisitors-General against heretical pravity throughout the entire Christian commonwealth, having before my eyes and touching with my hands the Holy Gospels, swear that I have always believed, do believe, and by God's help will in the future believe all that is held, preached, and taught by the Holy Catholic and Apostolic Church. But, whereas—after an injunction had been judicially intimated to me by this Holy Office to the effect that I must altogether abandon the false opinion that the Sun is the center of the world and immovable and that the Earth is not the center of the world and moves and that I must not hold, defend, or teach in any way whatsoever, verbally or in writing, the said false doctrine, and after it had been notified to me that the said doctrine was contrary to Holy Scripture—I wrote and printed a book in which I discuss this new doctrine already condemned and adduce arguments of great cogency in its favor without presenting any solution of these, I have been pronounced by the Holy Office to be vehemently suspected of heresy, that is to say, of having held and believed that the Sun is the center of the world and immovable and that the Earth is not the center and moves:
>
> Therefore, desiring to remove from the minds of your Eminences, and of all faithful Christians, this vehement suspicion justly conceived against me, with sincere heart and unfeigned faith I abjure, curse, and detest the aforesaid errors and heresies

> and generally every other error, heresy, and sect whatsoever contrary to the Holy Church, and I swear that in future I will never again say or assert, verbally or in writing, anything that might furnish occasion for a similar suspicion regarding me; but, should I know any heretic or person suspected of heresy, I will denounce him to this Holy Office or to the Inquisitor or Ordinary of the place where I may be. (quoted by Santillana, 1955, p. 312)

Madariaga, most probably, would have approved: here the Inquisition had worked at its best to "maintain the standards which it professed to serve: the purity of the faith" which Galileo had offended by his heliocentrist beliefs.

What I find to be missing in the Véliz-Madariaga discussion of the Inquisition is a sense of profound revulsion for a set of UP scripts which combine at the same time an official love for the Jesus of the Gospels, and for torture and murder for *Christian reasons*. As seen above, Galileo abjured his heliocentrism by "having before my eyes and touching with my hands the Holy Gospels."

Long term, the real tragedy was less the killing of 90, or 1,000, or even 10,000 people, but the *Anaconda Effect*: the climate of fear first, then of automatic self censorship by people who gradually, one generation after the other, incorporated more and more of the *Anaconda's* wishes until the great snake was seen as an integral part of the chandelier encouraging dedication to its interests. When that stage was reached, the Inquisitors could almost in good faith condemn Giordano Bruno and Galileo, and kidnap young Edgardo Mortara and make a priest of him, while proclaiming allegiance to the Gospels!

Said differently, what Véliz and Madariaga seem not to know, or not to care about, is that—in the development of a people's ethnopsychology—a moderate force applied constantly for several generations (and there were many generations between 1478 and 1834) has an infinitely greater impact than a great force applied for only a few years. Both Véliz and Madariaga are right in pointing to great horrors perpetrated elsewhere, be it in England or France; but none of those horrors continued systematically for so long, and in so many places.

The fact that the simple people loved the Inquisition—especially its *autos-da-fé*, and gladly informed the Inquisition—rendered the situation even more difficult for the intellectuals. Everything worked against them. The system (State, Church, Inquisition) was pro-active in its high praise of everything that was both Spanish and Catholic (theology, philosophy, the arts and literature), while non-Spanish but Catholic creativity could only be average, and the works of non-Catholics were at best fourth class, or worthless. So the Hispanic people, high and low, were told not only what to dislike and avoid, but what to love and seek out. And many concurred.

Back to Véliz and Madariaga, once the horrors of the Inquisition of the Spanish kings are minimized, or treated in a cavalier manner, all else becomes unclear in the history of both Spain and Latin America: *c'est la bouteille à l'encre* (it is the inkwell). Not only do we have *España, un enigma histórico* (*Spain, a historical enigma*), as per the title of the magistral opus by Sánchez-Albornoz, but also *Latin America, un enigma histórico.*

III. The *Vertical Component* of the Hispanic Character

1. As Seen by Madariaga

> There are two components in every human will: one, which might be described as "horizontal", acts in inter-human relations, works through human institutions and weaves the tissue of collective actions; the other, which we might call "vertical", seeks the summit of individual life, universal, diabolic or divine, and by its very orientation avoids the community and yearns to rise above the individual. The shade that separates the Spaniards from the other Europeans consists in this: that in his will the vertical component is far stronger than the horizontal. A number of features of the Spanish character which passed on to the Spanish American can be deduced from this... For the vertical component is religious and the horizontal component is moral, since the former yearns towards God and the latter spends itself among one's neighbours. Hence, perhaps, the Spaniard's fidelity to Catholicism which, in fact if not in theory, is the most religious and vertical, the least moral and horizontal of the Christian confessions.... The vertical component, as happens in artillery, by firing high, reaches the far distant horizon instantly and freely... Science, at any rate its technological and analytical aspects (which hold sway nowadays), belongs to the horizontal set of man's activities. Technology means control over matter. It requires patience and objectivity, horizontal virtues not prominent in the Spaniard. When Ortega, an advocate of horizontal virtues, complained of the lack of inventions in Spanish scientific history,

> Unamuno (a verticalist) retorted: "Let them invent!", meaning by "them" the other Europeans, and he praised above all scientific inventions the mystical intuitions of St. John of the Cross, the swiftest rising skylark in our skies. This shrinking from science, particularly in its practical and technical aspects, has been scored against Spain often enough. (1962, pp. 51-53)

So, here we have the vertical component with its praise for the mystical intuition, with its disdain for the horizontal interest in the relationship between man and objective reality; the vertical component with its search for the integral existence in which thought cannot be separated from the spontaneous vital process.

2. As seen by José Antonio Primo de Rivera

Moving now from the verticalism of the mystics to that of the politicians, the following—from the pen of José Antonio Primo de Rivera, Marqués de Estella, the founder of the Spanish fascist party, the Falange—deserves to be mentioned for its clarity and UP:

> *España puede tener un Estado fuerte, porque es, en sí misma, una unidad de destino en lo universal. Y el Estado español puede ceñirse al cumplimiento de las funciones esenciales del Poder, descargando, no ya el arbitraje, sino la regulación completa, en muchos aspectos económicos, a entidades de gran abolengo tradicional: a los sindicatos, que no serán ya arquitecturas parasitarias, según el actual planteamiento de la relación de trabajo, sino integridades verticales de cuantos cooperan a realizar cada rama de producción.*[166] (quoted by Cerezo Carrasco in "*Sindacalismo vertical*", 1952, p. 11)

How dangerously vertical, and mystical, is this "Spain as a unity of destiny in the universal"! As used by José Antonio, these words—unity,

---

[166] (Spain can have a strong State because it is in itself a unity of destiny in the universal. And the Spanish State can assume the essential tasks of the Power, performing, not arbitration but the complete regulation of many economic aspects of traditional entities of long ancestry: of the trade unions, which will not be parasitic architectures, as now in the present situation of the labor relations, but the vertical integralities of those who cooperate within each productive branch of the productive system.)

destiny, universal—imply a vertical relation to the highest, to the *supremo* (be he the King, José Antonio, or Francisco Franco) who knows if there is unity of destiny or not (and, obviously, there will be unity of destiny only if he is obeyed completely). Outside of this unity there are automatically only *arquitecturas parasitarias*.

3. As seen by López Ibor

> *Existen hombres de esencias y hombres de contingencias.... El español es, ante todo, un hombre desarollado, con preferencia, en las dimensiones verticales....*[*Los otros son*] *hombres contingentes y casi anedócticos, sin eje interno ni reciedumbre de alma....Español: hombre vertical. Hombre de esencias. Posición erecta, dura y dificil ante la vida, raíz de sus heroísmos y madre de sus desgracias.*[167] (1969, pp. 159, 162-63)

Seen by a vertical thinker—and many Hispanics were/are vertical thinkers—horizontal men are less than admirable because they are: "*hombres contingentes y casi anedócticos, sin eje interno ni reciedumbre de alma.*[168]

4. As Seen by Bunkley

> [Hispanic man] although he has tried to think in terms of the abstractions of the Occident, has never been able to remove his personality from his thought. Although his philosophy and theology have attempted to categorize existence in European terms, he has always thought in terms of an integral reality. Although he has sought to approach life rationally and logically, he has

---

[167] (There are men of essence and men of contingencies.... The Spaniard is, before anything else, a man developed with preference, along the vertical dimension.... [The others instead are] contingent people and almost anecdotal, without an internal axle, without strength of soul.... Spaniard: vertical man. Man of essence. With a straight position, hard and difficult toward life, the root of his heroism and the mother of his misfortunes.)

[168] Clearly, the way in which Madariaga and López Ibor set the game, they cannot but win. If the poles are vertical and horizontal thinking, and given that indeed the Hispanics are vertical thinkers, then all the others, automatically, are lesser beings who think horizontally, or at best diagonally.

> approached it spontaneously and vitally. The egocentricity, personalism, and integralism of Hispanic man have made possible his great art and his genial self-expression, but they have kept him from perfecting the political, social, and economic institutions of the modern world. They have created for him a history of Persons rather than of Ideas or Things.
>
> A panoramic view of Spain's history makes his personalistic character immediately evident. She produced no great Things. Her science was negligible. Her industry and commerce were backward. She produced no great Ideas. Her constitutions were borrowed. Her philosophy was either European or North American. She left no milestones of material or intellectual progress. But she produced great men and great personalities. Her history can be told either in terms of the success of her personal efforts or in terms of the failures of her material and intellectual efforts. (1952, p. 5)

5. *Vertical/Essential* People Are Intransigent

By the sheer nature of things, the vertical man of essence is intransigent and hates compromises:

> Argentina ranks first among all nations in the number of political parties and tendencies that during the twentieth century have adopted the label "intransigent". The Unión Cívica Radical (UCR) first gave currency to the term in its struggle for electoral reform. Then in the 1930s Leopoldo Lugones's Guardia Argentina adopted a program of "intransigent nationalism." Subsequently, the Radical party (UCR) split into two wings, one calling itself the Unión Cívica Radical Intransigente (UCRI), later simply the Partido Intransigente (PI). During the first resistance to military dictatorship, the deposed president, Juan Domingo Perón, also adopted a line of "absolute intransigence" for his political party. Peronist militants followed suit. In response to a second resistance to military dictatorship in the early 1980s they organized Intransigencia y Movilización Peronista (IMP) also known as Intransigencia Peronista (IP). Perón personified a peculiar Argentine syndrome: "I cannot strike a bargain with conservatives for a very simple reason: I aim to destroy them." Although intransigence antedates the formal appearance of the Argentine question, it made its debut as state terrorism as early

> as June 1956.... Whatever its social origins, intransigence is today deeply ingrained in the national character. What Marxists call the class struggle has been exacerbated by the presence of intransigence and the related posture of arrogance. Dubbed the peculiar Argentine malady, arrogance shows up politically in intransigent claims for one's social class and particular party. Because they are exorbitant, such claims cannot be met. Thus political stalemate plus intransigence have become an argument for lawlessness in Argentina. (Hodges, 1991, pp. 11-12)

6. The Inquisition, and the Emphasis on Dying Well

The Inquisition contributed in many ways to reinforcing the vertical thinking of the Spaniards and Latin Americans, as, for instance, described by Unamuno in his *Tragic Sense of Life*: "It was, in fact, the yearning for liberty, for inward liberty which, in the troubled days of the Inquisition, led many choice spirits to the cloister. They imprisoned themselves in order that they might be more free" (1913/1954, p. 289).

"It would be absurd to deny that Spaniards over the centuries have been preoccupied with the thought of death," wrote Bennassar in "To Die Well" of his *The Spanish Character* (1979, p. 237). Then, to make his point perfectly clear, Bennassar devoted three pages—out of the twelve of his chapter—to describe two *autos-da-fé* which took place in Valladolid on May 21 and October 1559:

> These ceremonies set in motion immense crowds whose size corresponded to the importance of the occasion. May 21, 1559: It was rumored that Doctor Cazalla, the Licentiate Herrezuelo, and persons of high rank would appear in the procession of the condemned; for the people, and above all for the humble classes, this was a primary source of attraction. The omnipotence of the Inquisition, which struck down the great of this world and did not halt at the gates of the palace, was made dazzlingly clear to their eyes..... During the night of May 20-21, two thousand persons kept vigil in the *plaza mayor* to assure themselves of good places. Men paid twelve, thirteen, and sometimes even twenty *reales* apiece for access to a balcony or a window overlooking the *plaza*. Beginning at one in the morning, masses were celebrated; the sound of prayer ran through the crowd. At dawn, at five o'clock, the auto-da-fé began....Then came the long, terrible reading of the sentence: fifteen persons

> were consigned to the stake, fifteen others were reconciled with the church. The people observed with emotion the reactions of the condemned. Agustín de Cazalla was especially demonstrative. He knelt before the princes and asked that he be allowed to confess his great sins....The drama was repeated on October 8, and from midnight on the *plaza mayor* was full. Some new elements attracted the multitude, and in the first place the presence of King Philip, absent from the auto-da-fé of May 21. (ib., pp. 242-44)

Bennassar then concluded with "a curious affair that reveals the small regard that men had for the lives of others and probably for their own.... Here we see a notary, a familiar of the Holy Office [Inquisition], unhesitatingly hiring two married peasants to kill a man; and they accept the job without a second thought, as if it were the most ordinary thing in the world. Thus the contemplation of death encouraged the absence of respect for life by developing a sort of exemption from risk" (ib., p. 248).

There was also the contribution of the *Spiritual Exercises*—developed by St. Ignatius of Loyola and popularized by the Jesuits—with their basic theme of: death unavoidable—judgment unavoidable—hell or heaven unavoidable.

Finally, an even more explicit link with death is provided by López Ibor who, to better explain the Spanish vertical way of thinking, narrated the story of a gravely sick twenty-year old girl, and of how both she and her mother spoke of death—be it in the hospital or at home—in a detached way, fully matter of fact. For them, and all other *vertical* people: "*No importa la muerte come contingencia, ante el hecho esencial de morir* [Death as a contingency is unimportant in the face of the essential fact of having to die]" (1969, p. 164). And López Ibor supplemented this comment with a quote from Unamuno on the proper way to die: "*sin hacer espectáculo de la muerte, como se mueren los verdaderos santos y los verdaderos héroes, casi como los animales se mueren, acostándose a morir* [without making any show out of death, as the true saints and the true heroes die, almost as the animals which lie down to die]" (ib., p. 165).

7. In Answer to Sarmiento: the Great *Vertical Men* of Unamuno and Madariaga

According to Véliz: "Sarmiento's blind love for Europe was only equalled by his contempt for Spain: 'I feel called to struggle against the

Spanish race, so incapable of understanding freedom, here as well as in Spain. The Spanish language is an insurmountable obstacle to the transmission of culture.... Spain has never had a thinker, not one notable writer, not a single philosopher.... Spain has condemned to backwardness the descendants of Europeans in America.' Such views reflect a silliness, ignorance, and pettiness, that is hard to associate with a man of Sarmiento's stature." (1980, p. 176).

However, on the matter of Spain not having had a great thinker or philosopher, López Ibor (1969) would have agreed with Sarmiento, and not with Véliz:

> *No nos engañemos. España no ha contribuido al mundo de ideas que se ha llamado moderno, en la proporción de otros pueblos.... He revisado, por ejemplo, la tabla cronológica de luchas decisivas de pensamiento que Heinrich Schmidt cita al final de su 'Philosophisches Wörterbuch'. No hay un solo nombre español. ¿Ceguera intelectual de un autor? No; fuerza es confesarlo, España, en las últimas centurias no ha producido más que epígonos. Frente a esta tremenda realidad cabe adoptar varias posturas. Una, la que propugnaban los hombres del 80: europeizar a España,...*[169] (pp. 62-64)

This last line, is pure Sarmiento. The trouble with Véliz's critique is that he neglected to mention those great thinkers, notable writers, and finally outstanding philosophers that Sarmiento should have known and admired. Véliz may have felt that this was not needed: those names were so obvious, but they were not obvious either to Sarmiento or to López Ibor.

However, they were known to Madariaga and Unamuno, even if Madariaga, as seen before, took his main choice, of the great mystic St. John of the Cross, from Unamuno's list of 1906. Other names listed by

---

[169] (Let us not delude ourselves. Spain has not contributed, to the world of ideas which is called modern, in the proportion of other countries. I have checked, for instance, the chronological tables of the decisive fights of the mind which Heinrich Schmidt mentions at the end of his *Philosophical Dictionary*. There is not one single Spanish name. Was this the intellectual blindness of the author? No; we are forced to admit it, Spain in the last centuries has only produced epigones. Facing such a tremendous reality, one must adopt various positions. One, proposed by the men of the 80s, is to Europeanize Spain,...)

Unamuno in that 1906 essay *Sobre la Europeización* were: Cervantes, Velázquez, Diego de Estella, Saint Teresa of Avila, and Saint Ignaz of Loyola.

The notable writer—sought by Sarmiento—is there; but I would not fault him for not being a devout admirer of the founder of the Jesuits and of the two mystics—St. John of the Cross and Saint Teresa of Avila—who, by mystically rising in the skies, leave the others alone in the fight against the UP of the Spanish Kings and their Inquisition.

Diego de Estella (Estella, 1524—Salamanca, 1578), that other great man for Unamuno, was a Franciscan friar, the author of the *Tratado de la vanidad del mundo* [Treatise on the vanities of the world][170] and of the *Cien meditaciones devotísimas del amor de Dios* [One hundred very devoted meditations on God's love]. Should we fault Sarmiento for not having thought of Diego de Estella when he complained of a shortage of great Spanish thinkers, writers and philosophers?

Madariaga, Unamuno, and Véliz, are entitled to their preferences (*de gustibus non est disputandum*)[171], however given the very difficult conditions in which Sarmiento found himself, his sharp words may not only be excusable, but praiseworthy. The yoke was so heavy that it demanded a major effort to shake it off, so strong that it lacked *l'esprit de finesse* demanded by those who came later and who had an easier task. Agreed, it is not true that "The Spanish language is an insurmontable obstacle to the transmission of culture," but if we replace "language" with "vertical thinking" we may not be very far from the obstacle which Sarmiento had in mind.

8. The *Vertical Component* as Medieval Thinking

In the words of Gurevich: "What mainly distinguishes the medieval world-view is its integrated structure; it is this that explains the specific non-differentiation, the inseparability of its various spheres. Hence also

---

170 This seems a text in line with the famous *De contemptu mundi* by Bernard de Morval (fl. 1150), in which we find the verse which inspired the title of Unberto Eco's *The Name of the Rose*: "*Stat rosa pristina nomine, nomina nuda tenemus* [Yesterday's rose endures in its name, we hold empty names]" (1929, p. 33).

171 In the same essay in which he praised St. John of the Cross, brother Diego de Estella, Saint Teresa of Avila, and Saint Ignaz of Loyola, Unamuno criticized France: "*Observad que el espíritu Francés no ha dado ningún gran místco, ningún verdadero gran místico puro* [Note that the French spirit has not produced any great mystic, any truly great pure mystic" (1906/1952, p. 916).

medieval belief in the unity of the universe....The language of mathematics was not independent; it was, rather, a 'dialect' of the wider language of Christian culture. Number was an essential element of aesthetic thought, a sacral symbol, a divine thought" (1985, pp. 9, 11). Gurevich could have equally-well said: "What mainly distinguishes the medieval world-view is its vertical structure".

In Spain, during the first centuries of the Reconquista: "*La vida se centró al cabo entorno a la fe y el apetito de salvación primó sobre todo apetecer.... Se avanzaba por la vida con el pensamiento puesto en la muerte* [Life was centered around the faith; and the hope of salvation primed above every other desire;... People moved in life with their mind focused on death]" (Sánchez-Albornoz, 1956/1991, pp. 319, 324).

9. Is this Uniquely Spanish?

No, this way of thinking is not uniquely Spanish; however, it is common to people who suffered, century after century, under the unity of power, be it of the Spanish kings and their Inquisition, or of the Tokugawa shoguns in Japan. In volume 3, I reported the following on Nishida Kitaro, which with minor changes could express the thoughts of a disciple of Unamuno:

> No one made a more profound attempt to emancipate the self and to actualize his individual aspirations beyond society than Nishida Kitaro (1870-1945). Unanimously, histories of Japanese philosophy single him out as modern Japan's most creative philosopher. His metaphysical system reveals many intellectual elements which he shared with less articulate Japanese....In 1899 he wrote to his closest friend: "What do you think is the means by which we might attain what you name the unity of thought? I believe the method of Zen is the shortest way to attain this end. If we could not get to this unity of thought through Zen ... no other means would do. Therefore, regardless of the actual outcome, I intend to discipline myself in the ways of Zen throughout my life." His diary and correspondence show how persistently Nishida meditated in Zen temples. "It has been already a few years since I started to sit at Zen. Progress is slow. I have gained nothing. I feel fully ashamed of myself.' Again he writes: 'Although I sit (Zen terminology indicating meditation), with the recurrence of such miscellaneous thoughts as the wish to be a

> college professor or to go abroad to study, etc. ... somehow I cannot purify myself." The intellectual's effort to free his inner self from the anxieties of social life is nowhere better and more honestly stated than in these writings of Nishida Kitaro....This category of pure experience is the essence of existence. "In order to comprehend the essence of existence and the genuine nature of the universe, we have to proceed from knowledge which cannot be questioned, knowledge which is free from all artificial hypothesis ... Science, for example, is based on hypothetical knowledge, and does not purport to be, or to attain, the profoundest explanation of real existence." (Arima, 1969, pp. 7-8).

After this, I wrote: "Basically, in the above, we encounter a set of aims which the unity of power loves completely: a) comprehend the essence of existence and the genuine nature of the universe, and proceed from knowledge which cannot be questioned; and b) despise science, because it does not purport to be, or attain, the profoundest explanation of real existence."

Here, I would first say: "This reads like Unamuno"; then I would add that if Mondino de Luzzi (see volume 1, chapter 13), Galileo and Newton had spent their best energies in comprehending the essence of existence, we would have no science.

IV. Key Terms: *Ser entero, hombre total, hombre integral, método de la pasión, arbitrariedad, complejo de inferioridad ante la ciencia y la técnica pero de superioridad como estilo de vida*[172]

Be it in Unamuno, in Sánchez-Albornoz, Madariaga, López Ibor, José Isaacson, we encounter a number of terms which are typical of much of the Spanish and Latin American thinking; terms like *ser entero, hombre total, hombre integral, método de la pasión, arbitrariedad, complejo de inferioridad ante la ciencia y la técnica pero de superioridad como estilo de vida.*

Already, according to Madariaga, passion is the psychological center of gravity of the Spanish people, as much as action is for the English, and thought is for the French (1969, p. 8).

---

[172] (Whole being, total man, integral man, the method of passion, arbitrariness, inferiority complex toward science and technology but superiority complex as to life style.)

Unamuno, had said exactly the same with his praise for

> *el método de la pasión, que es la arbitrariedad, a la cual no hay que confundirla con el capricho, como con frequencia ocurre. Una cosa es ser caprichoso, y otra, muy distinta, ser arbitrario.La arbitrariedad, la afirmación cortante porque si, porque lo quiero, porque lo necesito, la creación de nuestra verdad vital—verdad es lo que nos hace vivir—, es el método de la pasión. La pasión afirma, y la prueba de su afirmación astriba en la fuerza con que es afirmada. No necesita otra prueba.*[173] (1906/1951, p. 919)

This was forcefully presented by Unamuno in his drama *La venda* (*The blindfold*) of 1899 which one year later he transposed into a short story by the same title:

> *La venda* is a symbolic one-act play which was published in 1913 but was not staged until 1921. The plot is simple: María, who has gained her sight in the last two days, runs out into the street upon hearing that her father is dying. She had been born blind and had learned to move throughout the city with the utmost of confidence and facility. The world of sight is strange and foreign to her as she sees the streets for the first time. She is so thoroughly confused that she cannot find her way. Finally, María stops two strangers and asks them for a handkerchief to use as a blindfold; with it and a stick she immediately finds her way. She arrives in time to solace her dying father, but refuses to take the blindfold off. Marta, her practical, rational sister, tears off the blindfold for the sake of truth. The old man sees his daughter with sight for the first time and dies. The sight of her father's death is deeply imprinted in her mind. She closes her eyes and asks to see no more. Unamuno himself has indicated that María is the symbol of faith in man's life. Man is born with faith, lives with it, and learns to move through the obstacles of

---

173 (The method of the passion, i.e. of the arbitrariness, must not be confused with caprice, as is frequently done. It is one thing to be capricious and another, very different thing, to be arbitrary. Arbitrariness: the sharp affirmation to "why so?", "because I want it, because I need it", the creation of our vital truth—truth is what makes us live—all this is the method of the passion. Passion affirms, and the proof of its affirmation rests in the strength with which it is affirmed. It does not need other proof.)

> living with it, serenely and without question. However, science brings the sight of rational thought through which man begins to doubt until the blind faith is destroyed. But man cannot find his way through life with logic and reason; thus, it becomes necessary to re-create faith through the will of the *yo*. The conclusion is striking as reason tells man that death is the finality of human existence, but the blindfold created by the will asserts a blind faith in immortality. (Valdés, 1966, pp. 104-05)

The ending of the short story of 1900 is even stronger: Her brother touches María's shoulder and says:

María, eres madre, aquí te traen a tu niño, que olvidaste en casa al venirte;viene llorando...

— *¡Ah! Sí. ¡Angelito! ¡Qiere pecho! ¡Que le traigan!*

Y exlamó en seguida:

— *¡La venda! ¡La venda! ¡Tráeme pronto la venda, no quiero verle!*

— *Pero María...*

— *Si no me vendáis los ojos, no le doy de mamar.*

— *Sé razonable, María...*

— *Os he dicho ya que mi razón está en las tinieblas...*

La vendaron, tomó al niño, lo palpó, se descubrió el pecho, y poniéndoselo a él, le apretaba contra su seno.[174] *(1900/1973, p. 51)*

Less passionate people, who have a passion for clear thinking, will

---

[174] (María, you are a mother, here they bring your baby, whom you forgot at home when you came here. He is crying...

—Oh! Yes. Little angel! He wants to nurse! Bring him!

And then exclaimed:

—The blindfold! The blindfold! Bring me quickly the blindfold, I don't want to see him!

—But María...

—If you don't blindfold me, I will not nurse him.

—Be reasonable, María...

—I told you that my reason rests in darkness...

They blindfolded her, she took the baby, she touched him, she bared her breast, brought him close to her, and gave him her breast.)

ask what will happen in the coming days, months and years. Will María continue to wear her blindfold, and act blind? Will she continue to impose on others all the help that a blind person needs, even if she is now perfectly capable of moving around and doing all normal things in life? Will she always be a blind mother to her son, thereby limiting the help she can give him? And all this when medical science has given her sight?

Unamuno may well object to these non-passionate and non-arbitrary questions which, for him, are incapable of moving from the surface of the story to its profound meaning. Yet, if faith has to guide us better than science, then that superior guidance must not restrict itself to María's *hiperindividualismo*, but must extend itself to the others, at least to her baby.

Sánchez-Albornoz, too, had the greatest admiration for *el hombre entero*:

> *Ese hombre entero siguió siendo, como habia sido, menos propicio a deharse guiar por los consejos de la razón que por el impulso de las emociones y de los instintos; y más sensible a las ordenanzas de la moral....Por ello no fué anulada en España la espontaneidad por la racionalidad y por ello no se llegó en ella a la hiriente disyuntiva europea entre razón y vida, porque entre nosotros la razón vital triunfó siempre de la pura razón....Porque el* ***homo hispanus*** *siguió siendo el hombre entero a quien sólo interesaba la dignidad humana,...*[175]" (ib., p. 1299-1300).

And yet, as reported at the beginning of this chapter, "on March 24, 1976, a junta led by Gen. Jorge Rafael Videla seized power from President Isabel Peron. During the next seven years, at least 8,900 people disappeared, according to a 1985 government report, although human rights groups place the figure at around 30,000." Specifically, during the "Dirty War" of 1975-1978,

---

175 (This total man continued to be, as he had always been, less in favor of being guided by the advice of reason than by the impulse of emotions and instincts; more sensitive to the instructions of morality.... Because of this, Spain did not lose its spontaneity in exchange of rationality, and because of this she did not suffer of the European cutting disjunction between reason and life, because, among us, life's reason always triumphed over pure reason.... Because the *homo hispanus* remained always the total man whose sole interest was human dignity.)

> entire families disappeared. The bodies were covered with cement and thrown to the bottom of the river. The Plata River, the Paraná River. Sometimes the cement was badly applied, and corpses would wash up along the Argentine and Uruguayan coasts. A mother recognized her fifteen-year-old son, an Argentine, who appeared on the Uruguayan coast. But that was an accident — the corpses usually vanished forever. The corpses were thrown into old cemeteries under existing graves. Never to be found. The corpses were heaved into the middle of the sea from helicopters. The corpses were dismembered and burned. Small children were turned over to grandparents when there was mercy. Or presented to childless families. Or sold to childless families. Or taken to Chile, Paraguay, Brazil, and given to childless families. (Timerman, 1981, pp. 50-51)

Similarly, in the words of Donald Hodges (1991),

> The totalitarian controls imposed by the first military junta (1976-1978) aimed at intimidating the entire Argentine people. In the name of combating subversion any number of Argentines might reasonably be eliminated, affirmed General Videla. The obsession with rooting out subversion accounts for the thousands of disappeared persons, most of whom would be considered innocent by any standards other than those of the military assassins in power. Although some three thousand guerrillas met their death in the torture chambers of the armed services, to this figure may be added another seven thousand executions and from ten thousand to thirty thousand civilians who disappeared—a margin of error unimaginable unless it was deliberate.... Beginning in 1978 activists classified as irredeemable were permitted to have visits from their families, while false expectations were created that they would be freed. Subsequently, they were informed of the decision to transfer them to another camp. To assure their cooperation, they were injected with an ostensible tranquilizer that gradually rendered them unconscious. Unable to resist, they were loaded into trucks and transferred to their final destiny: to be hurled into the sea from military airplanes or to be murdered in open fields and special camps built for that purpose. The repression of virtually all civilian opposition to military rule was something new to Argentina. Torture was revived on a scale hitherto unknown for extracting information from the

guilty and instilling terror in the innocent. A clandestine state was erected parallel to the official one, with its own intelligence services, invisible "task forces," and some 340 secret detention centers. Prisoners were hooded or blindfolded during interrogation, their torturers operated under false names, and the authorities denied any knowledge of or participation in these clandestine operations. (p. 177)

In Spain, during the Civil War of 1936-1939:

On both sides the war was marked, especially in its early stages, by a ruthlessness which astounded the civilized world. Churches were burned or desecrated, and public religious observances forbidden throughout republican Spain; ten bishops and many thousands of priest, religious, and devout members of the laity were murdered in cold blood, for no political activity or crime. The Nationalists for their part organized mass executions (e.g., in Badajoz), and the Basque town of Guernica was subjected to terroristic air bombardment. (Carter, 1973, p. 1110)

In 1944, a spokesman for Franco's Ministry of Justice admitted that over 190,000 people were executed or died in prison. Even the most conservative subsequent investigations have reached figures in the tens of thousands. Torture accounted for the large numbers of suicides in prison. The authorities, feeling cheated by these "escapes" from their justice, often reacted by executing a relative of the prisoner concerned.... The sheer misery of life for the defeated in Franco's Spain accounts for the notable rise in the suicide rate. Mir[176] unearths many extraordinary cases of economic and sexual extortion by the powerful. Considerable cruelty was visited upon women under the rhetorical Francoist umbrella of "redemption" — rape, confiscation of goods and imprisonment — as retribution for the behaviour of a son or husband. Widows and the wives of prisoners were raped. Many were forced to live in total poverty and often, out of desperation, to sell themselves on the streets. Mir recounts many cases of soldiers billeted on poor families who took advantage of the unprotected women of the household. There was no short-

---

176 Conxita Mir's *Vivir es sobrevivir*. Lleida: Editorial Milenio.

> age of priests ready to defend the honour of male parishioners and to denounce their female victims as "Reds". Mir's work adds notably to the picture of Franco's Spain being created by important local studies of the repression. She shows that violence against the defeated was not limited to prison, torture and execution but extended to the psychological humiliation and economic exploitation of the survivors. (Preston, 2001, pp. 4-5)

"*Porque el* ***homo hispanus*** *siguió siendo el hombre entero a quien sólo interesaba la dignidad humana.*" Really?

No doubts, Unamuno's María was *ser entero, hombre total,*[177] *aficionado del método de la pasión, de la arbitrariedad, con su complejo de inferioridad ante la ciencia y la técnica pero de superioridad como estilo de vida*. And yet she is not a person we would like to have as sister, or daughter, or wife, or mother.

María is intransigent, and so were many Argentines. As reported above: "Argentina ranks first among all nations in the number of political parties and tendencies that during the twentieth century have adopted the label 'intransigent'" (Hodges, 1991, p. 11). Most of the Argentine militaries who commited those countless crimes must have considered themselves *hombres integros*, as did Tomás de Torquemada, the Grand Inquisitor.

This intransigence arises from a distorted polar vision: on one side the man of principles just discussed, on the other—by obvious contrast—the man without principles (liar, crook, unreliable, kowtowing, fawning). But this polarity is wrong, because the valid opposite of the *ser entero, hombre total, aficionado del método de la pasión, de la arbitrariedad,* is the man who thinks scientifically, i.e., the person who makes a serious effort to find valid evidence for his/her affirmations (a passion for evidence that has nothing to do with the *scientism* of which Unamuno loves to make fun). In essence, for the person who thinks scientifically, the rule is: *boldness in conjecture, but austerity first in refutation then in seeking evidence for the unrefuted.*

María may have been bold in discovering that, by remaining blind to the world, she had acquired a new inner sight, for instance in her relations to her father. But she did not evaluate the consequences of her

---

[177] I prefer to use here "hombre" as "human being"; and avoid gender differentiations.

stubbornness: her son will suffer for lack of assistance, and so will her family in having to provide for her and her son, given her "assumed blindness." María was unable to speak to her brother, to bring evidence for the soundness of her decision. She could only be arbitrary:

— *¡La venda! ¡La venda! ¡Tráeme pronto la venda, no quiero verle!*
— *Pero María...*
— *Si no me vendáis los ojos, no le doy de mamar.*
— *Sé razonable, María...*
— *Os he dicho ya que mi razón está en las tinieblas...*

We have here *el método de la pasión, que es la arbitrariedad*, the opposite of the Faraday/Pasteur method which, in this cases, would have consisted of a willingness by Unamuno-María to consider her duties as mother, to discuss in detail why a *razón en las tinieblas* was to be preferred to reason in the light.

Reading the ending of *La venda*, brings to mind a number of similar observations by Sánchez-Albornoz on the Spaniards:

> *Nuestro individualismo y el yo explosivo de los españoles. — [El español] se sintió vivir con tal ímpetu, violencia, intensidad, exaltación y dramatismo que no necesitó pensar para comprobar que seguía empuñando las riendas de la vida. Con frecuencia su yo explosivo se vertía en tan agitadas batallas contra el mundo o contra él mismo, que no le quedaba ocio para hallar en la callada reflexión el fluir de la sagre por su venas. Y así, hasta los misticos, ... — Ese hombre, así forjado, tuvo menos ocasiones de ejercitar la potencia de la razón que la energía de la voluntad*[178] (1991, pp. 1227, 1305, 1335).

---

[178] (Our individualism and the explosive I of the Spaniards. — [The Spaniard] felt that he was living with such a drive, violence, intensity, exaltation and dramatics that he had no need to think in order to prove that he was holding the reins of life. Often his explosive I was translated into battles so stormy against the world and against himself that the flow of blood in his veins did not give him the leisure to delve in quiet reflections. And so the mystics.... This man, so forged, had less occasion to exercise the strength of reason than the energy of the will.)

María-Unamuno[179] indeed, did not feel the need to think in order to prove that she was holding the reins of life; she had no time to spare for calm reflections; she, like the mystics, practiced the strength of her will and not of her reason.

Back to Argentina, there was also—even if in a different format—far too much *ser entero, hombre total, hombre integral*, especially in the military:

> The world of the Argentine armed forces is a closed, hermetic structure. Most of the officers' wives are the sisters or daughters of men in the military. Nearly all are related, and whenever there's a military regime, the civilians who participate are mostly relatives of the military or individuals who have frequented military circles, in precise anticipation of that moment when the armed forces will take power.... This pattern has separated the military from the most elemental currents of modern life, and has instilled in them a series of fantasies about the true meaning of the scientific, moral, literary, and religious elements that mankind has incorporated into its normal daily existence in recent decades. The ideology motivating the Argentine military stems more from a notion of the world they reject than from a world they would like to attain. (Timerman, 1981, p. 94)

V. The Spaniards as the Last of the Pre-DP Medievals[180]

- *En España perdura aun algo de la civilización medioeval, es decir, de la civilización individualista por excelencia.*

---

[179] I allow myself, here, to follow the example of Thomas Mann who in his essay "On the *Faust* of Goethe", wrote both of Goethe-Faust, and of Faust-Goethe.

[180] The Middle Ages is the conventional name for the period in Europe from the collapse of Roman civilization to the end of the 15th century: from ca. 395 to 1500. According to the GAM/DP theory this period should be divided into two parts: *pre-DP Middle Ages* from 395 to 1077 (Canossa), and *post-DP Middle Ages* from 1077 to 1500. In the case of Spain, one might have used the expression "pre-feudal", or simply "early medieval", and not the cumbersome "pre-DP medieval". However "early medieval" does not say much, and the positive elements of feudalism, qua feudalism, have been negated, with some reason. For instance, Menéndez Pidal noted that "it has been pointed out as a great misfortune for Spain that it never had feudalism, that is to say, a strong, enterprising nobility. But if it did not have a number of feudal states, it had a variety of kingdoms which were able in freer ways to develop their personality and spread their influence far and wide through the Mediterranean, through Africa, as an apprenticeship and trial for

[In Spain, something of the medieval civilization still lives, that is to say of the individualistic civilization par excellence.] (Galvez, 1930, p. 43)

- *Aquende el Pirineo perduraba intacto hasta nuestro días el yo voluntarioso del español del medioevo.*
  [On this side of the Pyrenees, still to this day, the self-willed I of the Spaniard of the Middle Ages continued intact.] (Sánchez-Albornoz, 1991, p. 1341)

Already in section III.7, I spoke of the vertical thinking of the Spaniards as medieval, as a continuation of the way people thought at that time. Hereafter I will discuss a few other features of the Spanish/ Latin American ethnopsychology which also are a continuation of the medieval, specifically the pre-DP medieval ethnopsychology.

1. The Many Types of Individualism

There are many kinds of individualism, such as the *universal* or *common individualism*, the *pre-DP (heroic or semi-heroic) individualism,* the *DP visitor individualism* (e.g., the *visitor communal individualism*), and the *DP skeptical/cynical individualism.*

Briefly, the *universal/common individualism* is that of the *challenged personalities* (with their high GxAxM) and of the *leaders* (be they, for instance, a king or queen by election or heredity). Clearly, Alexander the Great, Socrates, Julius Caesar, and Cleopatra were individualists, for one or both of these reasons.

---

the greatness when those kingdoms became reunited in the sixteenth century" (1946/ 1966, pp. 81-82). On the other hand, I suspect that much of what other authors (e.g., Véliz and Sánchez-Albornoz) have found missing in Spain—and have attributed to the fact that Spain did not experience feudalism—is in reality the DP which the rest of Western Europe began to have from 1077 onward. Sadly, Spain was not involved in the *War of Supremacy* between emperors and popes of 1075-1313, especially in its first phase of the *Investiture Controversy* of 1075-1122, as noted, for instance, by O'Callaghan (1975, p. 308).

If then the Spaniards and Latin Americans are indeed (or were until recently) the last of the pre-DP medievals, then I must change what I said in volume 1, and no longer call the Germans "the last of the medievals", but "the last of the feudal medievals" in which "feudal" means (beside the detailed description I gave in volume1) just what Menéndez Pidal said: "a strong, enterprising nobility."

The *DP visitor individualism* has been discussed throughout volumes 1, 2 and 3, while the *DP skeptical/cynical individualism* was only briefly touched on in volume 1.[181]

One of the most famous *pre-DP (heroic) individualists* was Homer's Achilles:

> Achilles' wrath, to Greece the direful spring
> Of woes unnumber'd, heavenly goddess sing!
> That wrath which hurl'd to Pluto's gloomy reign
> The souls of mighty chiefs untimely slain;
> Whose limbs unburied on the naked shore,
> Devouring dogs and hungry vultures tore:
> Since great Achilles and Atrides strove,
> Such was the sovereign doom, and such the will of Jove!
> (Beginning of the *Iliad*, in the verse translation by Alexander Pope)

Achilles is the greatest Greek warrior in the Trojan War. Attention is focused on him due to his refusal to continue fighting on the side of the Greeks because Agamemnon, commander-in-chief of the Greek forces, has taken Achilles' captive, Briseis. Disastrous consequences follow the quarrel: not only must the Greeks fight without Achilles' support, but they must endure the enmity of Zeus who has been per-

[181] There I compared my thoughts on the birth of the *visitor personality* during the times of the *Investiture Controversy* of 1075-1122 (part of the *War of Supremacy* between popes and emperors of 1075-1313) to the studies of Colin Morris (1972) on the origin of individualism during the same epoch. From the concluding pages of Morris' book, I report the following: "The discovery of the individual was one of the most important cultural developments in the years between 1050 and 1200. It was not confined to any one group of thinkers. Its central features may be found in many different circles: a concern with self-discovery; an interest in the relations between people, and in the role of the individual within society; an assessment of people by their inner intentions rather than by their external acts. These concerns were, moreover, conscious and deliberate. 'Know yourself' was one of the most frequently quoted injunctions. The phenomenon which we have been studying was found in some measure in every part of urbane and intelligent society" (p. 158).

suaded by Thetis (Achilles' mother) to turn the tide of battle so that the Greeks will see how much they need Achilles.[182]

2. Not Inclined to Solidarity with the Community

Another *pre-DP heroic individualist* is the paladin Roland of the *Song of Roland* (ca. 1100). Roland is courageous and brave—nobody surpasses him in these virtues, but he may be over-courageous and brave, especially in fostering the glory of his clan.

> The many forms and manifestations of the *démesure* hypothesis make testing it a particularly complex task. The idea that Roland causes a disaster at Roncevaux through self-destructive error is one of the oldest ideas in medieval literary studies.... The idea [of *démesure*] is made explicit at least as early as Gaston Paris's *Histoire poétique de Charlemagne* (1865): "[Oliver] asks Roland to sound his ivory horn.... Roland refuses out of an exaggerated sense of honor." In Louis Petit de Julleville's

182 This heroic individualist—even when doing exceptional deeds, of his own initiative—still belongs to an *affiliate clan*; and, in normal conditions, will give evidence of sharing most of the key characteristics of the *affiliated personality* as discussed in chapter 6 of volume 2, and in chapter 11 of volume1, where I reported the following from Le Goff (1988) on the *pre-visitor mentality*: "In medieval western Europe the individual belonged first and foremost to his family. The family... stifled the individual, forcing him to submit to collective ownership of property, collective responsibility, and collective action" (pp. 279-80). In turn, the individual was provided with the support and protection of the family: if not directly, then at least indirectly, through the revenge that the family would bring upon the offender, his relatives and descendants.(*) We find an echo of this medieval collegial responsibility in the ancient punishment, whether actually meted out or threatened, unto subsequent generations. Already in the Bible we find several examples of this collegial vengeance. For instance, after the prophet Elisha had cured the Syrian commander Naam of his leprosy, and refused any reward, his servant Gehazi ran after Naam and said that his master would accept a gift of money and garments which Naam gladly gave. Elisha found out the truth and said: "The leprosy therefore of Naam shall cleave unto thee, and unto thy seed for ever. And Gehazi went out from his presence a leper as white as snow" (II Kings, 5:27). On the other hand, Elisha was following God's example who had said to Moses that he would be "visiting(**) the iniquity of the father upon the children, and upon the children's children, unto the third and to the fourth generation" (Exodus, 34:7).

(*) The counterpart of this protection was the obligation to take arms to avenge an insult or tort against any other member of the clan.

(**) "To inflict punishment for (wrongdoing); avenge [*visiting* the sins of the fathers upon the children]" (*Webster's New World Dictionary*).

> edition (1878), it has taken on something like its modern form: "Roland destroys the entire rear guard, and himself, at Roncevaux, through his foolish obstinacy....[Such flaws] lend his character [the effect of] life and truth." (Cook, 1987, pp. 144-45)

At the critical moment, Oliver implores: "Roland, my compagnon, now blow your Olifant; Charles will hear it, and will come back with the army, and will help us with his barons." Roland answered "May please God that I should not dishonor my house, nor bring disrepute to fair France."

> The honor of the clan (mi parent) and of the tribe (dulce France) would be tainted by a call for help. A proud man must stand by himself alone. The "olifant," the horn, may apparently be used for gathering the host under its commander, but not for a call to the king for succor....
>
> [However], no shadow of doubt remains about the poet's meaning, about his conclusion. Knightly conduct does not imply folly, measure is better than proud stupidity, Roland has destroyed the army by his irresponsibility, thus bringing upon his head much greater shame than by the sounding of the horn. (Borkenau 1953/ 1981, pp. 423, 426)

The *pre-DP heroic individualist* knows everything about honor, his own and that of his clan, and possibly also of his country, but little of global strategy, of working with his superiors and peers at achieving a common long term goal, in the case of Roland of winning the war against the Saracens, not of simply winning a battle, or of showing heroic mettle. Roland was a great hero, yet subsequent, more enlightened ages have learned to praise men who pay less attention to their personal and family glory and more to the common good.

Moving to modern times, we can detect a residue of the Roland of the past in the following text by Ramón Menéndez Pidal:

> The Spaniard is inclined not to feel a sense of solidarity with the community except in so far as it will bring him immediate advantages, for he will always neglect indirect or future benefits. Hence he is rather indifferent to the welfare of the community or its problems, but in compensation he possesses a lively

> perception of his own individual case as well as that of his neighbour. This over-valuation of the individual has direct bearing upon the two cardinal principles of communal life: justice which regulates it and selection which divides it into hierarchies.... It is clear that there is in Spain a special weakness in the spirit of association. The benefits to be derived from co-operation are less clearly felt than the advantages of separate individual action, even though this, in the long run, shows less results. Communal life comes to be regarded as something hampering owing to the restrictions it imposes upon the individual, for every one wishes to work at his ease without having to take account of his neighbour. (1946/1966, pp. 44, 75)

When the link with the clan disappeared, when the individual no longer had to answer to parents or the extended family, and no longer had help to receive from them, then modern *hiperindividualismo hispano* developed, which was mentioned so often by Sánchez-Albornoz in his *España, un enigma histórico* (1959/1991), and was discussed above by Menéndez Pidal.

3. A Violent Passion for Honor

A residue of the heroic individualism of the past remained in the violent way in which honor was conceived and defended. As discussed by Bennassar:

> If there was one passion capable of defining the conduct of the Spanish people, it was the passion of honor. Rarely have such diverse sources been in such perfect accord as on this point. This passion has a long history. As early as the thirteenth century, the Castilian Code of the *Partidas* carefully defined honor as "the reputation a man has acquired by virtue of his rank, his high deeds, or his valor."...More and more, honor was defined as a code independent of all morality, a code to which social prejudice and pride, and even vanity was the only key.... Honor must be defended even at the cost of another person's life, and it matters little if that person is blameless....For centuries Spain was a land of chronic violence. Contained, indeed successfully repressed, in the age of the Catholic Sovereigns, then by Charles V and Philip II by judicial institutions of high quality, it surged up again from the seventeenth to the nineteenth

> century. I am not suggesting that the Hispanic conception of honor was responsible for all or even for the majority of the acts of violence of this period, but it certainly encouraged them.... For Lazarillo de Tormes, honor is but "a thirst for vainglory and a lie," "a priesthood without virtue." For Guzmán de Alfarache it is "a mask under which one may steal and lie, and a useless burden to boot." This message was not understood. A distorted ideal of honor long sterilized Spanish energies. (1979, pp. 213, 232-36)

4. Is the *Hombre Integro y Total* More Prone to Being Envious?

In the chapter "The Spaniards as UP *Hidalgos*" of volume 2, I devoted three sections to the high level of envy of the Spaniards, as discussed by Carlos García, Unamuno, Madariaga, Menéndez Pidal, Diaz-Playa, and López Ibor. In the third of these sections, the one devoted to an explanation, I wrote:

> I see envy as the direct consequence of the emphasis put on the pride in one's *linaje y honra*, and the corresponding lack of emphasis on personal development and success through initiative, creativity, risk, hard work, trial and error. Automatically, if the others are more and have more, it means one of two things: (1) they were better at birth (and this is automatically an insult for those less fortunate who did not deserve more at birth); or (2) they cheated, and have what they do not deserve. One way or the other, there is no reason why the others should be more, or have more; and precisely this "no reason" is at the source of envy.
>
> In other words, if the others have more, it is an insult to one's honor, it is an attack on one's *honra* which must be cleansed. Probably their lineage is fake, it was obtained by cheating, or they are not the legitimate branch, or worse.
>
> Worth stressing, an emphasis on *linaje y honra* means focusing on what one is at birth (e.g. *hijo de algo*, first-born, legitimately born). It also means paying close attention to the inheritance one will receive at the death of father and mother. And when one speaks of inheritance, one is immediately in the thick of the classic fights among siblings for who will receive or has received this or that part of the inheritance. There is later, the equally distorted time, before the death of the rich relative,

> when everybody is courting the old man or woman in the hope of receiving the highest share of the inheritance. And woe if one sibling received more than the other. This woe, also, is wrapped in envy, or causes envy.
>
> Part of the explanation is related to a disdain for work.[183] In a world in which *linaje y honra* are paramount, who has made his fortune through his industry is an ugly *nouveau riche*, a despicable *parvenu*. The only fine-smelling fortune is the inherited one. Indeed, who has inherited well must have done so because of his or her intrinsic value.
>
> The importance of hereditary wealth—linked to good lineage—goes in parallel with a dislike for working which goes back to the ancient Greeks, and before, when all practical work was banausic, to be left to slaves. (pp. 193-94)

In the above, several items bring to mind *el hombre integro y total*: what he is, or is not. By definition, the *hombre integro y total, "hecho de una sola pieza, incapaz de renunciamiento y de adaptiones* [made of one single block, incapable of renouncing and adapting)" (Gálvez, 1960, p. 44) is less interested—than those capable of renunciations and adaptations—in a personal development and in success through initiative, creativity, risk, hard work, trial and error. Automatically if he, in the end, is and has less than the others, his pride will be hurt, and one result will be envy. As discussed in volume 2, envy seems an older sin, which is less evident in modern times when the emphasis is on the self-made man, on doing without the help from one's parents, and possibly achieving more.

5. The Aristocratic *Soy Quien Soy*'s Contribution to Intransigence

Maravall, in *Teatro y Literatura en la Sociedad Barroca* (1972/1990), discussed at length the evil perpetrated in the Spanish world, at the time of the *Sociedad Barroca* [roughly during the *Siglo de Oro* of 1550-1650], by an aspect, of the aristocratic understanding of honor, which Maravall called "*la formula 'soy quien soy'* [the formula of 'I am who I am']".

---

183 Leaving aside such activities as fighting war, light intellectual pursuits, courting the prince, robbing one's peasants.

For Maravall:

> Honor is one of the most important themes of the propaganda of the baroque comedy in favor of the interests of the monarchic-aristocratic society....[Guided by the dictates of "I am who I am"] the victory over the self ceased to be the opposition and control of the evil tendencies always present in the human condition. Instead, and very different, it became the victory over the personal feelings: silenced in favor of the dictates of the social system. This, in turn, was based on the belief that the proper way of behaving consisted in performing the obligations of one's position in the aristocratic society. In this view, honor was dissociated from conscience—even when conscience was moved by impulses of natural justice—and, related, instead, to the notion of accomplishing what is due to one's aristocratic position:
>
> ... I am who I am
> and being who I am, I win
> over myself by remaining silent.
>
> So says a character of Lope de Vega, in *La Estrella de Sevilla*, when giving precedence to his noble status over the demands of his conscience.... [Thus] when a member of the aristocracy is asked—for convenience or for a sense of justice—not to act as he plans to do, his standard answer is that he has no choice but to follow the norm of "I am who I am", as stated hundreds of times by the key characters of the plays by Lope de Vega, Ruiz de Alarcón, Rojas Zorilla, Calderón de la Barca,[184] Moreto, etc.... From all the information that I [Maravall] have assembled, it is clear that the "I am who I am" is not a principle which leads to be true to oneself, as felt by everybody else. On the contrary, it is a principle which is always stated in

[184] Further on, Maravall listed a number of plays in which Calderón used the formula of *soy quien soy*: "*Amar después de la muerte, El médico de su honra, El mayor monstruo del mundo, A secreto agravio, secreta venganza, Darlo todo y no dar nada, etc*" (ib. p. 63). And, as expected—see chapter 10 "More on Lord Acton's Law of Power" in this same volume 4—the very person, in *El médico de su honra*, who explains himself through the formula of "I am who I am", is Don Gutierre, the absurdly evil "surgeon of his honor".

relation to the obligation to operate according to one's social status. (1972/1990, pp. 60-63)

Moving to Argentina's "Dirty War" of 1975-1978, one way to understand some of its horrors would be to see them as having been perpetrated—especially by the military officers—by the dictates of "*soy quien soy*" in which the person's rank ruled over the dictates of conscience in something reminiscent of the pre-Christian thinking when "classical philosophers regarded mercy and pity as pathological emotions—the defects of character to be avoided by all rational men. Since mercy involves unearned help or relief, it was contrary to justice. Therefore 'mercy indeed is not governed by reason at all,' and humans must learn 'to curb the impulse'; 'the cry of the undeserving for mercy' must go 'unanswered' (Judge, 1986:107). Judge continues: 'Pity was a defect of character unworthy of the wise and excusable only in those who have not yet grown up. It is an impulsive response based on ignorance'" (Stark, 1997, p. 212). Tragically, not a few Argentine officers—through their *"soy quien soy"*—may have felt that they had to avoid pity as "a defect of character unworthy of the wise and excusable only in those who have not yet grown up."

At one level, some of what we just discussed is nothing but part of the standard training in obeying orders; on the other hand, this training is never dissociated from the general philosophy(ies) of life at a given place and time; and the old Spanish *"soy quien soy"* may indeed belong to a group of similar formulae—two of them being "*sollen über sein* [must over being]" and "*Meine Ehre heisst Treue* [my honor is fidelity]", discussed in volume 1, which also have been known to obliterate the dictates of conscience.

VI. No, Gálvez[185]

1. In the Visual Arts

Was der Bauer nicht kennt, das isst er nicht.
[What the peasant does not know, he does not eat.]
(German proverb)

[185] Argentine novelist and biographer (1882-1962), "whose documentation of a wide range of social ills in Argentina in the first half of the 20th century earned him an important position in modern Spanish-American literature" (*Encyclopaedia Britannica,* 2002, *5*, p. 99).

In 1913, Manuel Gálvez wrote a volume of essays on Spain entitled *El solar de la raza* (*The birthplace of our people*). In it he recorded his impressions of Spain during his 1910 sojourn. "No other book by Gálvez received such unanimous praise from so many distinguished Spanish writers and critics, including Miguel de Unamuno, Ricardo León, Andrés González Blanco, and Ricardo Monner Sanz. León said that *El solar de la raza* represented 'one of the most ardent and moving apologies for Spain's national spirit that could be made'" (Lichtblau, 1972, p. 14). In the first chapter of *El solar*, Gálvez presented a summary of his key ideas.

Starting with the visual arts:

> Nothing speaks with more clarity and beauty of the soul of Spain than its art: perhaps the highest and noblest that has ever existed....I have never understood why people exalt the Greek sculpture. It is an objective art, materialist, of pure formal beauty. Those are faces without expression, those figures do not talk to the soul, they say absolutely nothing. They have in them no moral beauty which is the highest beauty, if not the only form of beauty. I would like to know which improvement can be derived by the contemplation of a naked man, or what benefit can our soul derive by admiring a well shaped foot. On the contrary, I believe that all this is low, superficial, miserable. And what is the use of the cold beauty of the *Apollo del Belvedere*: a naked man with elegant and effeminate forms? Let us compare this superficial, sensual art, only preoccupied with the form, with some Christs by Montañes.... What does the Apollo say to us? Nothing..... However, let us contemplate a Christ by Montañes.... If we are artists, if we have a passionate and sensitive soul, we will be moved by that image of human suffering. It is an art at the same time naturalistic and spiritual, human and mystic. It is an art which troubles us, which we will never forget and which inspires yearnings of moral improvements. The contemplation of such works should make us kinder, more compassionate and humbler....The formal beauty, moreover, is conventional and changes constantly. The *Venus of Milo*, which is considered the prototype of feminine beauty, is a disappointment to every sincere and intelligent person who visits the Museum of the Louvre. That woman with her broad hips is the opposite of our concept of feminine beauty. Today, a body like that of the Venus would be considered vulgar [sic]. (1913/1960, pp. 31-33)

2. On Social Sense and Religion

For Gálvez

> Spain is a country hard to grasp, and it can be understood only by those who know and love it well....But why is it so difficult to understand Spain? The answer—the way I see it—rests in the special sense of life which predominates among the Spaniards: their exacerbated individualism, their Catholic mentality....The Spaniards, better said, the Castilians have a different concept of life than that of our times which consists in making money and enjoying the sensual pleasures....Spain is the most personal and original of the European countries.... A modern German city is little different than a French modern city. But in Spain every city has it character, its soul.....Salaverría, a true Spaniard, said that: "The Spaniard is a unique being, different from all other men on the earth....For most people, the Spaniard is incomprehensible. He resists every confrontation and analysis; he is an exceptional being. He moves therefore in an exceptional way....The Spaniard does not possess the instinct of malleability. He is made out of a single block. He is incapable of renouncing and adapting. He does not know what ductility is. Endowed with sovereign pride, this pride makes him incapable of the most important feature of civilization which is to give up, to bend, to rectify." Indeed, the Spaniards do not possess [*sentido social*] the social sense, as said by a truly Spanish writer: Pío Baroja. But it is just this lack of social sense which is not pardoned. All vices are tolerated in the educated, friendly and sociable man. However, a saint will be odious to people because he has no social sensibilities. (ib., pp. 44-45)

We have here, from Salaverría and Gálvez, something similar to what I quoted before by Ramón Menéndez Pidal:

> The Spaniard is inclined not to feel a sense of solidarity with the community except in so far as it will bring him immediate advantages, for he will always neglect indirect or future benefits. Hence he is rather indifferent to the welfare of the community or its problems....It is clear that there is in Spain a special weakness in the spirit of association. The benefits to be derived from co-operation are less clearly felt than the advantages of

> separate individual action, even though this, in the long run, shows less results. (1946/1966, pp. 44, 75)

Now, according to Gálvez—and many others like him—the greatest saints are the mystics who cannot but lack *sentido social*, even if that makes them odious to the little people. In other words, the giants are fully authorized to ignore the others, given that the great mystics do so.

However, this too is a monstrous distortion given that Jesus had said "Thou shalt love thy neighbour as thyself" (Matthew, 20:19), and, equally, St. Paul: "For all the law is fulfilled in one word, even in this: Thou shalt love thy neighbour as thyself" (Galatians 5:14). The *Encyclopedia of Catholicism*, in its section on mysticism, is equally clear: "No Christian mystic would suggest that one can leave behind the demands of charity and service of others in pursuit of purely introspective and isolated contemplation.... Rather, the deep union [with God] of, say Bernard of Clairvaux, Francis of Assisi, or Catherine of Siena, led them in greater activity and service of others" (Waldstein, 1995, pp. 901, 902).

In the end: Yes, many Spaniards and Latin Americans are different from other people by not possessing the instinct of malleability. Endowed with sovereign pride, this pride makes them incapable of giving up, of bending, of rectifying (even if one might have used less unpleasant words, like "understanding the needs of the others" instead of that ugly "bending").

In the end, Gálvez is wrong by not being truly critical—as Ramón Menéndez Pidal and Pío Baroja were—of those who are made out of a single block, incapable of renouncing and adapting. Gálvez is also wrong in distorting the religiosity of the great mystics as if they were blessing those without *sentido social*. He inverted the order of cause and result, while Unamuno, instead, saw it clearly: "It was, in fact, the yearning for liberty, which, in the troubled days of the Inquisition, led many choice spirits to the cloister. They imprisoned themselves in order that they might be more free" (1913/1954, p. 289). In the harsh world of the Inquisition, without *sentido social* but the interests of the Spanish king, the cloister—with its reduced *sentido social*—may indeed have been the only solution to avoid following the rule of "Thou shalt love thy King above thyself." Within the walls of the cloister the mystic could, if not love his neighbor, at least practice the rule of "Thou shalt love God above everything and everybody else."

3. On Switzerland

> [According to a well-traveled Argentine physician] Spain was hopeless, and Switzerland the most admirable country of the world. I objected that the Swiss lack soul, grace and talent. They are like those honest and vulgar persons with a well-ordered life, who eat, work and love at fixed hours, are incapable of committing a crime or saying a lie, for sure, but also are incapable to dream and create. (ib., p. 56)

However, which is the country that gave birth to Henri Dunant, that true Good Samaritan who so generously helped the wounded of the Battle of Solferino, of June 24, 1859? And which is the country that gave birth to the other four founders of the *International Red Cross*: Moynier, Dufour, Appia and Maunoir? That country was Switzerland.

VII. No, Sánchez-Albornoz

1. On the Arts and Literature

"*[El español] ha sobresalido en las actividades que precisan exaltación brusca y violenta, tension de arco al disparar la flecha y concentración tan intensa como pasajera: el arte, las letras, la mística o la guerra* [The Spaniard excelled in the activities which need brusk and violent exaltation, the tension of the bow before shooting the arrow, and concentration as intense as transitory: art, letters, mysticism or war]" (1991, p. 1299).

But has Spain really excelled in comparison with England, France, Germany, Italy, or Russia? For instance, has Spain excelled in the number of Nobel laureates for literature, in the space devoted in encyclopedias, mentions in books like *Makers of Modern Thoughts*, edited by Bruce Mazlish, 1972 (where, out of 36 great persons—from Leonardo da Vinci to Norbert Wiener, none is a Spaniard)?

There is no intention on my part to denigrate the high contribution of Spain in the arts, letters, mysticism, and war, yet the amount and quality of that contribution does not seem to justify the large voids in other areas (e.g., science). Also I may be tempted to say that Spain's glory was achieved despite her people's fondness for brusk and violent exaltation, for concentration as intense as transitory.

2. On Mysticism

"The Spaniard excelled in the activities which need brusk and violent exaltation , the tension of the bow before shooting the arrow, and concentration as intense as transitory: .... mysticism" (1991, p. 1299). I had never thought of mysticism in this way, but it may be so.

Also, expanding on the above, I suspect that something is wrong with this continuous emphasis on mysticism—by so many Hispanic writers (Gálvez, López Ibor, Madariaga, Maeztu, Unamuno, Zum Felde, besides Sánchez-Albornoz)—as if it would be the alpha and omega of Christianity, and as if this very best would be a Spanish monopoly.[186] I have nothing against mystics, but—as far as I know—Jesus never said a word about them. He who was truly praised by him was the Good Samaritan who generously helped the robbed and wounded man on the way to Jericho.

Do mystics love their neighbors? Possibly, in their spare time. And is the number and quality of its mystics an indication of the greatness of a country? It may just be the opposite. Their number and quality may indeed be evidence that nothing good is around, not the state, not the church, not the local community, and that the only solution rests in a direct intimate discourse with God, since every other discourse is futile.

Unamuno said it clearly: "I shall certainly be reminded that the reform of the contemplative Order of the Carmelites which Teresa de Jesús undertook was a Spanish work. Yes, Spanish it was, and in it men sought liberty. It was, in fact, the yearning for liberty, for inward liberty which, in the troubled days of the Inquisition, led many choice spirits to the cloister. They imprisoned themselves in order that they might be more free" (1913/1954, p. 289). But having to imprison oneself in order to be free, having to devote all one's life to the high solitary dialogue with God, in order not to run into trouble for talking with other men, is this something of which a country can boast so loudly?

I even wonder if all this praise for mysticism is an indication of the lack of true religiosity. Having paid full homage to the unachievable summit, there may be less need to do those simple things that are demanded from a plain good Christian.

---

[186] There were also great mystics in other countries, and an interest in their works seems to run along national lines. For instance, Heidegger was interested in Meister Eckart, the greatest German mystic, and T.S. Eliot was interested in Julian of Norwich, the great English mystic.

VIII. So Much Violence and Torture, and So Few Great Thinkers

1. Violence and Torture

"Violence seems to be institutionalized in the organization, maintenance, and changing of governments in Latin America. The methodology of force is found in advanced and in backward countries, in Indian, mestizo, and white republics, in large states and in small ones, in urban and in rural areas, in agricultural and in industrial organizations, in the beginning of the twentieth century, in the present period, and in the early, middle, and late nineteenth century—in a word, whenever Hispanic culture is to be found in the Western Hemisphere" (Stokes, 1965, p. 140).

In his *The Spanish Character*, Bennassar discussed twice the causes of violence, first in his chapter "Honor and violence", specifically in the section entitled "Honor as a source of violence", and a second time in his chapter "To die well" where he concluded that "Thus the contemplation of death encouraged the absence of respect for life" (1979, p. 248).

Also, if "essential dissenters" can be tortured and killed by the Inquisition, they can also be tortured and killed by others equally convinced of the "essential truth" of their position. For *vertical people*, life may be a secondary contingency, especially the life of others.

2. So Few Great Thinkers

According to Marcellin Defourneaux: "No, the Inquisition did not close Spain off to European culture; the whole history of Spain in the eighteenth century proves the contrary. But it gave some Spaniards the feeling that they were shut up in an intellectual prison through whose bars they could catch a glimpse of liberty" (quoted by Bennassar, 1979, p. 252). Now, given that "the Spaniard is one of the most intelligent of European people; as well as one of those in whose midst intelligence is distributed most equitably" (Madariaga, 1962, p. 51), those Spaniards, who felt they were shut in an intellectual prison, must have had their reasons to think so, be it only because of the Inquisitional *Anaconda Effect*, which is definitely not conducive to great studies and intellectual questioning.

The Inquisition may not have closed Spain off to European culture, but as said by Sánchez-Albornoz: "No, one cannot deny the awful damage produced by the fear of the Inquisition in the devotion to learning by the Spaniards. It crippled the free flight of any philosophical thoughts which could in any way evolve into touchy religious debates. It contributed indirectly to bleed the rich potential of intellectual curiosity which the humanism and the geographical discoveries had created in the Spaniards, and to brake the vigorous unfolding of this potential, promising and even fertile as long as lasted the relative intellectual freedom of the epoch of Charles V, or as long as lived those who had been formed in that epoch" (1956/1991, pp. 1275-76).

Things were made more difficult by a strong remnant of pre-DP medieval vertical/ essential thinking. If everything that is not vertical/ essential is unimportant, and therefore not pursued, then much is lost, because always the big achievement is the tip of an iceberg whose lower part is constituted of small studies and small discoveries. And, indeed, Unamuno—with his famous "*¡Qué inventen ellos!* [Let them invent!]" (quoted by López Ibor, 1969, p. 65)—clearly showed his contempt for the lower side of the iceberg of knowledge. For him, "*la luz eléctrica luce aquí y corre aquí la locomotora tan bien come donde se inventaron y nos servimos de los logaritmos como en el país donde fueron ideados,... los romanos no entendieron de arte ni tuvieron un matemático, lo que no les impidió hacer su papel*"[187] (ib., pp. 65-66).

López Ibor, himself, said similar things when he wrote:

> *¿[El español] Complejo de superioridad o de inferioridad? Para mi, evidente complejo de inferioridad ante la ciencia y la técnica; además de él, en muchos españoles existe un complejo de superioridad como estilo de vida. Las páginas siguientes se hallan cuajadas de alusiones al mismo. Porque no cabe duda acerca de que nuestro estilo de vida tiene ya, como fuerza radical, una peculiaridad superior al de otros pueblos. La mayor originalidad de nuestro estilo de vida puede ser un aliento*

---

[187] (The electric light illuminates here, and the trains run here as well as where they were invented, and we use the logarithms as well as in the countries that invented them,... the Romans had no understanding of the arts, nor any of them was an eminent mathematician, and yet the Romans were able to be important in history.) With some caveats, this last comment on the Romans may be true; but are these the people to whom the Spaniards of the 20th century are to be looking for inspiration?

que hinche las velas de nuestra difusa conciencia de superioridad.[188] (1969, pp. 29-30).

This same *conciencia de superioridad* is at the base of the thinking of Unamuno, especially when, at the end of his 1906 essay "*Sobre la europeización*," he wrote: "*Sean, pues, aquí mis últimas palabras, mientras me preparo a pensar cómo pueda españolizarse a Europa,...* [Be these then my last words,[189] while I begin to think at how we can Hispanisize Europe]" (1951, p. 920). This desire by Unamuno, of making Europe Spanish-like, is a form of vertical/essential thinking as his and López Ibor's pride in an inferiority complex toward science and technology but compensated by a superiority complex in life style.

In turn, the three causes in question: Inquisition, Philip II, residual pre-DP medieval personality (hyperindividualism and vertical/essential thinking) are strongly related. It was indeed King Philip II who said that:

---

[188] ([The Spaniard,] a superiority or inferiority complex? For me, an evident inferiority complex toward science and technology; on the other hand, many Spaniards are endowed with a superiority complex in their life style. In the following pages there are, here and there, allusions to this. Because there is no doubt that our lifestyle, as radical force, is superior to that of other people. The main originality of our lifestyle may consist of the spirit which inflates the sails of our conscience of superiority.)

[189] He was referring to two sets of verses by Lord Tennyson from his *The Ancient Sage*:

> For nothing worthy proving can be proven,
> Nor yet disproven. Wherefore thou be wise,
> Cleave ever to the sunnier side of doubt,
> And cling to Faith beyond the forms of Faith.

And,

> For Knowledge is the swallow on the lake
> That sees and stirs the surface-shadow there
> But never yet hath dipt into the abysm,

One can fully understand why these verses fascinated Unamuno. On the other hand, the British were able, e.g. with Harvey's discovery of the circulation of blood, with Newton's interpretation of the composite nature of light, with the creation of much division of power (Magna Carta, Parliament) to prove a number of things worth proving, even if not those laying at the bottom of the abyss.

"*Veinte clérigos de la Inquisición mantienen mis reinos en paz*"[190] (quoted by Sánchez-Albornoz [1991, p. 1277], who then added "*Sí, pero en una paz que acabó aletargando la vida filosófica y cientifica de España por siglos*"). In turn, King and Inquisition blocked developments: Spain had no Renaissance, and no bourgeoisie for a long time (ib., p. 721), and the pre-DP medieval mentality persisted.

IX. No Shumway

1. Wrongly Faulting the Founding Fathers of Argentina

According to Nicolas Shumway's last words in his 1991 book:

> While Argentina's recurrent crises obviously have many causes and explanations, I can't help sensing that the competing myths of nationhood bequeathed by the men who first invented Argentina[191] remain a factor in the country's frustrated quest for national realization. (1991, p. 299)

Earlier, specifically in his criticism of the Generation of '37,[192] Shumway had written:

> More central to our purposes is the ideological legacy left by Argentina's early liberals, much of which now appears unfortunate. *Foremost in this regard is the relative modesty of their overriding goal*, of their chief guiding fiction: to bring Europe to the Southern Cone. *Rather than create something new, to build a new Jerusalem and be a light unto the nations*, they merely sought to re-create Europe and North America, not as a new idea but as a successful derivation. (1991, pp. 165-66; my italics)

---

[190] (Twenty fathers of the Inquisition keep my realms in peace. — Yes [noted Sánchez-Albornoz], but in a peace which, for centuries, lethargized the philosophic and scientific life of Spain.)

[191] Shumway refers specifically to "the rhetorical ghosts of Moreno, Hidalgo, Rivavadia, Sarmiento, Alberdi, Mitre, Andrade, and Hernández [who] continue to haunt the land" (ib., p. 299).

[192] Its members were: Esteban Echeverría, Juan Bautista Alberdi, Miguel Cané, Vicente Fidel López, Juan María Gutiérrez, Domingo Faustino Sarmiento (ib., pp. 129-30).

So, for Shumway, it would not have been sufficient if the early liberals had made Argentina into a peaceful and prosperous country like Canada or Sweden, no, they had *to build a new Jerusalem, a light unto the nations*! This request is absurdly *vertical*, as can be seen, in contrast, by the title of a great book by Henry Steele Commager (1977), entitled *The Empire of Reason: How Europe Imagined and America Realized the Enlightenment*.[193]

As explained by Commager, the overriding goal of the North Americans of those times was simple and humble: Realize the ideas of the European Enlightenment, with no thoughts whatsoever of building a new Jerusalem and being a light unto the nations. And they succeeded.

Similarly, for a long time during their study and admiration of French and English literature, neither Goethe nor Tolstoy placed any nationalistic *vertical* demands. "Tolstoy used to say later that, as a young teenager, he so idolized Rousseau that he wished he could have worn his portrait round his neck in a locket, like a holy icon. In other conversations, he would imply that this actually had been his practice" (Wilson, 1988, p. 36). No, both Goethe and Tolstoy simply admired and enjoyed the foreign authors without mental reservations. And yet they became respectively the greatest German and Russian writers.

On the other hand, Shumway is not alone in this demand for the highest goals and accomplishment, and similar requests have been made

---

[193] In the very first words of Commager: "The thesis of this book can be stated quite simply: The Old World imagined, invented, and formulated the Enlightenment, the New World—certainly the Anglo-American part of it—realized it and fulfilled it. It was Newton and Locke, and their eighteenth-century successors in Britain and on the Continent—philosophes like Priestley and Bentham, Hume and Monboddo in Britain, Montesquieu, Voltaire, Buffon, and Diderot in France, Christian Wolff and Lessing, von Haller and Goethe in Germany, Genovesi, Filangieri, Beccaria, and Tanucci in the Italian states, and sometimes supporting them, 'enlightened' monarchs like Frederick of Prussia, Joseph of Austria, Leopold of Tuscany, and Gustavus of Sweden—who launched the Enlightenment, gave it respectability and, somewhat tentatively, experimented with it. But it was Americans who not only embraced the body of Enlightenment principles, but wrote them into law, crystallized them into institutions, and put them to work. That, as much as the winning of independence and the creation of the nation, was the American Revolution.... Only rarely, and sporadically, could Old world philosophes turn their energies to the tasks of improvement and reform. But in America where there were no ancient tyrannies to overthrow, no barriers of tradition or poverty or ignorance to surmount, and few iniquities—except the prodigious iniquity of slavery—to banish, the philosophes could devote their energies to realizing the program of the Enlightenment" (pp. xi-xii).

frequently by Hispanic authors, for instance, for Argentina, by José Isaacson:

> *Queremos una Argentina peculiar pero integrada al mundo moderno* [We want a special Argentina, yet integrated in the modern world.] (1983, p. 58);

and for Spain, by Ortega y Gasset:

> *España es el problema y Europa la solución; pero no para repetir o imitar, sino para lograr que España sea* ***una posibilidad europea****, para llegar a* ***la interpretación española del mundo*** [Spain is the problem, and Europe is the solution; but not to repeat or imitate, but to attain that Spain be *a European possibility, to achieve the Spanish interpretation of the world*] (reported by Marías, 1985, p. 355; highlighted in the original).

In turn what Ortega y Gasset wanted had already been asked by Unamuno:

> *Tengo la profunda convicción de que la verdadera y honda europeización de España, es decir, nuestra digestión de aquella parte de espíritu europeo que puede hacerse espíritu nuestro, no empezará hasta que no tratemos de imponernos en el orden espiritual de Europa, de hacerles tragar lo nuestro, lo genuinamente nuestro, a cambio de lo suyo,* ***hasta que no tratemos de españolizar a Europa****"*[194] (1951, p. 918; my highlighting).

The trouble with demands like those of Shumway and Isaacson, Unamuno and Ortega y Gasset, is that no advancement will ever be enough, every step forward will be seen as irrelevant in comparison with the high goal; everything will seem miserably *horizontal* when luminous verticality is demanded, when it has to be peculiarly Argentine.

---

194 (I have the profound conviction that the true and meaningful Europeanization of Spain, i.e., our digestion of that part of the European spirit which can become our spirit, will not begin until we can impose ourselves on the European spiritual order, until we manage to have Europe accept us, the genuine us, in exchange of their contribution, *until we manage to Hispanisize Europe*.)

Seen differently, how can Shumway, Ortega y Gasset, and Unamuno demand so much when, as reported before:

- "[Spain] produced no great ideas. Her constitutions were borrowed. Her philosophy was either European or North American. She left no milestones of material or intellectual progress" (Bunkley, 1952, p. 5);
- "Let us not delude ourselves. Spain has not contributed to the world of ideas which is called modern in the proportion of other countries. I have checked, for instance, the chronological tables of the decisive fights of the mind which Heinrich Schmidt mentions at the end of his *Philosophical Dictionary*. There is not one single Spanish name" (López Ibor, 1969, pp. 62-63);
- "Spain has never had a thinker, not one notable writer, not a single philosopher" (Sarmiento, quoted by Véliz, 1980, p. 176).

Said differently, where were the Giants upon whose shoulders the Founding Fathers of Argentina should have stood to see further? And yet, according to Shumway, they should have seen further: "*Facundo* [by Sarmiento] remains a work of astounding and prophetic creativity. Yet for all its originality, there is no escaping an unhappy fact: *the Men of '37 were ultimately more concerned with re-creating Europe in the Southern Cone than in developing a new country that would blend the best of the Old and New Worlds*. (ib., p. 164; my italics)

Once more, this is an absurd *vertical demand*: which best of Old and New? The old of the Inquisition, which Shumway treated so cavalierly? The only time that Shumway wrote about the Inquisition was to say that "The higher officers of the Inquisition did issue edict after edict demanding that incoming books, bookstore stocks, and private libraries be regularly scrutinized by the Holy Office.... [And] the efforts of the inquisitors were frequently honored more in the breach than in the observance thanks to extensive smuggling of heretical works, often with the collaboration of lower Inquisition officials and members of religious communities" (1991, p. 14).

However this farcical Inquisition is not the one discussed by Boleslao Lewin in *La Inquisición en Hispano America*, when already at the beginning of his book, he reminds his readers that "*en el periodo anterior y posterior a Mayo 1810, la vigencia efectiva de la Inquisición, y de sus secuelas, constituía un motivo de cuidado constante para los argentinos, hecho que se refleja de diversos modos* [in the period before and after May 1810, the effective reality of the Inquisition, and pertinent outcomes, produced a sense of constant worry in the Argentines which manifested itself in various ways]" (1967, p. 14).[195]

Also, if today the works of Montesquieu and Voltaire are not commonly read, that was not the case for the Men of '37 who knew both authors well; and if today we can reflect calmly on the tragedy of Galileo, that was not so at that time, and Sarmiento himself spoke of him in his chapter on the Inquisition: "*Su prisión, su persecución y su retractación solemne, han enriquecido a la historia humana con una de esas protestas vengadoras que han salvado al mundo: e pur si muove!*[196] [His prison, his persecution and solemn retractation has enriched the human history with one of those avenging protests which have saved the world: But it does move!" (1883/1953, p. 114).

> In the New World, the Crown was absolute master, and exercised a virtually papal authority of its own. No cleric could go to the Indies without royal permission; there was no papal legate in the New World, and no direct contact between Rome and the clergy in Mexico or Peru; the Crown exercised a right of veto over the promulgation of papal bulls, and constantly intervened, through its viceroys and officials, in all the minutiae of ecclesiastical life. (Elliott, 1964, p. 91)

---

[195] This farcical Inquisition is definitely not the one described by Mack Smith for Sicily: "By trying to eliminate heresy, the Inquisition was helping to attack a dangerous source of independent and sometimes revolutionary thought in politics as well as religion. Along with Jesuit control of education, this contributed to keeping Sicily orthodox and obedient. Barely a glimmering of the 'scientific revolution' was allowed to penetrate, and literature had if possible to be preserved as a courtier literature, full of conceits and ingenuity and affectation, but preferably devoid of any stimulating content. The Holy Office censored books, and foreigners on arrival might have any dangerous literature confiscated. On a number of perilous occasions, for example in 1516, 1560, 1585 and 1647, the chief Inquisitors, who had the interests of Spain as well as of orthodoxy at heart, helped to subdue popular riots and political rebellion" (1988, pp. 166-67).

[196] How bitter and sad Sarmiento would have been if he had known of the total victory of the Inquisition over Galileo who never said such words. That "E pur si muove" was mentioned for the first time by the Abbé Irailh—in his *Querelles littéraire*s of 1761, more than one hundred years after Galileo's recantation: "*Au moment, assure-t-on, qu'il fut mis en liberté, le remords le prit. Il baissa les yeux vers la terre et dit, en la frappant du pied 'Cependent elle remue'* [It is related that as soon as he was freed, he felt the pangs of remorse. He lowered his eyes to the ground and said, hitting it with his foot, 'But it does move']". Poor Galileo, would never have dreamt to say such words: the Inquisition had totally defeated him.

Also, as said by Porras in his section on the Real Patronato in the *New Catholic Encyclopaedia*:

> The methods of control established by the kings—many of them not canonical—limited the action of the Church in America and hindered its full development during the viceregal period. The dioceses kept their original limits, which made spiritual government difficult. Permission was denied for building monasteries and other pious establishments; the ordination and profession of mestizos was prohibited (during the 17th century), contributing to the scarcity of secular and regular clergy and of nuns, which deprived the faithful of proper training. The anticanonical subjection of the religious orders to juridical decree ... made them virtually officials of the State. Because of this system, which prevailed for three centuries, when independence was achieved in the American provinces, the Church lacked the training necessary for establishing autonomy with the State. The new governments tried to obtain from the Holy See the same privileges of patronate that rulers of Castille had enjoyed.... The new republics systematically refused nominees who arrived from Rome. (1967, p. 1116)

Back to Shumway's criticism, the Men of '37 could not blend the best of the Old and the New World because, among others, they had no way to separate the bad from the good, no way to sift their Hispanic heritage and refuse those parts overly distorted by the unity of power of the Spanish kings. And in truth, this screening of the Spanish inheritance has not been done so far; and if Gálvez's *El solar de la raza* of 1913 was so highly praised in Spain, it is because it did not refuse anything Spanish. Everything Spanish was marvellous, including the paintings of Montañes which Gálvez found superior to anything else.

On the other hand, how could the Men of '37 filter French and English ideas, when even the great Italian poet Alessandro Manzoni found it difficult, in 1821, to make up his mind on Napoleon in his poem *Il Cinque Maggio:*

> *Fu vera gloria? Ai posteri l'ardua sentenza*
>
> Was it true glory? Up to future generations the difficult verdict

Furthermore, Latin America probably suffered more, psychologically, from the Inquisition than Spain itself which was closer to the rest of Europe and had fewer restrictions in commerce and travel, and had more relations with the Vatican. Indeed, it was the King of Spain, Joseph Bonaparte, who in 1808 put in practice the command by Voltaire of *Écrasez l'infame* (tread down the infamous) and suppressed the Inquisition in Spain (which was later restored by Ferdinand VII in 1814). For at least six years the Inquisition in Spain was both abolished and declared *infame*. In Spain, in those years, there were large movements of French and English soldiers and officers who had nothing good to say about the Inquisition. The Spanish intellectuals profited from the relatively enlightened court of the Bourbon kings in Madrid, especially during the reign of Charles III (1759-1788); something which, again, was not available to Argentine intellectuals.

For Shumway,

> Thus it was that although Echeverría, Alberdi, and Sarmiento found plenty to criticize in Europe and the United States, when it came time to give substance to their protestations of independence from European and North American culture, none of the Men of '37 recognized much in Argentina that could be defined as positive and unique. Indeed, the uniqueness of America, glorified in the Americanism of Artigas and Hidalgo's gaucho, was in their minds an obstacle to progress. *Nor did they envision any special mission or potential unique to their country*; transplanting Europe and imitating the United States were good enough. Consequently, it is hardly surprising that in devising a framework for their New Argentina, they repeatedly failed to break away from foreign models and create institutions peculiar to Argentina. (1991, p. 160; my italics)

This "*Nor did they envision any special mission or potential unique to their country*" parallels Shumway's previous demand that the Founding Fathers should have built *a new Jerusalem and be a light unto the nations*, as if both were feasible; and they were not. At a time when the Argentines were still so Spanish in their scripts, even if not in their political allegiance to Spain, to demand that they should have broken away "from foreign models and create institutions peculiar to Argentina" is demanding the impossible, the inconceivable, as evident from the following by Shumway himself:

> Confirmed what he already believed: Spain was the cradle of barbarism, a parent to be cast off and replaced....Alberdi goes on to say, however, that *Spain ceased being nature's tool (and in a sense ceased being part of Europe) when she tried to close Spanish America to the superior ways of France and England*, thus violating the law of cultural expansion. This anti-Spanish bias among Argentine intellectuals was never seriously challenged until the twentieth century when, as Marysa Navarro Gerassi has shown, Argentine writers like Ricardo Rojas, Enrique Larreta, Manuel Gálvez, and Carlos Ibarguren, in what would become known as the *hispanidad* movement, would seek to vindicate Argentina's Spanish heritage. (Shumway, pp. 138-39; my italics)

So, according to Shumway, Sarmiento's and Alberdi's criticism was biased: both were unjustly anti-Spanish. However, from everything reported so far, Sarmiento's and Alberdi's criticism of Spain was fully justified, and Gálvez's unconditional praise unjustified.

2. But the Inquisition of Rosas (1829-1852) Took Place

"The Men of '37 were ultimately more concerned with re-creating Europe in the Southern Cone than in developing a new country that would blend the best of the Old and New World" (ib., p. 164). But, was such a blend conceivable in the 19th century when the Inquisition still existed in Spain as late as 1834, and was very active in the Papal states as late as 1852, and when, in 1870, Pope Pius IX proclaimed papal infallibility? Was such a blend conceivable when, shortly after having won its independence from Spain in 1810, Argentina suffered the long dictatorship (1829-1852) of Juan Manuel de Rosas?

> Rosas divided society into those who commanded and those who obeyed. Order obsessed him, and the virtue which he most admired in a people was subordination. If there was anything more abhorrent to Rosas than democracy, it was liberalism. The reason why he detested unitarians was not that they wanted a united Argentina but that they were liberals who believed in secular values of humanism and progress. He identified them with freemasons and intellectuals, subversives who undermined order and tradition and whom he held ultimately responsible for

the political assassinations which brutalized Argentine public life from 1828 to 1835....While he controlled the legislature, Rosas also dominated the judicial power. He not only made law, he interpreted it, changed it and applied it. The machinery of justice no doubt continued to function: the justices of the peace, the judges for civil and criminal cases, the appeal judge and the supreme court, all gave institutional legitimacy to the regime. But the law did not rule. Arbitrary intervention by the executive undermined the independence of the judiciary. Rosas took many cases to himself, read the evidence, examined the police reports and, as he sat alone at his desk, gave judgment, writing on the files 'shoot him', 'fine him', 'imprison him', 'to the army'. Rosas also controlled the bureaucracy..... The Church was a willing ally, except for the Jesuits, who were re-admitted and re-expelled. *Portraits of Rosas were carried in triumph through the streets and placed upon the altars of the principal churches. Sermons glorified the dictator and extolled the federal cause. The clergy became essential auxiliaries of the regime and preached that to resist Rosas was a sin.* Political orthodoxy was conveyed by word as well as by deed, and the printing presses of Buenos Aires were kept fully employed turning out newspapers in Spanish and other languages containing official news and propaganda, for circulation at home and abroad. But the ultimate sanction was force, controlled by Rosas, applied by the military and the police. (Lynch, 1993, pp. 21-27; my italics)

Although Rosas was nominally no more than the governor of the province of Buenos Aires, his rule eventually extended to all the provinces. His rough-style propaganda was based on total intolerance; his supporters were expected to wear red ribbons with the motto "Death to the Savage Unitarists," the small but secretly active party of the May Association. Rosas distrusted and hated foreigners and opposed all liberal ideas of European or north American origin. *His solidarity with the Catholic church, whose ecclesiastical appointments he controlled*, was symbolized in the custom of "permitting portraits of Rosas the Restorer to be placed on altars alongside statues of Christ the Redeemer." A fierce parody of his regime is Echeverría's *The Slaughter House*, in which a child, an Englishman, and a defi-

> ant young Unitarist leader are the victims of a blood-thirsty, orgiastic horde of Federalists, including their terrorist special police, the Mazorca ("ear of corn; but also a pun: *más horca*,"more hangings").... According to Sarmiento, Rosas embodied the Spanish heritage, which would have to be eliminated if Argentina were ever to enjoy a free and democratic society.... (Earle, 1971, pp. 46-47; my italics)

> [As Sarmiento wrote in 1843] "Rosas is the political inquisition of old Spain personified....In Spain thought has been prohibited for three centuries and there has been a tribunal to persecute and to burn alive those who speak against, write against, or even those who are suspected of disloyalty to the king and to the dominant ideas; Rosas has created that same tribunal in his own person, to drown all murmur of disapproval, to suffocate every seed of liberty. Freedom of the press was unknown to Spain until 1833; Rosas has destroyed the liberty that we enjoyed since 1810". (quoted by Earle, 1971, p. 47)
>
> "....and the *Gaceta* [Rosas' official newspaper] agitated and encouraged its readers to hate the Europeans, to despise the Europeans who want to conquer us; Frenchmen are called lousy clowns; Louis Philippe, pigkeeper, and generally European politics are described as barbarous, disgusting, brutal, sanguinary, cruel and inhuman" (Sarmiento, quoted by Véliz, 1980, p. 175).

Having experienced this evil, how could Sarmiento and the other Founding Fathers of Argentina have filtered their Spanish inheritance and develop "*a new country that would blend the best of the Old and New Worlds*"? How could they have done it, when Rosas was forcing the Old on everybody, and preaching hatred for everything that was New?

3. But There Were Gauchos and No Farmers

Was the requested blend of the best of the Old and New World conceivable in the 19th century in a country without farmers?

> One of the most crucial social effects of Argentina's expansion turned out to be a nonevent: the country never developed a peasantry, at least not in the pastoral areas of the pampa and the

crucial coastal provinces. The Conquest of the Desert in the 1870s virtually eliminated the Indian population, and the land was promptly distributed in large tracts appropriate for raising cattle and sowing grain. In contrast to the policy in the Great Plains of the United States, Argentina did not give its land to family farmers or individual homesteaders. (Skidmore & Smith, 1997, p. 76)

Abundance of cattle and horses and seemingly limitless pastures produced a race of mounted nomads and hunters more resembling the Bedouins or the Indians than the society of cities, plantations, and mining camps of the Spanish empire. The abundance of the plains provided for the existence of a free and undisciplined people and all the bonds of hierarchy and authority were relaxed. (Ferns, 1869, p. 21)

Whether good or bad, the classical gaucho asserted his freedom from all formal institutions; he was indifferent to government and its agents, indifferent to religion and the Church. He did not seek land; he lived by hunting, gambling and fighting. The nomadism of the gaucho had many social implications. It prevented settled work or occupation. Property, industry, habitation, these were alien concepts. So too was the gaucho family....Thus, the gaucho was forcibly converted from a free nomad into a hired ranch-hand, a *peón de estancia.* This primitive society was not qualified for constitutional government or political participation. The *estancia* dominated economic and social life and became the model of government. *Estancieros* ruled their domains by personal authority and demanded unqualified obedience....This primitive political structure, founded on individual power, raised upon personal loyalties, cemented by the authority of the patron and the dependence of the peon, was finally built into the state and became the model of *caudillismo*....The year 1820 was a year of anarchy. Independence from Spain had culminated not in national unity but in universal dismemberment. After a decade of conflict between Buenos Aires and the provinces, between central government and regional interests, between unitarians and federalists, the framework of political organization in the Rio de la Plata collapsed. Independent republics proliferated throughout the interior, and when Buenos Aires sought to subdue them they fought back. Provincial caudillos - Estanislao López of Santa Fe, Francisco Ramirez of Entre Rios - led their irregular gaucho hordes,

> the fearsome *montoneros*, against the capital. On 1 February 1820 they defeated the forces of Buenos Aires at the battle of Cepeda and proceeded to destroy all trace of central authority. (Lynch, 1993, pp. 14-18; my italics)

4. But Spain was Disliked

Lastarria's[197] Judgment

In his *Literary Memoirs*, Lastarria made some interesting comments which were valid for the whole of Latin America:

> When the United States emancipated themselves politically, they did not emancipate themselves from English literature; the latter could be of use to them, and in fact, *was* useful in their new situation, for their feelings and ideas, interests and social needs continued to be British, with the sole difference being that their civic responsibility should be better served by the new republican organization, and could be so, since the latter was no radical new development, but progress, a natural evolution of that same civic responsibility.
>
> Thus we observe the needs of their new organization and the interests of their new political institutions were brilliantly served by politicians and men of letters of the caliber of Franklin, Washington, Adams, Hamilton, Jefferson, Madison, and Jay; how social interest in the emancipation of the spirit was brought about by Paine at first, by Channing and Emerson thereafter....Among the Chileans, Spanish literature had no representatives, and if one or another came from outside, they did not manage to form a literary school that could help our education, nor even the needs of the new age....We believe, as did this writer, that our Republic needed a nation of people; but to have it, it was not enough, in our judgment, to provide an industrial education to the new generation, *but it was indispensable to remake our civilization, disavowing the entire Spanish past*, and disciplining the current generation in the practice of freedom, through systematically and sincerely reformed political institutions. (1836-49/2000; my italics)

---

[197] José Victorino Lastarria (1817-1888), one of nineteenth-century Latin America's intellectual giants, who wrote on literature, history, philosophy and politics, was not from Argentina but from Chile.

Belgrano's[198] Comments

> Since I was in Spain in 1789 at a time when the French Revolution was causing a change in ideas, particularly among the men of letters with whom I associated, the ideas of liberty, equality, security, and property, took a firm hold on me, and I saw only tyrants in those who would prevent a man, wherever he might be, from enjoying the rights with which God and Nature had endowed him.... (ca. 1812/1965, p. 276)

Von Humboldt's[199] Observations of 1799-1804

> The result has been a jealousy and perpetual hatred between the *chapetons*[200] and the Creoles. The most miserable European, without education, and without intellectual cultivation, thinks himself superior to the whites born in the new continent. He knows that, protected by his countrymen, and favored by chances common enough in a country where fortunes are as rapidly acquired as they are lost, he may one day reach places to which the access is almost interdicted to the natives, even to those of them distinguished for their talents, knowledge, and moral qualities. The natives prefer the denomination of *Americans* to that of Creoles. Since the peace of Versailles, and, in particular, since the year 1789, we frequently hear proudly declared: "I am not a Spaniard, I am an American!," words which betray the workings of a long resentment. In the eye of law every white Creole is a Spaniard; but the abuse of the laws, the false measures of

---

[198] "Manuel Belgrano (1770-1820) was born in Buenos Aires, educated at the Universities of Salamanca and Valladolid in Spain, and made a name for himself in the last years of colonial rule as secretary of the *consulado*, which was set up at Buenos Aires in 1794, and as an economist. He edited the *Correo de Comercio* in 1809, played a prominent part in the revolution of May 1810, was a member of the governing junta which was then established at Buenos Aires, and finally became a general" (Humphreys & Lynch, 1965, p. 275).

[199] "Alexander von Humboldt (1769-1859) spent nearly five years in Spanish America, from 1799-1804, exploring, collecting, observing. No one did more to reveal a New World to the Old: his writings remain an indispensable source of information on almost every aspect of Spanish American life during the closing years of the colonial period" (Humphreys & Lynch, 1965, p. 269).

[200] South American nickname for a European-born Spaniard. Criollos or creoles were the Spaniards born in America.

> the colonial government, the example of the United States of America, and the influence of the opinions of the age, have relaxed the ties which formerly united more closely the Spanish Creoles to the European Spaniards. (1811/1965, p. 270)

Echeverría

Anti-Spanish sentiment became even more virulent among the Men of '37, symbolized by a striking tendency, still common in modern Argentina, to exclude Spain whenever the term *Europe* is invoked. *Europe* in Argentina came to mean northern Europe, the source of modern (non-Spanish) culture.The impulse behind this peculiar usage is well seen in the Men of '37. Echeverría, for example, argues that *Spain left Argentina a tradition "that denies analysis and choice; in a word, the suicide of reason"* (*Dogma* 191). He later deplores "the rancid Spanish enlightenment, which with its books and its concerns sowed such bad seed in American soil" (*Ojeada* 121). (Shumway, 1991, pp. 136-37; my italics)

X. A Divided Nation

Things were particularly difficult in Argentina (in reality all over Latin America) also because of the sharp division between a small number of intellectuals living in coastal cities, and the large mass of people living in the interior. Sarmiento observed sharply:

> The Argentine Revolutionary War was twofold: 1st, a civilized warfare of the cities against Spain; 2nd, a war against the cities on the part of the country chieftains with the view of shaking off all political subjection and satisfying their hatred of civilization. The cities overcame the Spaniards, and were in their turn overcome by the country districts. (Sarmiento 1845/1961, p. 63)
>
> We may add, that throughout the revolution, Cordova was the asylum of all fugitive Spaniards.... Had revolutionary ideas, such as are found in Rousseau, Mably, and Voltaire, happened to spread over the pampas and descend into this Spanish catacomb—if we may say so—what response would they have been likely to find.... (ib., p. 99).
>
> *In 1810, Buenos Ayres was filled with partisans of the revolution, bitterly hostile to anything originating in Spain....* The *Social Contract* flew from hand to hand. *Mably and Raynal*

> *were the oracles of the press; Robespierre and the Convention the approved models. Buenos Ayres thought itself a continuation of Europe*, and if it did not frankly confess that its spirit and tendencies were French and North American, it denied its Spanish origin on the ground that the Spanish Government had patronized it only after it was full grown. (ib., p. 102)
>
> On one side Cordova, Spanish in education, in literature, and in religion, conservative and strongly opposed to all innovations; and on the other, Buenos Ayres, revolutionary by nature, ready for any change and progress. *These were the types of the two parties that divided every city; and I doubt if there is another such phenomenon in America; that is, two parties, conservative and revolutionary, retrograde and progressive, each represented by a city having its own peculiar form of civilization, and receiving opinions from entirely different sources: Cordova, from Spain, the Councils, the Commentators, the Digest; Buenos Ayres, from Bentham, Rousseau, Montesquieu, and French literature in general.* (ib., p. 106; my italics)

XI. The Natural, but by Far Incomplete, Solution: Learn from Europe

1. European Books

"Hispanic man was agonized by his failure" (Bunkley, 1952, p. 8)[201]

---

[201] "During the Age of the Enlightenment, Spanish history manifested two important trends. The first of these was the artificial attempt to 'twist the habits of the people' to fit the civilization of the time. The new Bourbon royal family and its court attempted to introduce certain French customs, political reforms and economic reforms to the Hispanic world. Spanish thinkers and artists turned to French, German, and English models. They wrote neo-Classic literature, and they painted neo-Classic paintings. They compiled encyclopedic knowledge, and worked within a framework of a rationalistic philosophy. *But in none of these fields did they excel. They produced nothing great and nothing original, and Spain gave the world no universal figure during the Enlightenment.*

The second trend was one of bitter self-criticism. Seeing the wide gap between the Spanish way of life and the way of life that was bringing Western civilization to new heights of political and economic achievement, *Spanish writers and thinkers violently attacked their own tradition and asked why it was unable to measure up to that of its neighbors and rivals. Hispanic man was agonized by his failure*" (Bunkley, ib., pp. 7-8; my italics).

Not only did the Latin American intellectuals know, on their own, of the evil done in America by the UP of the Spanish kings, but they knew about it by the extensive literature that the Enlightenment produced in France, England, the USA, Italy, and Germany. This literature decried not only what the Inquisition had done, but it said directly, or implied, that the Spaniards were worse than ugly barbarians who could only be justified by their ignorance. As a result, it was natural for the Latin American intellectuals to say: "Not me; I do not belong to that evil group; I am different, and my mind and hands are clean." The books by the *philosophes* came to Latin America:

> *Los destellos de luz que en tanta copia despidieron por aquella época los Estados Unidos de América y la Francia dieron una dirección más feliz a las ideas. A pesar de la vigilancia de la Inquisición, penetraron en las posesiones españolas las producciones inmortales de algunos filósofos; buscábanse con tanto más ardor cuanto más perseguidas eran; estudiábanse en la soledad; y comenzaron a germinar en varias cabezas los principios luminosos de los varones ilustres, que tanto honor hicieron a su especie y tanto bien.*[202] (García del Río, who had been secretary of Bolívar, quoted by Sarmiento, 1883/1953, pp. 131-32)

2. Initial Euro-Argentine Political Solution

From 1810 to the Constitution of 1819, these men tried to create a government based on law, a nomocracy,[203] a political form in which the personality was fitted into a pattern of political behavior created by an abstract system of law. These leaders

---

[202] (The sparks of light which came in such profusion at that time from the United States of America and France gave a wonderful direction to the ideas. Notwithstanding the vigilance of the Inquisition, the Spaniards could get hold of the immortal productions of some philosophes. They searched with the greatest enthusiasm those works that were most persecuted; they studied them in secret; and there began to germinate in various minds the luminous principles of the famous men who brought to the human race so much honor and so much good.)

[203] "The form of government in which an abstract system of law is above and beyond the personality. To this system, whether it be constitutional or traditional, the personality owes its loyalty. It limits and directs the political life of the individual.... The ideas of Moreno and Rivavadia did not center around *who* ruled but on *how* they ruled" (Bunkley, 1991, p. 19).

> chose conventions. They elected representatives. They promulgated constitutions. They passed, amended, and repealed laws. They dissolved assemblies, and they originated new ones. With great persistence, they continued to seek a formula from the French Revolution that would suit their circumstances. In vain they tried one solution after another. Mariano Moreno translated Jean Jacques Rousseau's *Social Contract*. This was his Bible, and through it he saw the answer for Argentina's trouble. "They sought then to legislate according to the principles of the *Social Contract* and thus to erect insurmountable barriers to the national illiteracy, to the incapacity of the half-breed and the Indian to deal with representative government, to the barbarity of gauchocracy. (Bunkley, 1991, pp.19-20)

The Founding Fathers may have aimed too high, not too low!

3. Travel to Europe

> The number of upperclass Latin Americans sent to Europe for study or mere travel grew at the end of the [XIXth] century to proportions never achieved under Iberian rule; on their return to capital cities or family properties in the interior the gap between civilization and backwardness was both disturbing and unforgettable. The steamship and the telegraph augmented the flow of information, technological, economic, literary, from the advanced and modernizing centers of the North Atlantic basin: European newspapers and periodicals regularly received found their way into the interior as well as the Latin American capital cities and served as a constant reminder of the superiority of white peoples over those in the backwash of change. In searching for the causes of backwardness, the Latin American elite conveniently pinpointed the second factor in racism, the "apathy, indolence and improvidence" of the masses.... The human resource endowment was inadequate.... To the Latin American elite in the last quarter of that century, the only road to progress was simply to substitute local manpower by massive immigration or failing to attract it, to hope that a long process of "whitening" might bleach out racial deficiencies. The vision of progress via immigration was coupled with racial prejudice and racial pessimism, although a few devoutly believed that well-planned prac-

tical education in science and technology might modify the incapacity of the masses to change. (Stein & Stein, 1970, pp. 183, 185)

4. Claraval's *Mal de Europa*

Gervasio Claraval is the main character of Gálvez's novel *Men in Solitude* of 1938:

> Claraval was no antiEuropean. He left this to the patriotists and the criolists. Nobody was burning inside with so much passion for everything European. Going there for a long vacation, calmly travel through the classical countries, this was his dream. He loved Paris far more than his brother-in-law, but not the Paris of the boites, of the onerous or easy loves and of the socializing among compatriots, but the Paris of the intelligence, of the spirit and the freedom. However if Claraval loved Europe for itself, even more he loved it for the contrast with our country [Argentina]. Europe for him was not only a set of features essential for life which we did not possess, but also the only way to evade. In Europe he saw the only refuge against the infinity of our evils which he thought were irremediable: the isolation and the solitude, the materialism and the impudent bad manners, the uniformity of events and of the spirit, the lack of respect for the respectable, the rabble and the stupid jokes, the lack of hierarchical feeling, the incapability of an intellectual life, of a life of the spirit, the huge pressure of society on the souls. He had already been there, when yet unmarried. Only after his marriage, there grew in him what he called "*el mal de Europa*" (the European sickness), an infirmity which, in the last five years had at the same time embellished and poisoned his hours. (1938/1994, pp. 19-20).

Like many *porteños* [i.e., from Buenos Aires], Claraval seeks his real identity in European cultures. To mask his loneliness he longs for residence abroad, where as a foreigner he may live unencumbered by the social and cultural pressures encountered at home, while at the same time absorbing a rich intellectual experience. Surrounding Claraval is a group of malcontents who likewise flee from the banal reality of their existence, finding an escape in politics, religion, illicit love, frivolities, or extended trips to Europe. To the basic theme of inner solitude Gálvez adds secondary ones that he associates with modern Argentine society: the religious poverty of Buenos Aires, the tedium of life resulting from the mechanical, formalized behavior forced on the *porteños*, dissimulation and hypocrisy in personal relationships, arrogant pretension, and emotional isolation. (Lichtblau, 1972, p. 89)

5. As Seen by Véliz

The republican nationalism of nineteenth-century Latin America was uncompromisingly outward-looking, avid to learn and imitate anything coming from France and Britain and equally vehement in its rejection of everything related to its own Hispanic past. José Victorino Lastarria, the eminent Chilean writer and man of affairs, calls the three hundred years elapsing between Columbus and Bolívar "a black Winter." The Argentine poet Esteban Echeverría, writing in the first half of the nineteenth century, is more explicit "we are independent but we are not free; the arms of Spain no longer oppress us, but her traditions weigh us down,....the social emancipation of Latin America can only be achieved by repudiating the heritage bequeathed by Spain." Had Echeverría written with complete candor, he would have added, "and by embracing the liberal heritage of France and Britain." (Véliz, 1980, pp. 164-65)

Learning from Europe was a natural solution, but by far incomplete.

XII. Was *Hispanidad* the Answer?

For Shumway,

the Generation's reverence toward northern Europe and the United States contrasts sharply with their facile dismissal of Spain. Spain, for the Men of '37, was the cradle of barbarism, the backward daughter of Europe, the poor relation to be avoided at all costs. Such a matricidal attitude could hardly build national self-confidence. Indeed, one of the most important intellectual currents of this century, the *hispanidad* movement that begins in the early 1900s and figures prominently in the rise of Peronism, was specifically targeted against the anti-Spanish bias of nineteenth-century liberals. (1991, p. 161)

However, could the *hispanidad*, of *El Solar de la Raza* by Gálvez, be of any help to Argentina? I doubt it; and even less that of Ramiro de Maeztu, the author of *Defensa de la Hispanidad* (1934).

In the words of Navarro Gerassi:

> Of all Spanish writers of the generation of '98, the one who had the highest influence on the Argentine nationalists of the right was the antiliberal Ramiro de Maeztu... The help he gave to Primo de Rivera made him the Spanish ambassador to Argentina in 1927.... Maeztu remained two years in Argentina, and upon his return he published his most famous book, *Defensa de la Hispanidad*, a fiery attack against republican Spain. According to Maeztu, democracy is a product of importation from abroad. It was the result of liberalism, the ideology which caused the decadence of Spain. In order to bring back the nation to its previous greatness, the Spaniards had to remain faithful to the founding principle of the Hispanidad, which began on October 12, 1492,[204] and which transformed Spain into "the arm of God on earth." This principle had been applied during the XVIth and XVIIth centuries. The central idea at the base of the concept of Hispanidad was Catholicism which had penetrated the ethics, laws, politics and arts of Spain....The idea of an evangelizing mission by Spain under the banner of the sword and the cross was not original of Maeztu. The first who had expressed that idea was Marcelino Menéndez y Pelayo in his *Historia del los heterodoxos españoles* (1880-1881), however Maeztu was the one who developed it and baptized it *Hispanidad*. Maeztu did not want only a renaissance of Spain but also of Latin America. The *Hispanidad* of Maeztu was a call to arms for the Latin Americans to join together under the banners of religion, authority and spirituality—embodied in the new Falangist Spain which Maeztu hoped would soon be realized—to fight against the great enemies of the Continent: the Russian Communism and the North American imperialist economy. Both were enemies because they were anti-Catholic, and Catholicism, the essence of *Hispanidad*, had to be defended before everything. (1968, p. 124-125)

[204] Discovery of America when one of Columbus' sailors first saw land from the prow of the "Pinta."

Is this the kind of thinking that should have guided the Founding Fathers of Argentina? Hate against democracy and total priority to Catholicism? Spain and Latin America allied as "*el brazo de Dios en la tierra* [the arm of God on earth]," "*bajo el estandarte de la espada y la cruz* [under the banner of the sword and the cross]"? "Spain and Latin America *bajo el estandarte de la espada y la cruz*" is certainly not what Véliz wants, yet he complained that:

> Lastarria and Echeverría were not alone in their admiration for Europe beyond the Pyrenees. Their views reflected a general intellectual climate that negated the nexus with the Iberian metropolis and its traditions and nurtured the opaque myth of the Black Legend, *the new obscurantism that turned Latin Americans against their own cultural heritage and led them readily to exchange three centuries of living history for the exotic novelties imported from Paris and London*. (1980, p. 165; my italics)

Then the question is the degree of approval or disapproval of Maeztu's *Hispanidad* given that—contrary to Lastarria and Echeverría—*Hispanidad* was true to those three centuries of living history, and did not include a trace of those exotic novelties from Paris and London. Said differently, should all the past be accepted with admiration and love only because it is "living history", and the foreign contributions branded and refused as "exotic novelties"? Certainly not. What is needed is a major effort at separating, all along those centuries, the wheat from the tares, both north and south of the Pyrenees.

XIII. Conclusion

Having been critical of much that has been said on the causes of Argentina's difficulties, any conclusion of mine must be preceded by a reminder of what I said in the introduction, that—similarly to the China specialists who hate to deal with the excruciating practice of *footbinding* , which darkened the lives of more than half of the women in China in recent centuries—the Argentina specialists hate to deal with the *mindbinding* practiced by the Inquisition (specifically by the UP of the Spanish Kings and their Inquisition), which darkened the lives of so many Spaniards and Latin Americans.

This said, my conclusion is that Argentina inherited from Spain much that can be labelled *UP hidalgo mentality*, in particular: an above average dose of vertical thinking and derived intransigence, a shortage of *sentido social* and abundance of hyperindividualism, a fondness for being *ser entero, hombre total, hombre integral incapaz de renunciamientos y de adaptaciones, incapaz de plegarse y rectificar,* a love for *el método de la pasión* and for *arbitrariedad.*

This combination of features explains why Argentina is a divided nation.This combination of mental features explains why that which was learned from Europe could only be applied in a limited way at home, why it was so difficult to develop a new country on the basis of the best of the Old and the New World, and why so many of the subsequent reforms failed. This combination of mental features was not conducive either to a good design or a proper manning of national and provincial institutions, as demanded by Karl Popper when he wrote that:

> You cannot construct foolproof institutions, that is to say, institutions whose functioning does not very largely depend upon persons: institutions, at best, can reduce the uncertainty of the personal element, by assisting those who work for the aims for which the institutions are designed, and on whose personal initiative and knowledge success largely depends. (Institutions are like fortresses. They must be well designed *and* properly manned). (1961, p. 66; italics in the original)

XIV. Recommendations

Recommendation No. 1

The Argentines should be courageous, and carefully study the *mindbinding* that the Inquisition imposed on the Hispanic people along their history. It should be a long multi-year study and general debate.

Recommendation No. 2

Independently of the above, scholars should study the Argentine vertical thinking, and the derived emphasis on *esencias y intransigencia.*

Recommendation No. 3

The literature in Spanish (from both Spain and Latin America) should be examined, in parallel, for the presence of *insular, vertical,* and *visitor* works.There is some truth in what Gálvez said at the beginning of

El solar de la raza: "*Nosotros, a pesar de las aparentes diferencias, somo en el fondo españoles*" [In spite of apparent differences, deep down, we are Spaniards] (1913/ 1960, p. 19). And there is some truth—after a major correction—in what Julián Marías said in *España inteligible*: "*Si se elimina el ingrediente español en los países hispánicos, se volatiliza toda continuidad historica entre ellos, desaparecen sus raíces compartidas, y con ello toda conexión social que pudiera llegar articularlos en un mundo coherente*" [If we eliminate the Spanish component in the Hispanic countries, every historical continuity among them evaporates, their common roots disappear, and with it all social connections that could integrate them in a coherent world] (1985, p. 414). The correction I have in mind is to replace his "historical continuity" with "selective cultural continuity."

Once this correction has been made, then the question remains, what kind of *ingrediente español*? The Spanish/Gálvez ingredient which gives priority to Montañes and cannot find anything good in ancient Greek sculpture? The mystic component *sin sentido social*? The arbitrary thinking of Unamuno? The Spaniards of the Inquisition, of Philip II, of Francisco Franco? No!

If, instead, by *ingrediente español* we mean the humor and wisdom of *El Ingenioso Hidalgo Don Quijote de la Mancha*, then the answer is a clear "yes". I am talking here, for instance, of Don Quixote's "Learn, Sancho, that nobody is more than another unless he does more than another" (1, xviii), and of his prior answer on the lineage of Dulcinea: "She is not of the ancient Roman Curii, ... nor of the Moncadas and the Requesenes of Catalonia;... but of El Toboso of La Mancha, a lineage that though modern may yet give birth to the most illustrious families of future ages" (1, xiii).

And there is also a clear "yes" to the criticism of Calderón de la Barca of *El médico de su honra* and his king, a clear "yes" to the poetry and wisdom of Jiménez's *Platero y yo (Platero and I)*, to the thinking and deeds of Mario in Delibes's *Cinco horas con Mario (Five hours with Mario*). The answer therefore should be: this Spain "yes", and that Spain "no".

There is no doubt in my mind, that the Latin Americans should cultivate their *españolidad,*[205] and should know and love many Spanish and Latin American authors. However this love must discriminate: this

---

[205] *Hispanidad* would have been a better word, but it brings immediately to mind the antiliberal *Defensa de la Hispanidad* by Ramiro de Maeztu.

work "yes", that other "no"; *visitor* texts, "yes"; *insular* and *vertical* texts, "no". And clearly, the first thing to refuse is any command that says that: "*La tradición hay que aceptarla en su integridad. Hay que aceptar a España entera, desde sus tiempos primeros hasta ayer, hasta hoy mismo. Todo lo que tiende a excluir, a reducir, a recortar, a sembrar recelos, a entontencer a los españoles, no es falangista.*[206] [The tradition must be accepted in its entirety. One must accept the whole of Spain, from its first times till yesterday, till this very day. Everything which tends to exclude, to reduce, to trim, to seed misgivings, to befuddle the Spaniards, is not Falangist"] (Tovar, 1953, p. 3). My recommendation is the opposite: be selective, do not accept the whole, exclude some, reduce, trim, seed misgivings.

Obviously, my recommendation to be selective has nothing to do with the efforts of Zum Felde who opposed *Hispanidad* because of its neo-imperialist program[207], while accepting everything else Spanish: "*Nadie más amador de España que el autor de estas aclaraciones; nadie más fervoroso gustador de las esencias de su genio profundo.... Sus pintores, sus poetas, sus capitanes, sus místicos, y hasta sus toreros.*[208]" (1943, p. 50). Here I would definitely suggest some discrimination: not all painters, not all poets, very few captains, few mystics, and not many toreros; and much prudence with vertical expressions like "*las esencias de su genio profundo.*"

I would recommend the same approach, for instance, with Julián Marías's *Ser español: Ideas y creencias en el mundo hispánico*: accept half of those ideas and beliefs and refuse the other half. But who is to decide which half to accept, and which to refuse? The answer is simple: each reader, on the basis of his or her personal judgment. The important thing is less what has been accepted or refused, but an individual process of selection which can be explained and defended. In

---

[206] This brings to mind the condition that Sarmiento had criticized so strongly (as quoted before): *Existo; ¡luego no pienso!*

[207] "We don't like *hispanidad*, which is a Spanish affair, but *americanidad* which is our affair: *americanidad* with a content of *hispanidad*, which is a completely different matter. Because this *hispanidad*—when included as an element of our *americanidad*, thanks to our historical heredity or living tradition—is what has a legitimate tendency to remelt with the other, equally lawful, elements of our historical reality, in an integral synthesis of our personality in gestation" (Zum Felde, 1943, p. 49).

[208] "Nobody loves Spain more than I; nobody appreciates more than I the essence of her profound genius... Her painters, poets, *capitanes,* mystics, even her toreros,..."

other words, the motto should be "A wise person makes distinctions or says nothing", and makes distinctions which are neither arbitrary, nor those of old when everything marked *nihil obstat* and *imprimatur* was OK, and nothing was OK if it had been placed on the *index librorum prohibitorum*.

Possibly, after twenty or fifty years, there may evolve a consensus as to which poets, novelists, or mystics deserve the highest praise by *visitor* Argentines, but that should not be the goal. The goal, instead, is a prosperous and peaceful Argentina, possibly also "a light unto the nations", peopled by just, fair, and generous Argentines.

In the end, my recommendation is nearly identical to Shumway's demand of building "a new country that would blend the best of the Old and New Worlds", provided that the blend be made in a long-term *visitor developmental process*, by *visitor personalities* who understand their past (recommendation No. 1), who have significantly reduced their passion for vertical thinking and derived dangerous emphasis on *esencias y intransigencia*, who are not intransigent, not *hechos de una sola pieza, incapaz de renunciamientos y de adaptaciones, sin sentido social* (recommendation No. 2), who have discovered a *visitor españolidad* through a critical screening of Spanish and Latin American literature (recommendation No. 3).

17

In Supping with the Tsar the Russian Church Lacked a Long Spoon[209]

Latest figures on religious affiliations[210]

***Russia*, in 1997:**	
Russian Orthodox	16.3%
Nonreligious	75.4%
***United States*, in 1995:**	
Christian	85.3%
Nonreligious	8.7%

1. The Spoon Was Too Short — Part 1

Practically from its beginning, but especially from the time of Tsar Peter the Great, the Russian Orthodox church was short of intellectual weapons with which to fight for its independence from the state. It is true that the Church had a fair amount of liturgical power over believers, but this power tended to melt away when confronted with the new ideas that came from the West.

It is remarkable how little interest has been shown by the Russian Church (in contrast not only to the Catholics and Prot-

---

209 *He should have a long spoon that sups with the devil.* (English proverb)

210 Source: *2002 Britannica of the Year*, Chicago: Encyclopaedia Britannica.

> estants but even to the Greek Orthodox) in theological problems. The duty of the Russian Christian has been rather to observe the ceremonies and to obey the rules, than to think about the doctrines of the faith. It is also important to note that the Russian Church did little to transmit secular learning. *Whereas the Catholic Church took over the literature and the worldly knowledge of Rome and Greece, and passed it on to medieval European society, and thus also to its Protestant successors, hardly anything of this reached Russia.* (Seton-Watson, 1967, pp. 30-31; my italics)
>
> *The Russian monasteries were totally devoted to the contemplative life and, unlike their counterparts in western Europe, they played no active part in public life or scholarship. Orthodoxy preached humility and, more than any other Church, it made a cult of passive suffering* (the first saints of the Russian Church, the medieval princes Boris and Gleb, were canonized because they let themselves be slaughtered without resistance). The second consequence of this mystical approach was the burden that it placed on ritual and art, on the emotional experience of the liturgy, as a spiritual entry to the divine realm.... The Russian Church is contained entirely in its liturgy,[211] and to understand it there is no point reading books: one has to go and see the Church at prayer. The Russian Orthodox service is an emotional experience.... The liturgy has never become the preserve of scholars or the clergy, as happened in the medieval West. This is a people's liturgy. (Figes, 2002, pp. 296-97; my italics)

The liturgy may be a wonderful emotional experience, but by itself it is a weak weapon in any *war of the scripts*, be it against Tsar Peter the Great's modernizing efforts, or later against the Enlightenment, Socialism and Communism.

Even the above listed 16.3 percent of affiliation with the Orthodox Church is, probably, far more the result of the failure of Communism

---

[211] "There was no Russian Bible—only a Psalter and a Book of Hours[*]—until the 1870s" (Figes, p. 58).

[*] The *Psalter* is a liturgical book containing the psalms and biblical canticles used in divine services. The *Book of Hours* is a liturgical book containing the prescribed prayers for the canonical hours of the day; such a book is also called a *Breviary* or the *Liturgy of the Hours*.

than of a true victory of the Russian Church. Let us not forget that Christianity won over the Paganism of Greece and Rome because of its intellectual/moral message, as contained for instance in the Four Gospels and in the letters of St. Paul, and not for its liturgy (which did not exist for a long while). Similarly, Martin Luther won against Rome because he fought a mighty intellectual *war of the scripts* which he began by nailing his *Ninety-Five Theses* on the door of All Saints Church at Wittenberg, on October 31, 1517. And the great papal victory at Canossa in 1077, was prepared by the treatise *Against the Simonists* of 1058 by Humbert of Moyenmoutier, Cardinal Bishop of Silva Candida, and by the *Dictatus papae* of 1075 by Pope Gregory VII.

In the end, the Russian Church supped with the State with too short an intellectual spoon, and the results were nearly fatal.

2. The Spoon Was Too Short — Part 2

Also:

> It is hard to overstress the importance of the fact that Russia received its Christianity from Byzantium and not from the West. It was in the spirit of the Byzantine tradition that the Russian Empire came to see itself as a theocracy, a truly Christian realm where Church and state were united. *The god-like status of the Tsar was a legacy of this tradition*[212]....*Unlike the Western Church, Byzantium had no papacy to give it supranational cohesion.* It had no lingua franca like Latin—the Russian clergy, for example, being mostly ignorant of Greek—and it was unable to impose a common liturgy or canon law....*For the Western reader, accustomed to conceive of religious differences in terms of doctrine and moral attitudes, it may be difficult to understand how rituals can define a national group.* But rituals are essential to the Orthodox religion—indeed, the very meaning of the concept 'Orthodox' is rooted in the idea of the 'correct rituals'. This explains why Orthodoxy is so fundamentally conservative—for purity of ritual is a matter of the utmost importance to the Church—and why indeed its dissenting movements have generally opposed any innovations

---

[212] The reader may want to refer to my volume 1, chapter 14 "The Low Creativity and High Unity of Power of the Byzantine Civilization", and to the two sections on Byzantium of the present volume 4, chapter 10 "The DP Christian Iconography versus UP Christian Iconography."

in the liturgy, the Old Believers being the most obvious case in point. (Figes, 2002, pp. 300-01; my italics)

Being illiterate, the average nineteenth-century Russian peasant knew very little of the Gospels, for there was no real tradition of preaching in the countryside. Even peasant readers had little means of access to the Russian Bible (which did not exist in a complete published version until the middle of the 1870s).[213] The

---

[213] The reader may remember what I wrote in chapter 11 of volume 3, "The Liberating Power of the English Bible under DP". The complete title was "The Strong Long-Term Civic Liberating Power of the Vernacular Bible (in English) in a DP Climate versus the Weak Long-Term Civic Liberating Power of the Vernacular Bible (in German) in a UP Climate." How much stronger would now be a contrast with the Bible in Russian!

Early in that chapter, I reported the following comment on Christopher Hill's (1994) *The English Bible and the Seventeenth-Century Revolution*:

What caused the English civil war? What brought Charles I to the scaffold? Answer to both questions: the Bible. To sustain this provocative thesis, Christopher Hill's new book maps English intellectual history from the Reformation to 1660, *showing how scripture dominated every department of thought from sexual relations to political theory*. (John Carey in the *Sunday Times*; my italics)

However the answer to both questions is incomplete. The answer should have been: "the Bible in a DP climate," because also Germany had its vernacular Bible at that same time, but there the Bible did not have the same civic impact. Under the UP of the German princes, the Bible had a strong religious and linguistic/literary impact, but a small civic influence. (Therivel, 2003, *3*, p. 280)

Further on, I quoted Hill directly:

The availability of the Bible in English was a great stimulus to learning to read; and this in its turn assisted the development of cheap printing and the distribution of books. *It was a cultural revolution of unprecedented proportions, whose consequences are difficult to over-estimate. Direct access to the sacred text gave a sense of assurance to laymen which they had previously lacked, and so fortified long-standing criticisms of the church and its clergy*" (Hill, 1994, p. 11; my italics).(*)

(*) "Scholars estimate that by 1640, one-quarter to one third of Englishmen, perhaps even 40 percent, were voting in elections for Parliament. Then there was the evidence of petition signing .... All of which argues for a considerable public political awareness... Roughly a quarter of England's two million households owned a Bible—the most influential indictment of pharisees, courtiers, and tyrants ever printed [provided one is allowed to notice it]" (Phillips, 1999, pp. 47-48).

"Ivan Roots's epigram that the Reformation started in the ale-houses of England, expresses a necessary truth. Where else could ordinary people meet for discussion? The printing explosion during the relative freedom from censorship of Edward VI's reign enabled protestantism and religious discussion to establish themselves in England" (Hill, 1994, p. 15).

Lord's Prayer and the Ten Commandments were unknown to the average peasant. (ib., p. 320).

The teaching of the Orthodox church was inherently favourable to autocracy. Its emphasis was on ritual and on the other world, on the individual's improvement of his own soul. It was not felt to be the duty of the Church to interfere in the things of this world. This again is in marked contrast to the Western Churches, both Catholic and Protestant. Both of these have usually taught that if a ruler acts contrary to Christian faith or morality it is the duty of Christians to oppose him. 'Render unto Caesar...' applies to secular government but not to matters of morality. But there is a strong trend in Russian religious thought that the affairs of this world are no concern of the Christian, that a ruler, even if he be Antichrist himself, may dispose of the believer's body in this world, but if the believer leads a Christian life and performs the ritual he will inherit eternal life. *The passivity of the Church in political matters benefitted the autocracy indirectly, no less than its active support of many rulers helped it directly. Indeed, in the purely political field, religion was much more frequently 'the opium of the people' in Russia than ever in the western countries of which Marx was thinking when he invented the phrase.* (Seton-Watson, 1967, pp. 13-14; my italics).

3. The Devil at Work

In 1721 Tsar Peter the Great issued the *Spiritual Regulations*, abolishing the Moscow patriarchate and placing the church under a 'spiritual college' or synod, containing married clergy as well as bishops. Deprived of its leadership, the Russian church was rendered subject to the secular authorities to an extent that had never been the case in Byzantium. (Ware, 1993, p. 164)

The *Spiritual Regulations* were a sort of constitution of the Church. They were anything but spiritual. It was the Regulations which established the subordination of the Church to the Imperial state. The Church was governed by the Holy Synod, a body of laymen and clergy appointed by the Tsar to replace the Patriarchate, which was abolished in 1721. The duty of the clergy, as set out in the *Regulations*, was to uphold and enforce the Tsar's authority, to read out state decrees from the pulpit, to carry out administrative duties for the state, and inform the po-

> lice about all dissent and criminality, even if such information had been obtained through the confessional.[214] The Church, for the most part, was a faithful tool in the hands of the Tsar. It was not in its interests to rock the boat. (Figes, 2002, p. 292)
>
> It was particularly galling that the synod's principal Civil Servant and supervisor was to be a layman who might not even be Orthodox by faith. Peter's original instructions required no more than he be 'a good man...to be chosen from among the officers': he was to be 'our eye and personal representative for the affairs of state'. Significantly, his title *Oberprokuror* was borrowed from the German language and the Protestant milieu. Not surprisingly, 'Peter listened to no part of what I told him more attentively', noted an Anglican divine in 1698, 'than when I explained the authority that the Christian Emperours [sic] assumed in the matters of religion and the supremacy of our Kings'. Such supremacy of kings was much in keeping with the Russian ruler's plans. Thus the kind of spiritual, even administrative, autonomy which the patriarch of Constantinople might enjoy within the confines of the Islamic Ottoman empire—his role as administrator of the Christian population or *millet* was clearly defined and generally respected—was not conceded, in an overtly Christian society, to the synod of the Russian church. (Hackel, 1990, pp. 520-21)

Hackel reported an engraving showing the famous equestrian statue of the Tsar, to which he added this caption: "The Russian Emperor and the Synod. The figure of Peter the Great, sculpted by E. Falconet in 1782, appears to dominate the buildings of the Holy Synod in St. Petersburg [shown behind the statue]. This interrelationship corresponds to the

---

[214] "In contrast to the Catholics, the Orthodox confess, not to a priest, but to the icon of Christ with a priest in attendance as a spiritual guide. The icon is the focal point of the believer's religious emotion—it links him to the saints and the Holy Trinity—and for this reason it is widely seen by Russians as a sacred object in itself" (Figes, 2002, p. 299).

Also: "With its lack of codified morality, the Russian Orthodox Church in particular placed great emphasis on purgation through confession, a concept that became integral to Russian attitudes over centuries. In bringing Raskolnikov to the understanding that peace of mind could come only through public confession of his crime, Dostoyevsky demonstrated his insight into the Russian psyche. In its own time, [Solzhenitsyn's] *One Day* performed a similar function for the nation" (Burg & Feifer, 1972, p.175).

emperor's own intentions in respect of church-state relations. These intentions were to prevail from 1721 to 1917" (p. 521).

We are here in a strong UP world, miles apart from the DP world of the Magna Carta of 1215 whose first article says: "In the first place have granted to God, and by this our present chart confirmed for us [King John] and our heirs for ever that the English church shall be free, and shall have its rights undiminished and its liberties unimpaired"; we are even further away from the famous First Amendment of 1791 to the American Constitution, the one establishing "a wall of separation between Church and State", which says: "Congress shall make no law respecting an establishment of religion, or prohibiting the free exercise thereof; or abridging the freedom of speech, or of the press, or the right of the people peaceably to assemble, and to petition the Government for a redress of grievances."

> In a lengthy preamble [to the *Spiritual Regulations*], Feofan Prokopovich (and, through him, Peter) explained the decision to abolish the one-man rule of the Patriarchate and replace it with collective administration:
>
> "From collegiate government in the church there is much less danger to the country of sedition and disorder than may proceed from rule by a single spiritual ruler. For the common people do not understand the difference between the spiritual power and that of the autocrat. Instead, dazzled by the splendor and glory of the highest clergyman, they think that he is a second sovereign equal to or even greater than the autocrat, and that the spiritual power is of another and better realm. If then there should be any dispute between the patriarch and the tsar, they might take the part of the patriarch in the belief that they were fighting for God's cause."[215]

---

[215] In reading the "Preamble", I felt that its author(s) was speaking more of the Pope than of the Patriarch of Moscow, probably a reflection of the feelings of Prokopovich, an orphan in childhood who "was educated by his uncle, a learned monk and rector of the academy in Kiev, and went on to Jesuit colleges in Poland and then to a special college in Rome. There, he studied theology, took Catholic orders and, in 1700 at the age of twenty-two, witnesses the coronation of Pope Clement XI. The effect of his three years in Rome, however, was to instill in Prokopovich a permanent dislike of the Papacy and the Roman church" (Massie, 1981, pp. 811-12).

We are reminded of Luther's journey of 1511 to Rome where "his earnestness was shocked by the levity of the Roman clergy and by the worldliness so evident in high places" (Rupp, 1978, p. 189).

> For the next two centuries, until 1918, the Russian Orthodox Church was governed by the principles set down in the *Spiritual Regulations*. The church ceased to be an institution independent of government; its administration, through the office of the Holy Synod, became a function of the state. (Massie, 1981, p. 814)

The Tsar was so powerful that he could spell out his intentions: "If then there should be any dispute between the patriarch and the tsar, they might take the part of the patriarch in the belief that they were fighting for God's cause."

4. Give the Devil his Due

> In preserving the predominance of Orthodoxy in Russia, Peter demanded of the church that it make itself useful to society. In his view, the most useful thing that Russian priests could do, besides saving souls, was to teach. There were no schools, and priests were the only channel by which enlightenment could come to the Russian peasantry scattered across the immense land. But, for this purpose, the clergy seemed a woefully inadequate instrument. Many priests were hopelessly ignorant and unshakably lazy. Some were as superstitious as their parishioners. Few had any knowledge of how to preach, and therefore such education and morality as they did possess could not be transmitted. Attempting to overcome this deficiency, Peter sent a number of country priests to Kiev and other theological schools to learn not just theology, but how to speak in public. Beyond what might be called the innocent ignorance of the Russian clergy, there was another failing which drove the Tsar into a rage. This was the widespread superstition of the Russian people and the playing on this trait by certain unscrupulous people, including members of the clergy. (Massie, 1981, p. 807)

5. The Tragic End

> The most striking feature of the *Ecclesiastical Regulations* was that it met no opposition, either within the church or among the people. In large part, this was because Peter had not tampered with the elements which mattered most in the Russian church, the sacred ritual and dogma. Who administered the church

> was of overriding concern to Peter; the form of the liturgy and sacraments did not interest him, and so he did not touch it. In time, however, the assumption of state control over the church had an injurious effect on Russia. Individual parishioners could seek salvation and find solace from life's burdens in the glory of the Orthodox service and its choral liturgy, and in the warm communality of human suffering found in a church community. *But a tame church which occupied itself with private spiritual matters and failed to stand up against successive governments on behalf of Christian values in questions of social justice soon lost the allegiance of the most dynamic elements of Russian society.* The most fervent peasants and simple people seeking true religion gravitated toward the Old Believers and other sects. *Students, educated people and the middle classes disdained the church for its conservative anti-intellectualism and slavish support or the regime.* The church, which might have led, simply followed, and ultimately the entire religious bureaucracy established by Peter followed the imperial government over the cliff; the Holy Synod was abolished in 1918 along with all the other governing institutions of the imperial regime. (Massie, 1981, p. 815; my italics)

In the end, tragically, in supping with the Tsar, the Russian Church lacked a long spoon.

6. Strong Rulers Have Always Liked Their Church to be Collegial and Conciliar, not Monarchical

As reported above, Peter the Great, in his Preamble to the *Ecclesiastical Regulations* for the Russian Orthodox Church praised the collegial government of the church against the one-man rule of the Patriarchate. This praise is consistent with the Eastern thinking on both collegial and conciliar government of the Church.

For instance, in the words of Kallistos Ware,

> Whereas the Western church during the Middle Ages grew increasingly monarchical, Orthodox Christians in Byzantine as in modern times have always seen the church as conciliar. As the Italo-Greek Barlaam the Calabrian, sent by the Byzantine emperor [Andronicus III] to negotiate reunion with the West, told Pope

> Benedict XII at Avignon in 1339: "There is only one effective way of bringing about union: through the convocation of a general council to be held in the east. For the Greeks anything determined by a general council has the authority of law." At each true council, on the Orthodox understanding, the miracle of the Pentecost is renewed; the Holy Spirit descends, the many become one in mind and heart, and the truth is revealed. (1993, pp. 136-37)

Far be it from me to disagree with this Orthodox understanding of when and where the Holy Spirit descends, making the many of one mind and heart. On the other hand, every lay ruler, be he a Byzantine emperor or a Russian tsar, will always favor a conciliar church and oppose a monarchical church. Among the members of a council, there will be a number of clerics whom the emperor can influence or corrupt so that the council will do his bidding; not so in a monarchical church. In other words, councils are easier to control than a pope. In addition, it is far easier to exert pressure on a council held in one's territory than in one held far away, thereby explaining the demand of the emperor's emissary for a "convocation of a general council to be held in the east."

It is very important to recognize, that if the Western Church had been conciliar in 1077, 1176, and 1215, there would have been no popes like Gregory VII, Alexander III, and Innocent III, and correspondingly no Canossa, no Legnano, no Runnymede, and no Western civilization significantly different from the Eastern civilizations.

The more conciliar the church, the more monarchical the State over the Church, as was already the case with Constantine the Great:

> The emperor completely controlled the bishops himself. His dominance over them, embodied in the statement 'my will must be considered binding'—which some earlier Christians would never have accepted—was a prime example of that monarchical control of the church described as Caesaropapism. Athanasius declared that the Fathers never sought imperial sanction, but in Constantine's time they had to, and they did. And it was he himself who chose every bishop when a vacancy arose. (Grant, 1993/2000, p. 159)

Emperor Constantine had practically hijacked the church and enslaved it to the state, so strongly that Kallistos Ware, early in his chapter on Eastern Christendom could write that:

> Our modern distinction between church and state would have been meaningless to the Byzantines [and later to the Russians for a long time]. They did not envisage church and state as two contrasted entities, subsisting in co-operation or conflict, but they thought of society as a single, integrated whole. The emperor had religious as well as civilian responsibilities. He was concerned with the entirety of his subjects' lives, and therefore among other things with the way they worshipped God. (1993, p. 133)

The emperor or tsar may have had religious as well as civilian responsibilities, yet—with few exceptions—his civilian interests had top priority, and the church was regularly used for un-Christian activities.

Sadly, but when supping with great rulers, a collegial/conciliar format nearly always proves to be a "short spoon". When, instead, supping among equals, the opposite is true: a collegial church—or better many different churches—will foster initiative, constant renewal, the valid participation of all members, as proved by the statistics I reported at the beginning of this chapter.

18

The DP Origins of *Courtly Love*

This chapter is divided into the following sections:

1. Introduction
2. What is Courtly Love?
3. More in Praise of Courtly Love
4. The Three Phases in the Development of Courtly Love
5. Causes
   Phase 1 – Provençal
      A. Basic DP
         a. The various kinds of powers, and related scripts
         b. Additional DP: the *juvenes* theory of Duby
         c. The triangular DP of Troubadour-Lady-Lord
      B. Arab/Islamic influences
      C. The High Creativogenicity of Detours Abroad
   Phase 2 – Champenois
      A. Political-Sociological Influences
      B. Major Cultural Influences
         a. Ovid, Chrétien de Troyes and Andreas Capellanus
         b. Ibn Hazm
      C. The High Internal DP of *The Art of Courtly Love* by Capellanus

1. Introduction

Compared with this revolution the Renaissance is a mere ripple on the surface of literature. (C. S. Lewis, 1937, p. 4)

Most critics since the sixteenth century have shared the conviction that modern European poetry begins in twelfth-century Provence and that the troubadour concept of love is utterly different from that which was current in antiquity. (Boase, 1983, p. 667)

The above praise for *courtly love* by C. S. Lewis may surprise at first, but it is my hope that, at the end of this and the next chapter, many readers will agree with him, not by lowering their opinion of the Renaissance, but by appreciating at its proper value the positive revolution brought about by *courtly love*. In turn, and this explains my DP interests in it, an act of creation so important and radically new in the history of the World could only have been the result of several forms of division of power (and many assistances) which I plan to discuss hereafter.

2. What is Courtly Love?

As summarized by various authors:

- "Courtly love is the name given to a conception of love elaborated in the courts of southern France at the end of the eleventh century.... First, medieval love poets did write within a literary tradition inspired by a particular ideal of 'true love': Provençal poets spoke of *verai'amors*, *bon'amors*, and *fin'amors*, words that find their counterparts in other Romance languages. Second, their use of the adjective 'courtly' establishes a link between Provençal poetry of the early twelfth century, northern French romances of the late twelfth century, German minnesinger poems and romances of the thirteenth century, Galician-Portuguese lyrics of the same period, Italian literature of the early fourteenth century, English literature of the late fourteenth century, and Catalan and Castilian literature of the fifteenth century.... In essence, courtly love was an experience of contradiction founded on the precarious coexistence of erotic desire and spiritual aspiration; 'a love,' according to F. X. Newman, 'at once illicit and morally elevating, passionate and self-disciplined, humiliating and exalting, human and transcendent.' These paradoxes, and the social and literary conventions that embodied them, provided the aristocracy with an outlet for fantasy, *tempered the arrogance of men trained as warriors, and inculcated ideals of courtesy and courtship whose civilizing influence has only recently begun to wane*" (Boase, 1983, pp. 667-68; my italics).

- "Courtly love was the theme of the troubadours' lyric poetry. Their conception of love was novel to a world that had looked upon licit love as a domestic comfort, upon passionate illicit love as a madness or as a masculine form of enjoyment and pleasure. Their cult of the beloved was novel to a world that regarded woman as man's help-mate, the bearer of his children or the object of his lust and sensuality. For the troubadours love was a noble and ennobling passion even though illicit and adulterous. The beloved held a place of superiority over the lover: his desire, anxious and fearful spurred him on to rise in worth to her and to attain the union he sought.

  The literary innovation of the troubadours had an enormous influence on the subsequent literature of France, Italy, England, Spain and Germany and on those countries touched by their civilization and culture....What is it then that makes Courtly Love to be Courtly Love? What is it in the troubadour conception of love that distinguishes it from that of the ages that had preceded them? The first essential characteristic is twofold and inseparable - the ennobling force of love and the elevation of the woman loved to a place of superiority above the lover. These two elements, the cult of the beloved and the ennobling power of love, are inextricably linked one with the other. The lover grows more virtuous through his love of the beloved and it is love which raises him in the scale of worth and virtue" (Denomy, 1944, pp. 175-76).

- "Let us then look at some of the more important 'rules' as they acquire form and authority in the verse of these singing poets, whose ranks contain high nobles as well as more lowly entertainers....The first requirement of the new lover suddenly smitten is that he leave behind his male arrogance and place himself humbly in his lady's service. No longer is she the passive prey to his desires, but an ideal to which he must aspire from below, the focus of his every thought and energy.... There is much talk of love-service in these poems, and one Provençal text mentions specific stages through which the toiling lover must pass in his quest for his lady. First he is the aspirant to her favours, silently adoring from afar. After a while, emboldened by his passion, he will become a suppliant, somehow acquainting her with his ambitions. If fortune and the lady favour him he may then become her recognised suitor. The ultimate success will be his when he is taken as an accepted lover; but this is a triumph rarely treated in song, on the principle perhaps that passion fulfilled makes dull poetry.

Our courtly lover, then, enlists in his lady's service, binding himself in all honour to carry out her least command, and pledging his unswerving loyalty, even to the death. In return he is entitled to expect his beloved not to let her eyes wander elsewhere. It was an expectation all too liable to be deceived, as many poems testify.

The astonishing reversal of roles brought about by courtly love (and that quite early in the twelfth century) is now apparent. The amorous relationship has taken on the outward form of the feudal pact between lord and vassal; but now the lady has become the lord, a fact strikingly illustrated by the troubadour practice of transferring to her the masculine form of address (*midons*, 'my lord')" (Owen, 1975, pp. 25, 29-30).

- "The essential characteristics that set aside the courtly conception of sexual love and distinguish it from that of previous ages is the cult of the beloved on the one hand—the elevation of the lady loved far beyond the level of ordinary mortals in physical, moral and intellectual worth—and secondly, the ennobling force of that love whereby the lover grows in virtue, worth and esteem" (Denomy, 1946, p. 133).

- "First developed in the castles of Provence, courtly love flourished within an aristocratic milieu and expressed the ideals and values of an aristocratic class. Some of the troubadours, like Duke William IX of Aquitaine, were powerful lords; others were humble knights or minstrels turned poets who depended on the generosity of wealthy patrons for their living.... While making the lady the object of his secret passion, he gave her an aggrandized self-image by addressing her as his lord.

Playing the role of the lover, the poet gave voice to the aspirations of the courtly class. Love made him noble, for only the noble in spirit could engage in courtly love. This new kind of love incorporated the concept that nobility was based on character and actions rather than on wealth and rank; in so doing, it appealed to the poor but aspiring knights. *It also celebrated woman as an ennobling spiritual and moral force, thus expressing a new feminism that contradicted both the antifeminism of the ecclesiastical establishment and the sexual attitudes endorsed by the church. Rather than criticize romantic and sexual love as sinful, the troubadours praised it as the highest good, a force that enabled the individual to realize his greatest potential. Love inspired the lover to be temperate because he did not desire other women, and to be courteous, generous, and brave in order to win a good reputation and impress his lady. It thus acted as a refining, socializing force. Although courtly love was mainly an aesthetic and erotic ideal, it acted as a moral force, ennobling those who followed it....*

Courtly love appears in the works of most major medieval authors, including Chaucer, Gower, Dante, Marie de France, Chrétien de Troyes, Gottfried von Strassburg, and Malory" (Bornstein, 1983, pp. 668-69; my italics).

- "A major new feature of the value-system associated with courtesy was that it applied to both sexes. Women had been virtually invisible in the *chansons de geste*. In the courtly romance, on the other hand, as in the court itself, women had important roles to play. Marie de Champagne was the patron of Chrétien de Troyes, while Marie de France composed romances herself. The texts expressed ideals for ladies as well as for knights, and they had much to say about relations between the sexes. Prominent among those ideals was that of courtly love.... The courtly romance appealed to the nobility all over Europe. In Italy, for example, the fusion of *cavalleria* and *cortesia* may be illustrated from Ludovico Ariosto's rewriting of the story of Roland in his epic poem *Orlando Furioso,* first published in 1516. From the start Ariosto declares his intention to write about both love and war, knights and ladies *(Le donne, i cavallier, l'arme, gli amori*) and also about courtesy" (Burke, 1996, pp. 16-17).

3. More in Praise of Courtly Love

As said by C.S. Lewis:

- "Our code of etiquette, with its rule that women always have precedence, is a legacy from courtly love, and is felt to be far from natural in modern Japan[216] or India.... French poets, in the eleventh century, discovered or invented, or were the first to express, that romantic species of passion which English poets were still writing about in the nineteenth. *They affected a change which has left no corner of our ethics, our imagination, or our daily life untouched, and they erected impassable barriers between us and the classical past or the Oriental present*" (1937, pp. 3-4; my italics).

And, as said by D. D. R. Owen:

- "We may wonder how much these rules [by Andreas Capellanus] were taken to heart and how far courtly love was truly practised. But there is no doubt that, whatever its fictional trappings, it came to be a real force in society and had a refining influence on the attitude of man to woman. And whether a cause or a product of the move towards female emancipation in the twelfth-century courts, or more likely a little of each, it brought a bright polish to both literature and life" (1975, p. 38).

In reality, courtly love was much more than a simple refining influence, far more than a bright polish on literature and life, it was a fundamental revolution nearly as important as the DP which began at Canossa in 1077. To understand that revolution, let me briefly remind the

---

[216] Fosco Maraini described, in *Meeting with Japan*, the encounter, in the years immediately after the end of WWII, between the Japanese women and the American soldiers who "seemed to the Japanese female to be the fulfillment of her wildest dreams.... Conditioned to obedience by centuries of submission or to the sadness of petty revolts promptly crushed between the paper of the *shoji* and the silk of the *tamoto*, now suddenly walked abroad in a heaven of understanding, gentle affection, and warm-hearted confidence. Here was a man who treated her like a queen! Not only did he open the door to let her pass, not only did he offer her his seat, not only did he hold her hand in the street, and help her up a step that was slightly steeper than usual, but he wanted to stay with her at all times and share his pleasures and amusements with her; and, finally, he opened his mind to her without restraint, not withholding anything of himself" (1960, p. 91).

reader of the universal macho attitude that pervaded the rest of the world. For instance in ancient Rome, man behaved toward woman

> with an attitude of superiority toward the love object, who was often treated like a slave. The attitude emblematic of the Roman lover was not holding his beloved by the hand or around the waist or, as in the Middle Ages, putting his arm around her neck; the woman was a servant,... The Roman way was the way of the seraglio. A small amount of sadism was permissible on the pretext of making her obey.... Petrarch's praise of passion would have scandalized the ancients or made them smile. The Romans were strangers to the medieval exaltation of the beloved, an object so sublime that it remained inaccessible. (Veyne, 1978, pp. 204-5)

The reader may want to relate this by Veyne to the long string of Roman macho quotes I reported in chapter 2 of volume 2.

For Japan, the reader will remember what I wrote on the training of little girls in chapter 9 of volume 3, specifically:

> A girl is taught that she must talk very little, but listen sympathetically to the conversation of her superiors. If she has a brother, she must look up to him as her master, even although he be younger than herself. She must give way to him in every detail. The baby boy places his tiny foot upon his sister's neck, and she is thenceforth his slave. If he is sad, her one care must be to make him happy. Her ambition is to imitate as nearly as possible the behavior of her mother towards her own lord and master. (Menpes, 1905, quoted by Moloney, 1962, p. 217)

Even in a Kabuki "romantic love story" with a happy ending, as *Kuruwa Bunsho (Tales of the Licensed Quarter)*, the notion of woman serving—and not being served—remains unchanged even if she is a first class geisha tending an impoverished lover (the play was patterned on a real-life famous and accomplished courtesan who died in 1711).

At the beginning of the last act, Izaemon, the son of a wealthy merchant is by now totally impoverished, and disowned by his parents, for having spent extravagantly on the courtesan Yugiri. Izaemon has learned that Yugiri is very ill, and even if so impoverished that he wears only a 'kamiko' a paper garment that speaks openly of his poverty, he asks to see Yugiri once again. He is admitted, however he is quickly faced with the fact that Yugiri attends to a new wealthy patron. He also discovers that Yugiri truly loves him. Izaemon becomes upset by jealousy, and Yugiri makes every effort to pacify him:

> Yugiri comes to him. She is gentle and understanding and humours her lover as a mother might treat a small child. Izaemon is cold and angry, abruptly rejecting all her peace offerings. Yugiri patiently lets his jealousy work itself out and then they are reconciled. At that moment a train of porters bearing heavy boxes approaches down the *hanamichi*. They bring the money sent by Izaemon's mother who has relented and wishes him to ransom Yugiri. The maids of the house hasten to dress her in a wedding robe and prepare a feast. Izaemon puts on a silk coat worthy of a wealthy merchant's son and the curtain is drawn on a triumphant pose. (Halford & Halford, 1956, p. 212)

This is not a case of Chinese footbinding, nor of ancient Roman macho behavior, but neither does it include a vision of woman as something high, inspiring and ennobling. Beautiful and accomplished Yugiri becomes a regular wife, probably humble and submissive as those women who followed the standard path to marriage and motherhood.

As seen before, courtly love "celebrated woman as an ennobling spiritual and moral force, thus expressing a new feminism that contradicted both the antifeminism of the ecclesiastical establishment and the sexual attitudes endorsed by the church. Rather than criticize romantic and sexual love as sinful, the troubadours praised it as the highest good, a force that enabled the individual to realize his greatest potential. Love inspired the lover to be temperate because he did not desire other women, and to be courteous, generous, and brave in order to win a good reputation and impress his lady. It thus acted as a refining, socializing force. Although courtly love was mainly an aesthetic and erotic ideal, it acted as a moral force, ennobling those who followed it" (Bornstein, 1983, p. 669).

The change is radical, of enormous importance, also because it created a new valid highly-creativogenic DP: between man and woman.

4. The Three Phases in the Development of Courtly Love

Before moving to a discussion of the causes at the origin of courtly love, let me briefly sketch its first three phases (place, times, key author(s) and works):

Phase 1 – Provençal

This phase developed and grew in the south of France, in Provence. Its key exponent was the troubadour[217] William IX, duke of Aquitaine and count of Poitiers (r. 1086-1127).

Phase 2 – Champenois

This phase developed at the court of Marie de Champagne, in the north of France. Its key exponents were:

- Chrétien de Troyes, the author of *Érec et Énide* (ca. 1170), *Le Chevalier de la Charette (Lancelot)* (1177-1181), and *Conte du Graal (Perceval)* (before 1181)
- Andreas Capellanus (fl. ca. 1170-1190), the author of the witty, mock-serious treatise about the love-dialectic, *De amore libris tres* (Three books about love, also known as *De arte honeste amandi* (The art of loving honorably).

Phase 3 – Early European

Courtly love then expanded all over Europe. Key representative countries, works and authors were:

- France - *Le Roman de la Rose* by Guillaume de Loris who began it in 1237, and Jean de Meun who finished it before 1280;
- Italy - *Vita Nuova* (c. 1293) by Dante; *Filocolo* (1336-1338) and

[217] "Both troubadours and trouvères were composers of lyric songs in strophic form. The former composed in Old Occitan (Old Provençal), while the latter used Old French. Their songs were destined to be sung, but there is not much information about the conditions of public performance. Usually the same person composed both words and music; but the composers were not usually the performers (jongleurs). The most typical and productive period for the troubadours was about 1150-1200, while the most illustrious of the trouvères belong to the thirteenth century" (Akehurst, 1989, p. 214).

Filostrato (1340) by Boccaccio; *Canzoniere* (first edition 1360) by Petrarch;

- England - *Troilus and Criseyde* (compl. 1385) by Chaucer.[218]

The German case is special, and is dealt with in the next chapter.

5. Causes

According to the GAM/DP theory, the origin of something so radically new and important as courtly love should be found in a situation of high DP both at the level of the powers that be (to use the language of I Corinthians 13:1) and of the scripts. Indeed, the time of the reign of William IX (r. 1086-1127) is that of the Investiture Controversy of 1075-1122, and his dukedom was a neighboring country of the highly cultured Arab/Islamic Spain.

In the words of Roger Boase (1983): "It has been argued that [courtly love] was an importation from Muslim Spain, a product of sociological factors, a consequence of the interaction of Christianity and a pre-Christian matriarchal culture, a vehicle for Cathar doctrines, an ideology influenced by Neoplatonism, a secular response to Marianism, a misreading of the ironies of Ovid's *Ars amatoria*, and a tradition evolving out of folk ritual" (p. 667). However, with the years, several of these

---

[218] Sadly, but with the return of the UP (e.g., that of the *signori* in Italy, as discussed in chapter 2 of volume 2; that of Napoleon as discussed in chapter 5 of volume 2) some of the ground won by women (in reality by women and intelligent men) was lost. Estelle Freedman (2002) reported this phenomenon as seen from the results:

> Reviving classical notions about women's irrationality, Renaissance depictions viewed woman as the flesh, sexually uncontrollable and liable to temptation by the devil. This association of women, sin, and the devil laid the groundwork for the accusations of witchcraft that targeted women during the Inquisition and the later witch hunts. From 1500 to 1800 four times more women than men were persecuted as witches, with estimates of those killed ranging from a hundred thousand to more than a million.
>
> The narrowing of European women's sphere affected their public influence. For centuries nuns had ministered to the poor and sick in their communities. Increasingly, however, the church required nuns to remain in the convent rather than the world at large. (p. 35)

With some research one could expand on the notion of a clear long-term parallelism between the level of the division of power, in a given civilization, and the positive feminine condition (e.g., agreement that women and men are inherently of equal worth; educational, professional, economic, political, sportive, creatively, sexual equality between women and men).

interpretations have lost importance, as for instance that of the interaction of Christianity and a pre-Christian matriarchal culture, or that of being the vehicle for Cathar doctrines, or as a secular response to Marianism,[219] and a tradition evolving out of folk ritual. Instead the sociological factors, the importation from Spain, and the reading of Ovid have won more explanatory powers. *Of these three, the one I see as the most important are the sociological factors if read primarily as division of power in its two forms of actual power and scripts, because this DP is the precondition for the other two, the importation from Spain and the reading of Ovid, which, instead, under UP would never have happened.* I will therefore first discuss the basic initial DP, and later the secondary DP brought by the Arab/Islamic importations, and by reading Ovid.

Phase 1 – Provençal

A. Basic DP

a. The various kinds of powers, and related scripts[220]

- "*Occitania was a highly decentralized patchwork of duchies, counties, principalities, and towns* that included Aquitaine, Gascony, Béarn, Foix, Toulouse (Languedoc proper), Limousin, Auvergne, Venaissin, Provence, and such border areas as Poitou, Roussillon, Nice, and even Catalonia, *each with its idiosyncratic dialect and, of course, with constantly changing borders and regimes.... Though doubtless a powerful lord, William IX was no king of France or of England;*

---

[219] "The fall of the old civilization and the coming of Christianity did not result in any deepening or idealizing of the conception of love.... That Christianity in a very general sense, by its insistence on compassion and on the sanctity of the human body, had a tendency to soften or abash the more extreme brutalities and flippancies of the ancient world in all departments of human life, and therefore also in sexual matters, may be taken as obvious. But there is no evidence that the quasi-religious tone of medieval love poetry has been transferred from the worship of the Blessed Virgin: it is just as likely—it is even more likely—that the colouring of certain hymns to the Virgin has been borrowed from the love poetry. Nor is it true in any unequivocal sense that the medieval church encouraged reverence for women at all" (C. S. Lewis, 1937, p. 8).

R. H. Green, in the *New Catholic Encyclopedia*, concurs in not seeing a derivation from the worship of Mary. On the contrary a certain language of human love in the devotion of the Blessed Virgin "is more likely to have been influenced by the literature of courtly love than to have been among the causes of the secular phenomenon" (1981, p. 395).

[220] Throughout this section, all italics are mine.

his repeated failures as a crusader could hardly inspire the development of a native Occitanian chanson de geste. Moreover, Occitania possessed no genuine—or aggressive—power center, as did northern France or Castile in Christian Spain. (Toulouse, at a later date, would most closely approach this ideal, especially subsequent to the diminution of Catalan influence, but the power of its extraordinary counts would be crushed by France and the papacy.)

In addition, feudal structures in the Midi were far less developed than in the north. By 1100 the social status and the political power of women were relatively more elevated in Occitania than in France or England: the *midons* (lady) of so much Provençal poetry was by no means entirely a lyric fiction. Women could inherit property and, in some cases, exercise effective suzerainty. *(Midons* derives from the Latin masculine *meus dominus,* my lord.) Some centers of learning existed in the south but, by and large, were of a character different from those established north of the Loire....

Meanwhile, though the aristocratic worldliness just described tended to undermine ecclesiastical authority, it ought not to be construed as an explicit attack on established orthodoxy. *Conversely, twelfth-century Occitania witnessed the spread, at first among the middle and the lower classes, of what would eventually be branded as outright heresy, the Albigensian, or Cathar, movement*" (Uitti, 1988, pp. 159, 163-164).

- "*Most of the protectors of the troubadours had been, if not heretics, at least liberally disposed toward the heretical party*, in whose downfall they consequently found themselves involved" (Berry, 1970, p. 263). "[The adherents to the mystical doctrines of the Cathari] were strong in Provence, and can sometimes be localized in those very castles where the troubadours lived" (Salter, 1970, p. 665).

- "*The fabric of the Church at that time was torn by heresy, by anti-clericalism*, by laxity of morals. That may explain the conscious, deliberate and carping denunciation of abuses extant in the clerical and lay world expressed for example, in the *sirventes* of the troubadours" (Denomy, 1945, p. 180).

- "William IX of Aquitaine (Guillaume de Poitiers, 1071-1126) is known today primarily for his poetry. He is the first of the troubadours whose works have come down to us, the first poet in whom we find evidence of the 'courtly love' tradition that soon dominated Western literature....*The chronicles show, first of all, that William's political and military achievements were mediocre at best.* The first of the two crusades he undertook (1101) was a major disaster. While crossing Anatolia most of the army he had raised was massacred in an ambush. (It is true that his second crusade, in Spain in 1120, was a smashing success.) *At home, William's devious efforts to increase his holdings resulted in persistent local warfare and conflict with the church,* rather than in any significant territorial gains. In 1097 he temporarily annexed Toulouse after its lord had left for the First Crusade and while the land was under the protection of the church. *Finally, his flagrant adultery and open devotion to promiscuity led him into further conflict with the church, and he was excommunicated in 1114. The excommunication was not lifted until 1117.*

The accounts of William's relationship with the church, the many anecdotes illustrating his worldliness, *and the chroniclers' evaluations of his personality show that William drew attention to himself less by his official acts in battle or politics than by his ostentatious violation of church-sponsored standards of conduct....*

The historian William of Malmesbury reports that William sang of how he planned to set up an Abbey of Whores in which various known prostitutes would hold the positions of authority. In the same passage, this historian describes William's violent behavior toward Peter, bishop of Poitiers, when Peter excommunicated him because of his famous affair with Maubergeonne, viscountess of Châtellerault. [He had repudiated Hildegard of Anjou and married Maubergeonne taken away from the viscount of Châtellerault]....

The William of the burlesque poems is the William in whom the chroniclers, judging from the clerical perspective, saw only unbridled sensuality, a foolish disregard of the judgment to come, and absence of purpose, a man who made a joke of everything and whose wit masked superficiality. The remaining poems are serious, indicating that William found spiritual and intellectual pleasure as well as physical pleasure in his life....

All but one of the serious poems are about love. They praise noble love between two people who are worthy of it as the source of 'joy,' the highest earthly good. This love both demands and mysteriously confers spiritual strength and refinement, making the lovers more noble than they were before. William's portrayal of love is consistent with what we find in later 'courtly love' lyrics" (Lipp, 1989, pp. 630-632).

First developed in the castles of Provence, courtly love flourished within an aristocratic milieu and expressed the ideals and values of an aristocratic class: "Certain factors in its make-up are easily definable. Feudalism, with its binding obligations of loyal service to a lord upon whom all fortune depended provided the general background and even, perhaps, some of the poetic vocabulary—the troubadours spoke often of the adored lady as 'my lord' (*midons*). It may also be true that the conditions of life in the castle civilization of southern France at the time were favourable to the development of courtly love. *Rich but isolated centres of refinement and culture, the castles contained many men but few women*; to the lady of the castle the knight and squire may have felt themselves feudally inferior. The adulterous basis of such love is understandable in the light of medieval marriage conventions, and the attitude of the church to marital passion. Marriage was usually a business contract, involving property and military power but little sentiment; no woman had freedom either to choose her husband or to prevent the annulment of the marriage for political or dynastic reasons if, as often happened, it was not a success. The teaching of the church upon the relationship between husband and wife was ambiguous as far as passionate love was concerned; it could have done little to encourage the growth of romantic devotion. For these reasons the idealization of passion could not be based upon the marriage state; poets looked beyond it" (Salter, 1970, pp. 664-65).

b. Additional DP: the *juvenes* theory of Duby

As summarized by Diane Bornstein (1983): "Georges Duby's discussion of marital practices in twelfth-century France provides a sociological explanation for the popularity of the game of courtly love. In order to curb the proliferation of potential heirs and to avoid divided inheritances, the heads of households pursued increasingly restrictive matrimonial strategies. They tried to marry off all their daughters to create a widespread network of alliances. However, they wished to marry only their oldest son, unless the others could be matched with heiresses. *This practice resulted in a large number of bachelors, who formed turbulent bands that roamed the country in search of adventures, wives, or the protection of a lord....* The real aim of a *juven* was to obtain a wife, by honest or dishonest means, who would bring him stability and turn him into a *senior*. The game of courtly love allowed the *juvenes* to engage in fantasies of abduction" (pp. 31-32).

Moving to Duby himself (1978): "The new situation prompted the chivalric idealogy to extol the life of adventure more resoundingly than ever as a mock compensation for the frustration of 'youth'. Here what is called 'courtly love' came into its own.... *The code of the young extolled such triumphs over the control exercised by the elders, who alone had the power to bestow women.* But now this dream of abduction was carried inside the very house that provided for the young men, into the house of the patron who gave them shelter, who 'retained' them, and who every night went to bed with his own wife. The favors of the lady thus became the stake in the competition among the bachelors of the court, a game that was similar in every point to the tournament, for it was aimed at attaining a mock capture that derived much of its excitement *from flouting the strict prohibition of adultery and was tantamount to a kind of revenge against the common seigneur. The game of love was thus the expression of profound hostility to marriage....* By exhibiting his largesse to the point of letting his lady pretend that she was gradually giving herself, he was able to gain an ever stronger hold over the young men of his household, to domesticate them in the proper sense of that term" (pp. 13-14).

c. The Triangular DP of Troubadour-Lady-Lord

It may be true, as Duby said, that by exhibiting his largesse to the point of letting his lady pretend that she was gradually giving herself, the lord was able to gain an ever stronger hold over the young men of his household and to domesticate them. Yet, at the same time he too was domesticated, and his lady empowered. By witnessing—albeit accepting—the praise and love/sexual allusions of the troubadours, he acknowledged—even if only as a game—the high role and importance of his lady and of womanhood.

Also, it would have been normal for the lady to demand a certain continuation of the politeness of courtly love in private life. Regular life is not the same as the stage where the actor quickly abandons his role and fully resumes his normal self. The triangular game troubadour-lady-lord was always a power game with broad implications, especially cultural ones, as proven by that extraordinary institution of the *courts of love*, those literary salons where people read poems and romances, debated questions of love, and played word games of flirtation.

On the troubadour-lord side, each actor had to play the game with finesse, fully recognizing the power of the other. A troubadour who overstepped his role risked either beating (if of low social class) or expulsion (if of high class). Conversely a troubadour could depart and move to another court if he was not treated with the respect due to him, and that departure would stain the reputation of the misbehaving lord or lady. The game had to be played correctly, and each player had to train him/herself to win glory and reputation.

In the introduction, I spoke of both DP and assistances, and much of the above can be read as assistances. For instance, as reported above, "Rich but isolated centres of refinement and culture, the castles contained many men but few women; to the lady of the castle the knight and squire may have felt themselves feudally inferior" (Salter, 1970, p. 664). The knight and squire yes, but less or not at all the wandering troubadour, to whom a rich and cultured castle, ruled by an intelligent and refined couple, or by a great lady alone, was a major material as well as spiritual assistance.

B. Arab/Islamic Influences

Many of the conventions of courtly love can be traced to Ovid, but Ovid's comic, satirical spirit has little in common with the elevated, refined spirit of courtly love. A source that has been proposed for this difference in tone is the Arabic poetry of Muslim Spain. There were periods in Spain, particularly during the eleventh century, when the Muslims and Christians lived side by side and benefitted from each other's culture. *A group of wandering poets came into existence who passed from court to court and sometimes visited the courts of neighboring Christian countries.* The situation closely resembled the one that developed a century later in southern France. Contacts between these Spanish poets and the troubadours of southern France were frequent. The Spanish poets used metrical forms and themes that were similar to the ones later used by the troubadours, and an influence seems possible, particularly since Mozarabic poetry provides many of the motifs that cannot be traced to Ovid (Bornstein, 1983b, pp. 671-72).

- "The most important single source of influences, however, was the poetry of Muslim Spain which experienced its greatest flowering in the 11th century, roughly one hundred years before courtly poetry developed in southern France. Hispano-Arabic poetry shows striking similarities in poetic forms and literary clichés; and many authors now accept as a fact that the troubadours were indebted to the Muslim Spanish poetry to a similar extent as the German minnesingers, in turn, were influenced by the Provençals. *In either case, what happened was not bookish imitation, but an adaptation animated by spontaneous interest*.... Hispano-Arabic love poetry with features foreshadowing the later troubadour lyrics is known to have been produced since about A. D. 820. The poetic forms known as *muwassaha* and *zagal* reportedly were invented in Spain at the end of the 9th century. Andalusian love poetry reached its greatest height in the 11th century" (Moller, 1958, pp. 139, 141).

It is important to mention that also Dante was influenced by Arab/Islamic authors:

- "The Spanish Arabist, Father M. Asin Palacios, in a work published in 1919, came to the conclusion, after collating parallels between Islamic descriptions of journeys to the Other World [by Mohammed] with the *Commedia*, that Dante's subject-matter was drawn to a great extent from Islamic sources.... Some years later an unsuspected key link was found in the translation made by an Italian, Bonaventura da Siena, into Latin and Old French of a lost Castilian version of an Arabic *mi'raj*—a popular text describing this very journey of the Prophet to the Other World. Enrico Cerulli, the scholar who found and published the translations, was able to prove that the work was current in early fourteenth-century Italy and was mentioned by followers and imitators of Dante as the *Libro della Scala*" (Anderson, 1980, p. 277).

- "If we compare Guillaume's [William's], Marcabru's and Rudel's forms of poetry with the forms current in contemporary Muslim Spain, as well as in the East, we cannot fail to find considerable analogies which can only be explained by imitation or adaptation, not by independent invention. Subsequent Troubadours were original enough to elaborate on what they preferred in the work of their predecessors or contemporaries, according to their moods and to the moods of their respective audiences" (Nykl, 1946, p. 379).

- "Of particular interest for the early development of Provençal lyrical poetry is the evidence which links William IX and his father to the Spanish frontier. In 1064 William VIII of Aquitaine, accompanied by his vassals from Limousin and Poitou, was among those who participated in the 'crusade' which captured Muslim Barbastro. *The count, however, was apparently himself captivated by his Muslim prisoners and, in particular, by the female lute-playing singers. He was even reputed to have shared out these female singers among the nobles of his lands in southern France.... William IX, the troubadour, married the young widow of the king of Aragon, and her retinue would almost certainly have included some jongleurs or female singers similar to those who had been captured at Barbastro....* William IX was too young to participate in Alfonso VI's conquest of Toledo, but in 1125 he accompanied Alfonso I of Aragon (the 'battler') in a campaign during which the Christians thrust down as far south as Granada and brought back with them several thousand Mozarabic families who were resettled along the Ebro valley.

Given the circumstances, therefore, there was certainly no lack of contact opportunities with the lyrical and musical traditions of al-Andalus, and this lends credence to the theory that the troubadours imitated a Spanish-Arabic model.... *It has even been plausibly argued by some scholars that, just as William VIII is reputed to have conversed with his female captives in broken Arabic, so William IX's fifth poem illustrates that he was acquainted with the spoken Arabic of the peninsula.*

Even if the arguments in favour of Spanish-Arabic influences are accepted, however, it is important to note that, in passing from Spain to Provence, the poetical art-form underwent important changes. In the feudal society of Provence the lyric-songs, from being popular, became the preserve of noble and princely circles, and subsequently the new poetic form, elevated to the art of courtly love, was reimported into Spain with resounding success during the late twelfth and thirteenth centuries" (MacKay, 1977, pp. 93 - 94).

In his detailed study of the various theories on the origin of courtly love[221] Roger Boase said the following on the first theory, that of the Hispano-Arabic origin:

- "Courtly Love was either imported into the south of France from Muslim Spain, or was strongly influenced by the culture, poetry and philosophy of the Arabs....

(a) Scholarship and culture of the Islamic world

---

[221] *The Origin and Meaning of Courtly Love: A Critical Study of European Scholarship* of 1977, in which he analyzed the following theories of the origin of courtly love: Hispano-Arabic, Chivalric-Matriarchal, Crypto-Cathar, Neoplatonic, Bernardine-Marianist, Spring Folk Ritual, and Feudal-Sociological.

It was from the Arabs, from the tenth century onwards, that Christian Europe became better acquainted with Hellenic culture. *Europe, which was still emerging from barbarism, assimilated the literary, scientific and technological achievements of a civilization which had already reached its maturity....*

(b) Chivalry of the Arabs

The institution and the ideal of chivalry was initially unconnected with the feudal system; it was introduced into Europe from Moorish Spain or from the Middle East, together with Arabian prose romances, through cultural, commercial and military contacts, and as a consequence of the Crusades....

...(g) The concept of love in poetry

Some, if not all, of the essential features of Courtly Love can be discerned in Hispano-Arabic (and even Middle Eastern) poetry; the insatiability of desire; the description of love as exquisite anguish; the elevation of the lady into an object of worship; the poet's submission to her capricious tyranny; and the emphasis on the need for secrecy....

The cultural supremacy of the Islamic world in the period immediately preceding the rise of the troubadour lyric is indisputable, and the importance of Arabic scholarship as a medium for the transmission of Greek classical texts is widely acknowledged" (1977, pp. 62-63).

C. The High Creativogenicity of Detours Abroad

Back to the GAM/DP theory, the reader will remember—from volume 1—the importance I gave to the reading of foreign texts as a major kind of DP. There I discussed the foreign readings of such great masters as Goethe, Pushkin, Tolstoy, Dostoyevsky, Dante, and Lady Shibiku Murasaki. To this let me add the following on Kant:

> English and French writers were among his favourite authors. In his study of physics he was most indebted to Newton. Frequently he refers to Locke, and repeatedly to Berkeley; but owing to the critical spirit of Hume and the impulse which he had received from him, Kant placed him higher than any other Scotch or English metaphysician. ....
>
> Among French authors, he had read Montaigne while still a student, and knew many passages of his "Essays" by heart. Montesquieu's political views and his knowledge of various na-

> tions made him a great favourite.... But of all French writers he preferred Rousseau, whose views of human nature, of education, and of the freedom and rights of man, deeply interested Kant and made a lasting impression on his mind. The only picture which adorned his house was one of Rousseau. His "Emile" so engrossed the philosopher's attention that for several days it kept him from his usual walk. This author and Montesquieu had most influence in moulding his social and political views. (Stuckenberg, 1882, p. 147)

In that same chapter, I also discussed the great contribution to creativity of the immigrants. In each case, the *detour abroad* weakened the UP of the national scripts and started (better said reinforced in challenged minds) a *war of the scripts* in which each thinker could use his/her critical mind first, then build a creative work from the ideas stimulated by the dissonance between national and foreign scripts.

The literary detour abroad provided its readers with a freedom denied to them by their national literature, even by its best authors, because after a while much of what those authors had said had been incorporated within the general culture. For instance, it is nearly impossible nowadays for any Western reader to react to Hamlet's "To be or not to be" as a personal discovery. That sentence has long belonged to everybody: from educated parents to school teachers. Instead, with a lesser known or totally unknown foreign text, things are different: the new author belongs to the reader alone. To such a reader, the foreign author says things that nobody else has said before. The reader can then reject what is read out of hand, or accept it. *One way or the other, the reader is in command.* The reader's enthusiasm is genuine, not the parroting of what was said in class.

If, moreover, those poems were the forbidden fruits of a superior civilization, as was the case for Islam in the 11th and 12th centuries, then clearly they drew attention and had an impact.

Using my terminology, the new Arab/Islamic ideas adopted north of the Pyrenees, the new Arab/Islamic poems read with pleasure, were *detours abroad* far more important than the actual voyages by William IX to the Near East and Spain; more important than Kant's and Tolstoy's reading of Rousseau.

My key point is that each of the main causes which have been proposed for the origin of the courtly love (political-religious-social, Islamic influences; and the influence of Ovid) is in itself a major creativogenic form of DP, being both fuel and match for a great fire, or with a different simile being both bricks and inspiration for the construction of a beautiful *chateau*.

This high praise of mine for what happened then, can be better understood if we imagine what would have happened under UP. Under lay UP (royal or imperial) or ecclesiastical UP (episcopal or papal or "brotherly" as was later the influence of Savonarola), the reading of Arab literature would have been strongly restricted if not totally forbidden, and all intellectuals, big and small, would have been put in the service of the UP. Under UP, all Arab literature would have been on the imperial or papal Index.[222]

Clearly, the political/ecclesiastical power had to be weak and divided to permit those detours abroad. It is here that those strange relations between William IX and the episcopacy come to merit, and also the criticism of the Cathars against the Catholic orthodoxy. If William IX and the other troubadours had truly feared excommunication, they would have stayed miles from any possible Arab/Islamic influences.

What was said about the Arab/Islamic influence is equally true for the influences, to be discussed shortly, of Ibn Azm and Ovid. Also the reading of these two authors were major detours abroad: the immersion in the thoughts and feelings of two other civilizations (much different, and yet strongly related to Western thinking).

Phase 2 – Champenois

A. Political-Sociological Influences

Here, the major political-sociological-economic influences were Marie de Champagne (1145-1198) and the county of Champagne she ruled, first as wife of Henry the Liberal of Champagne, then—widowed in 1181—as regent for her young son from 1181 to 1187, then again from 1190 to 1197 during his participation in the Third Crusade. Her link with Provence and its poetry was direct: she was the first child of King Louis VII of France and Eleonor of Aquitaine the daughter and heiress of William X, duke of Aquitaine and count of Poitier, son of William IX.

---

[222] Emperor Frederick II paid dearly for flirting with the Arab civilization (and his opposition to the Papacy), by being branded an heretic and excommunicated, and as such placed by Dante in one of the burning tombs of Circle VI. Indeed, the first critique, leveled against him—by Pope Innocent IV, in 1245—was that Frederick "was the lover of Saracen company" (Abulafia, 1988, p. 369).

Marie de Champagne was the patroness of the most important Champenois works about courtly love. "Living at a time of great prosperity for Champagne, derived principally from its fairs, and when the counts of Champagne were among the king's most powerful vassals, Marie seems to have made the court of Troyes a cultural center. In his *Chevalier de la charrette*, Chrétien de Troyes claims he obtained the subject matter of his story from her and attributes to her at least some of its thematic and structural orientation. Gautier d'Arras composed his romance *Eracle* for her, and several lyric poems are associated with her. In his *De amore*, Andreas Capellanus, chaplain at Marie's court, mentions her several times as an authority on matters of love casuistry" (Burgess, 1987, p. 135).

B. Major Cultural Influences

What made Phase 2 distinct and so important were the influences of Ovid and Ibn Hazm, stronger than before and impinging on the ideas and feelings of Phase 1.

a. Ovid, Chrétien de Troyes and Andreas Capellanus

Ovid – "Western vernacular poetry reveals Ovid's prodigious influence. The complex medieval conception of romantic love, for example, contains a heavy Ovidian component, centering on the mutuality and transforming power of the erotic experience. Such stories as those of Echo and Narcissus, Philomela, and Pyramus and Thisbe appear in Old French, German, and Netherlandish versions, and later in such works as the *L'amorosa Fiammetta* of Boccaccio, *The Legend of Good Women* of Chaucer, for whom Ovid was "Venus' Clerk," and the *Confessio Amantis* of John Gower. This vernacular Ovidian vogue passed undiminished into Renaissance poetry and art.

Although the Latin elegiac comedies were not translated, they do appear in the same contexts as the vernacular adaptations, along with the fabliaux. Chrétien de Troyes translated the *Ars amatoria*, and, though the work is lost, its influence is apparent in his Arthurian romances" (Dwyer, 1987, p. 314).

Chrétien de Troyes – "[His] training was that of a clerk, schooled in Latin and familiar with the seven liberal arts. At the start of his second romance, *Cligés*, Chrétien describes himself as a clerkly artisan working in the vernacular, claiming among his accomplishments the text of *Erec et Enide* as well as the translation into French of such Ovidian poems as *The Commandments of Ovid; The Art of Love; The Metamorphosis of the Hoopoe, Swallow, and Nightingale*; and *The Shoulder Bite* (probably the story of Philomela and the account of Pelops from the *Metamorphoses)*" (Freeman, 1983, p. 308).

Andreas Capellanus – "The codification of courtly love is exemplified in the *ars amandi* (art of love), the medieval adaptations of Ovid's *Ars amatoria* (Art of love). The most important one, from the point of view of both its popularity during the Middle Ages and its influence on later scholarship, is the work by Andreas Capellanus [of about 1174]. *The popularity of Andreas' work is shown by the number of surviving manuscripts and by translations into the vernaculars. Twelve complete manuscripts survive* (twice as many as we have of Chrétien de Troyes's *Chevalier de la charrette*) besides several that contain extracts.... Book 1 of Andreas' work on the nature of love and Book II on how love may be retained, are based on Ovid's *Art of Love*: Book III, on the rejection of love, is based on Ovid's *Remedies of Love*. The work is not a translation but a free adaptation of Ovid's ideas to a medieval setting" (Bornstein, 1983b, pp. 670-71; my italics).

"The first and by far the longest book defines the nature and goal of love and then offers a series of dialogues between men of the upper nobility, gentry, and the bourgeoisie and their female counterparts.... The second book deals with the retention and loss of love.... The third book retracts all the advises and repeats all the clichés of medieval antifeminism" (Jackson, 1983, pp. 88-89).

b. Ibn Hazm

"A rival claim has been made for the origins of courtly love in Arabic mystical philosophy; even if this is judged controversial, it is highly likely that the literature of Muslim Spain had a direct influence upon southern French poets. A treatise, the *Tawq al-hamamaw* (*The Dove's Neck Ring*) written in 1022 by the Andalusian religious philosopher Ibn Hazm (994-1064), contains in detail most of the ideas which recur in all medieval treatments of courtly love from the troubadours onward. The *Tawq*, which draws upon many earlier eastern textbooks of love, and ultimately upon Platonic philosophy, contrasts strongly with Ovid's manual in that it seeks to discover spiritual significance in passionate relationships....

The contribution of Andreas Capellanus, writing late in the 12th century, was a textbook codification of the whole doctrine of courtly love, much on the lines of Ibn Hazm's *Tawq*.... The initial definition given by Andreas could have come from the *Tawq*: 'love is a certain inborn suffering.'[223] He stresses that love develops nobility, but also that it is the prerogative of the leisured and cultured classes" (Salter, 1970, p. 665).

"In or about 1022, Ibn Hazm wrote *The Dove's Neck-Ring*, which in scope, realism, and beauty is vastly superior to *The Art of Courtly Love*, a stilted[224] command performance, composed more than a century and a half later by Andreas, a chaplain at the court of Marie of Champagne. In Muslim Spain, love poetry was still popular after the coming of the Almoravides in 1086, but slowly waned in the 12th and 13th centuries" (Moller, 1958, p. 141).

In the words of Diane Bornstein: "Many of the conventions of courtly love can be traced to Ovid, but Ovid's comic, satirical spirit has little in common with the elevated, refined spirit of courtly love. A source that has been proposed for this difference in tone is the Arabic poetry of Muslim Spain.... In Mozarabic poetry, we find two different attitudes toward love: a sensual tradition, which may have been influenced by Ovid, and a spiritual tradition, which seems to be based on the work of Plato as it had come down through the commentary of Arabic scholars. The *Tawq al-hamama* (The Dove's Neck Ring), written about 1022 by the Andalusian Ibn Hazm, illustrates the spiritual tradition" (1983b, p. 671).

C. The High Internal DP of *The Art of Courtly Love* by Capellanus

a. Denomy's Islamic theory of the Christian evolution of courtly love

---

[223] In full: "Love is a certain inborn suffering derived from the sight of and excessive meditation upon the beauty of the opposite sex, which cause each one to wish above all things the embraces of the other and by common desire to carry out all of love's precepts in the other's embrace" (ca. 1174/1959, p. 28).

[224] Stilted or not, Capellanus's work had probably a higher impact on Western civilization than Ibn Hazm on the Arab/Islamic civilization. Why so? The answer, in a nut shell, is the high division of power (persons, institutions, scripts, mentalities) in the West, and the relatively high unity of power in the Arab/Islamic countries, especially from around the 14th century onward.

Alex Denomy (1953) has proposed an "Islamic" theory of the ennobling (Christianization) of courtly love based on 1) the thoughts of Avicenna on the morality of sexual love, 2) the clash between these thoughts and Christianity, and 3) the reconciliation of this clash by the doctrine of the "double truth" of the Latin Averroists. What Denomy said is important to understanding both the condemnation of Andreas Capellanus as heretic (see next section b) and the important high internal DP of his *The Art of Courtly Love* (see section c).

On Avicenna

- "Courtly love is wholly unconcerned with Christian morality and in that sense it is neither moral or immoral: it is amoral, that is completely divorced from the Christian concept of the morality of sexual love. No Christian teaching could ever give rise to the troubadour conception of sexual love; paganism, while recognizing the delights of sensual desire and indulgence, did not recognize the dogma that carnal desires, fanned by sensual delights, ennobled man. *The conception that such love ennobles is to be found in Arabian philosophy and in the mystical writings of Avicenna....* For Avicenna and the Muslim mystic, as for the troubadours, the morality of sexual love rests not on religious and legal grounds but just on whether it tends to ennoble man or to degrade him, whether it tends to make him increase in merit and worth or to reduce him to the level of the animal. For Avicenna and the Muslim mystic as for the troubadours that love is pure and true, the source of nobility and progress in virtue which is based on the freedom of the rational soul of man from solely carnal ephemeral delights. That love is impure and false which is grounded on the subjection of man's heart and mind to sensual, fleeting indulgence. For the troubadours there was to be found in Avicenna and Muslim mysticism a norm of morality that they made for their own, a norm unavailable and unknown elsewhere" (pp. 155-56; my italics).

On the clash with Christianity

- "For the Christian, the morality of sexual love rests on his status as a supernatural creature with a supernatural end. Whatever therefore in that love tends to avert him from that end which is eternal beatitude in the contemplation of God is immoral: whatever in that love tends to further him toward that end is moral and good. For the courtly lover, the morality of sexual love rests on his status as a rational creature with a natural end. Whatever tends to avert him from that end which is progress towards natural worth and virtue is immoral; whatever tends to further him towards that end is moral and good. *The chasm that lies between Courtly Love and the Christian morality of sexual love is immediately apparent. No Christian teaching could ever give rise to such a conception of sexual love with all its implicit sexual desires and delights as the source of natural good and virtue.* Not even perverted and bad Christians could fail to recognize its sinfulness even though they practiced it. How, then, could men as troubadours conceive of love as the source of good which as Christians they knew to be sinful and immoral?" (ib., p. 156; my italics).

On reconciliation through the doctrine of the double truth

- "The only reasonable basis for the acceptance of Courtly Love with all its implications while remaining Christians is the doctrine of the so-called 'double-truth', itself a product of the conflict between philosophy and theology among the Arab philosophers" (ib., p. 156). Basically, this doctrine sees religion and philosophy as separate sources of knowledge giving from time to time contradictory truths without detriment to either. "Therefore the philosopher can and does speak otherwise than as a Christian: he accepts as necessary the results of philosophic speculations but as a Christian he believes as true what Revelation teaches" (ib., p. 156).

b. The clash between Orthodoxy and the Double Truth, and the condemnation of Andreas Capellanus as heretic

- "[The theory of the double truth] is heretical and its adherents, the Latin Averroists,[225] were condemned at Paris in 1277 on the grounds that they held 'those things may be true according to philosophy but not according to faith just as if they were two contradictory truths'. It is significant that the *De Amore libri tres* of Andreas Capellanus, the text-book of Courtly Love, was condemned at the same time and on exactly the same ground.... The troubadours and Andreas Capellanus must have been acquainted with the attitude of mind engendered among the Arabian philosophers when confronted with the problem of how to think as philosophers and how to believe as followers of Mahomet" (ib., p. 156).

---

[225] "The theory of the double truth was definitely not formulated by Averroës, but rather by the Latin Averroists" (Rosenthal, 1978, p. 539).

"On rational grounds Andreas proved in the *De Amore*[226] that sexual love is the most desirable thing in this world, that it is the source of all that is good and noble here below. He taught Walter that by his very nature man is compelled to love and that he must do so if he would rise in natural virtue and attain to the natural felicity of this world. In the *De Reprobatione* Andreas repudiated this teaching entirely and proved—significantly enough in regard to just those parts which would later form the matter of propositions to be condemned in 1277—from divine authority that such love is sinful before God and before man, that it is the source of all criminal excesses....

Thus, there is built up an opposition between the *De Amore* and the *De Reprobatione*, between the rational, natural teaching of the former and the theological, supernatural teaching of the latter. In them is seen to be erected an opposition between nature and grace, between reason and faith, between philosophy and theology. What Andreas teaches to be true according to nature and reason, he teaches to be false according to grace and to divine authority. Thus emerges in his book the doctrine of the so-called double truth wherein contradictory propositions are held to be true at the same time. It is not surprising, therefore, that Stephen Tempier included *nominatim* [by name] the work of Andreas in his condemnation of those 'who say that those things may be true according to philosophy but not according to Faith as if there were two contrary truths and as if the truth that lies in the sayings of the accursed gentiles were against the truth of Holy Scripture'" (Denomy, 1946, pp. 148-49).

c. The high internal DP of *The Art of Courtly Love*

What is fascinating, in terms of the GAM/DP theory on one side, and of the development of courtly love and its literature on the other, is the high dialectical DP which exists between the first two books on *De Amore*, of Andreas Capellanus's *The Art of Courtly Love*, and the third book on *De Reprobatione Amoris*. With the text by Capellanus, the DP does not take months or years to become evident, as was the case for the debates and fights of the *Investiture Controversy*. With Capellanus,

---

226 To be clear in what follows, and basically in line with Denomy, I call *De Amore libri tres*—or *The Art of Courtly Love*—the whole text, *De Amore* (On love) the first two books, and *De Reprobatione Amoris*—or only *De Reprobatione*—(On the reprobation (or rejection) of love) the third and last book.

the DP exists, and is very obvious, within the same book, a relatively small one of little more than two-hundred pages. Correspondingly, Bishop Stephen Tempier was not the only one who was faced with this double truth: every reader had the same problem. And every reader had to make up his/her mind: this or that truth, both, none, when and how? How? William Anderson, in *Dante the Maker,* gave us his interpretation:

- "The world-wise tone of Ovid imbues the most important treatise on love associated with the ideal of amour courtois, Andrew the Chaplain's *De amore*. In this late-twelfth-century work, the author instructs a young nobleman called Walter in the ways of love, with the ultimate object of warning him off it. *Few readers paid attention to its final pages,* so gripped were they by the witty examples of speeches by lovers of different and equal ranks and by the way it presented a world of rules and behaviour, not so much at variance with the world of conventional Christian morals as existing in another continuum from it. Andrew, through one of his speakers, even describes an afterworld in which ladies receive rewards, corrections, or torture according to the measure in which they showed kindness to their lovers in life. This book was translated into Tuscan and was known in Dante's Florence where it exercised a strong influence on lovers and poets, especially in its insistence on discretion and secrecy: an influence to be detected in Dante's use of the ladies who were "a screen of the truth" to disguise his love for Beatrice in the *Vita Nuova*" (1980, p. 97; my italics).

What Anderson just said, that "few readers paid attention to [Capellanus's] final pages" is interesting, because while a *fox-trapper* like Boccaccio must have been one of the many who did not pay attention to those final pages, a *hedgehog-universalist* like Dante, most probably, read those final pages with much attention.

Accordingly, while Anderson may be correct in relating Capellanus' insistence on discretion to Dante's use of "a screen of truth," Dante so completely purified the idea of discretion as to make it an example of the highest and spiritual love.

Phase 3 – Early European

In Phase III the works of the authors of the previous two phases become better known and influential, as can be seen by the following on Boccaccio and Chaucer. In his discussion of the literary background of the young Boccaccio, Thomas Bergin wrote: "As one might expect, Ovid is all-pervasive, but there are recognizable echoes of Virgil, Statius, Lucan, and most of the Latin authors that circulated in the late Middle Ages.... [Also] undoubtedly Boccaccio was familiar with both the old Provençal *partimen*, in which love problems were debated in verse, and the illustrative dialogues set forth in the essay on love of Andreas Capellanus" (Bergin, 1981, pp. 86-87).

Chaucer himself, wrote in *Troilus and Criseyde*:

> Go little book, my little tragedy,
> May God grant thy maker, before he dies,
> The power to compose a comedy!
> But, little book do not envy
> Any other poem; but be subject, as am I,
> To poesy itself, and kiss the footsteps
> of Vergil, Ovid, Homer, Lucan and Statius.
> (Book V, lines 1786-1799).

Two other sets of footsteps might also have deserved a kiss: those of Boccaccio and of Andreas Capellanus.

As to the first, the story of *Troilus and Criseyde* is based on Boccacio's *Il Filostrato* even if the poem by Chaucer has more individual characterization and humor. The Trojan prince Troilus falls hopelessly in love with Criseyde who gives him permission to serve her in the courtly love tradition. "Chaucer makes his Troilus more innocent, more courtly than Boccaccio's Troilo. He makes Criseyde, on whom he lavishes all his art and tenderness, more delicate, timid, loving and vulnerable than Boccaccio's heroine, and without her ready wantonness" (Brewer, 1973, p. 350).

"Some critics consider *Troilus and Criseyde* Chaucer's finest work, greater even than the far more widely read *Canterbury Tales*.... At the end of the poem, when Troilus' soul rises into heavens, the folly of complete immersion in sexual love is viewed in relation to the eternal love of God" (Lumiansky, 1978, p. 64). This last comment brings to mind the third book by Capellanus on *De Reprobatione Amoris*, as discussed by Elaine Tuttle Hansen in *Chaucer and the Fictions of Gender*:

- "In the third book of *The Art of Courtly Loving,* Andreas Capellanus explains why it is useful for noblemen like Troilus to learn how to love women. Andreas instructs his male readers: 'Read this little book, then, not as one seeking to take up the life of a lover, but that, invigorated by the theory and trained to excite the minds of women to love you may, by refraining from so doing win an eternal recompense and thereby deserve a greater reward from God.' And Andreas concludes: 'For God is more pleased with a man who is able to sin and does not, than with a man who has no opportunity to sin.' This passage suggests quite clearly how the apparently conflicting points of view in *The Art of Courtly Loving* seriously cohere, and why Andreas's interest in courtly rules does not sit as oddly with his antifeminism as scholars have often thought; it also reads like a gloss on Troilus's experience, in Chaucer's telling. 'Invigorated by the theory' [*eius doctrina refectus]* of Love at the very beginning of the poem, the hero is then 'trained [by teacher Pandarus] to excite' the mind of Criseyde to love in Books II and III, and Troilus learns his lesson well. But his fidelity to love, however praiseworthy, cannot in itself serve to designate proper manliness or moral worth; it is, as Andreas says, indulging so far and then refraining from what is so delightful that matters" (pp. 180-81).

6. Conclusion

Many different forms of division of power contributed to the origin and devolopment of courtly love. Each was a conflict between two or more opposing entities, persons, and cultural scripts. The direct and indirect participants were confronted with a clash of ideas whose sparks stimulated new ideas first, then—through quality elaboration—new creative works of the highest order.

There was the division of power between nearby lords competing for supremacy, for influence, for display of wealth and culture, for having the best troubadours at their court. There was the division of power between lay and religious rulers as shown by the excommunication of Duke William IX by his local bishop. Also, "The intrusion of churchmen into conjugal matters caused resentment on the part of the husbands. William of Aquitaine.... makes fun of women who fall under the sway of priests and monks.... I interpret [this poem by him] not as a precursor of the chivalric debates between priest and knight that were to be popular a hundred years later, but as an angry expression of the animosity husbands felt against spiritual directors who challenged their power and encouraged their wives to be frigid" (Duby, 1983, p. 159). We encounter here another form of DP in the triangle of *man/husband-wife-confessor* in which the confessor gives power to the woman by the sheer act of refusing herself when she (and the confessor) felt that that was the proper way to behave as a woman and as a Christian. Deliberate frigidity gave woman power the way the celibacy of the clergy gave power to the clergy and the church, as discussed in my volume 3.

Then, there was the division of power between two Christian religions: Catholics and Cathars, and there was the overall religious and cultural division of power between Christianity and Islam, between Provence and Islamic Spain at a time when Islam was by far superior in nearly all domains.

Only so many forms of division of power and derived intellectual freedom can explain "the letter sent back by the Countess of Champagne on May 1, 1174" reported by Capellanus in the Seventh Dialogue of Book 1:

> We declare and we hold as firmly established that love cannot exert its power between two people who are married to each other. For lovers give each other everything freely, under no compulsion of necessity, but married people are in duty bound to give in to each other's desire and deny themselves to each other in nothing. (1959, pp. 106-07)

A direct consequence of this ruling is found, ten pages later, when a man of the higher nobility speaks with a woman of the same class:

> I admit that I have a wife who is beautiful enough and I do indeed feel such affection for her as a husband can. But since I know that there can be no love between husband and wife (the decision of the Countess of Champagne supports this statement), and that there can be nothing good done in this life unless it grows out of love, I am naturally compelled to seek for love outside the bonds of wedlock. (ib., p. 116)

Here is the new thesis (shortly thereafter contradicted by apellanus himself with the antithesis of his third volume) which will find its answer when marriages are no longert arranged for dynastic or monetary interests, when Romeo and Juliet will be allowed to marry, when romanticism will be an important ingredient at the time of courtship and after marriage. Christopher Brooke had noted this dialectic from the point of view of the historian of marriage:

> The knightly adventures dwell in a world of fantasy. They are commonly performed to do honour to a lady—and even when they are not, it is often implied that they ought to be, or that the audience will expect them to be.... We recognize that there is much of literary tradition and of fashion in them. Yet we cannot evade the power of the sentiments they express, or imagine them wholly without influence. Real life is full of paradox, and the romances reveal some of the paradoxes of medieval society with a good deal of force. But there is more to it than that, for the romances were used by their authors—and especially by the more intelligent and reflective of the poets—to search out the paradoxes and the meaning of the strange world their characters inhabit. (1991, p. 179)

Paradoxes and a strange world, this is DP; normality and conventionality are either UP or its cousin.

7. Appendix – On the Magnificent Results of the DP Revolution of the 11th and 12th Centuries

Already in volumes 1 and 3, I noted that I was not alone in seeing a fundamental change in the ethnopsychology of Western Europe at the time of the DP *War of Supremacy* of 1075-1312 between emperors and popes, and I paid hommage to Colin Morris, Alfred Crosby, Carlo Cipolla, and Harold Berman. To these names I am now pleased to add that of Denis de Rougemont, who in his chapter on "Love" in the *Dictionary of the History of Ideas* said that:

• "With the twelfth century came a complete change. As the witty sally of the French historian, Charles Seignobos, put it: 'Love is an invention of the twelfth century!' Love, which for us denotes sentimental feeling or passion, took on this meaning only with the poetry of the troubadours, written and sung, which appeared suddenly in southwest France (Poitou, Limousin, and Languedoc), and spread over the whole continent with surprising rapidity. This love resembled nothing that the ancient or Christianized world knew; it seemed to fall from the sky. La *cortezia*, with its fixed and refined forms and its absolutely novel doctrine, could not possibly have been only the more or less accidental discovery of a few pious musical members of the church at Saint-Martial de Limoges, and/or minstrels with little education, which is the 'prudent' thesis of most nineteenth- and twentieth-century specialists of the *Trobar*. But if they are right, how could this poetry have conceivably transformed our ways of feeling, our customs, and our arts for centuries? Would it not be, on the contrary, the sign of a more general revolution operative in the Western mind at this time?...Since the beginning of the eleventh century, heterodox religious movements proliferated in Italy, the German Rhineland, Flanders, the north and then the center and south of France—Reims, Orléans, Poitou, Périgord, Aquitaine—all these movements being more or less associated with Manichaeism. They confronted the Church with a purified spirituality; all condemned marriage—which Pope Gregory VII had just forbidden to priests—and all declared the soul to be divine, and judged the body to be so vile that nothing it did could be conducive to well-being" (1973, pp. 97-98).

With Rougemont, we are back to the ferments which preceded the *Investiture Controversy* of 1075-1122. Among these we must include both the Patarine movement of Milan of the first half of the 11th century, and the writings of the moving spirit of those time: Humbert of Moyenmoutier, Cardinal Bishop of Silva Candida, especially his treaties *Against the Simonists* of ca. 1058.

But why am I always stressing the humiliation of Emperor Henry IV at Canossa at the hands of Pope Gregory VII as the beginning of Western Civilization, and not all that preceded and prepared it? Because all those ferments would have come to nothing without Canossa. The history of every nation, of every civilization is full of similar ferments, of similar promising beginnings, and yet after fifty or one hundred years they disappear crushed by the unity of power.

So the problem is less one of birth but of survival[227] and then of expansion. And the same is true for the courtly love which might easily have disappeared leaving few traces behind, if the DP had not continued. And where that DP did not exist, as in Germany, courtly love did not take hold and was soon replaced by *Minne*, as discussed in the next chapter.

---

[227] With pride, Sánchez-Albornoz, told his readers of the Magna Carta, granted by King Alfonso IX of Leon in 1188, a Spanish Magna Carta older than the English of 1215 (1991, pp. 795-798). Yet, this magnificent precedent did not last, mainly because the Church was not involved in its making and was not its major beneficiary, as was the case with the English Magna Carta, as discussed in chapters 11 and 23 of volume 1.

19

The *Ritter*'s Reaction to Courtly Love

This chapter is divided into the following sections:

1. Introduction
2. *Minne* and *Minnesinger*
3. The German Criticism of the French Model of Service to the Lady
    A. Wolfram von Eschenbach
    B. Walther von der Vogelweide
    C. The Author of the *Nibelungenlied*
    D. Gottfried von Strassburg
    E. In the End, is Gottfried's *Tristan* Minne?
4. The Retrograde Courtly Love of Castiglione's Courtier
5. In the Opposite Direction: The DP *Minne Omega* in Dante's *Vita Nuova*
6. *Minne/Kultur* versus *Amour Courtois/Zivilisation*
7. Conclusion
8. Neidhart von Reuental's Courtyardly Love

1. Introduction

> The literature of Germany in the High Middle Ages is largely a reaction to contemporary literature in France.
> (Jackson, 1968, p. 55)

The world of the poetry of courtly love is divided into two provinces: that of *amour courtois* (French, Italian, English) and that of *Minne* (German). In turn the German province of *Minne* can be divided into two parts: *Minne A*, Lady oriented, along the French model, and *Minne B*, realistic, girl/wench/woman oriented. A notable feature of *Minne B* is its strongly critical attitude toward the French *amour courtois* of high praise for, and full dedication to, the Lady.

Basically, the poetry of *Minne B* is *Ritterminne*, i.e. the poetry which sings the loves of the *Ritter* shaped by the partial UP of his lord. This love for a girl/wench/woman is much different (more conventional) from that of the *amour courtois* of the *troubadour* or the *chevalier* shaped by the high DP discussed in the preceding chapter.

2. *Minne* and *Minnesinger*

In German, courtly love is called *Minne*, and *Minnesinger* is "the generic name by which the German lyric poets of the 12th and 13th centuries are known. Strictly speaking, and in the usage of Middle High German poets, the term *Minnesang* denotes only songs of love (*Minne*) written in the service of the ideal of *amour courtois*, or courtly love [Minne A], but it has come to be applied to the whole body of lyric poetry of the period, *Spruchdichtung* (political, moral and religious poetry) as well as *Minnelyrik*. The songs of *amour courtois*, like the concept itself and the court culture from which it sprang, came to Germany either direct from Provence or through northern France"[228] (Taylor, 1973, p. 533).

This description by Ronald Taylor is correct when referring to what I call *Minne A*; for instance to the courtly love of the poems of Reinmar von Hagenau, as discussed hereafter.

---

[228] "The lyric poets of Provence inspired those of Germany almost as soon as they inspired those of northern France, with the result that the latter exerted less influence in Germany" (Jones, 1968, pp. 16-17). "It [*Minne A*] is a reaction, sometimes little more than a feeble reworking but more often an attempt to express the thought and style of French literature in new terms" (Jackson, 1968, p. 55).

However, there is a more Germanic *Minne*, more Ritter: the *Minne B* of the poems of Walther von der Vogelweide, Wolfram von Eschenbach, Gottfried von Strassburg, and the anonymous author of the *Nibelungenlied.* While *Minne A* still pertains to the DP world discussed in the previous chapter, *Minne B* is a product of the partial UP-Ritter world discussed in chapter 25 of volume 1.

3. German Criticism of the French Model of Service to the Lady

A. Wolfram von Eschenbach

In the words of W. T. H. Jackson: "It is clear that German authors regarded the love they found in the works of their French models as characterized by love-service.... *This element of service, the most telling evidence of the influence of feudal society on the love concept, was regarded as a thoroughly bad characteristic by the greatest of the German authors. It is in this sense that they recognized 'courtly love' and in this sense that they rejected it.* There is no German version of Chrétien's *Lancelot.* All the poetry of these authors is a reaction to courtly love, and, in the best authors, it is a rejection. *The ideal of love based on service and admiration of specific qualities is to be replaced by mutual attraction. Each in his own way, Wolfram [von Eschenbach], Gottfried [von Strassburg], and Walther [von der Vogelweide] told of a love that was shared and in turning his back on the concept of service opened new ways for the consideration of the love phenomenon*" (1968, pp. 75-76; my italics).

"Wolfram's criticism of the idea of love-service is quite explicit. He has no time for the kind of man who overpraises his lady and regards her as the only one worthy of consideration. He does not consider that it is a knight's duty to maintain such a belief, nor should the lady expect it. If she wants him, she should do so on the basis of his knightly conduct in general rather than his service to her in particular. *Thus Wolfram clearly states that service to the lady is not the most important part of knightly duty and that knightly combat is not intended for the winning of ladies but for higher objectives*" (ib., p. 63; my italics).[229] But what can these higher objectives be? For a ritter this means probably serving his lord, and, time permitting, the emperor and the church.

---

[229] Wolfram von Eschenbach is the author of the poem *Parzival*, written between 1200 and 1210, based on an unfinished romance of Chrétien de Troyes. Wolfram uses Parzival's dramatic progress from folk-tale dunce to wise and responsible keeper of the Grail to present a subtle allegory of man's spiritual education and development.

"Thus Wolfram makes clear his criticism of the ideas of love which he had received from his sources. He rejects the two basic constituents of courtly love, the right of a woman to demand unquestioning service in whatever triviality she fancies, and the need for a knight to gain honor for his lady by embarking on a series of otherwise pointless adventures, adventures which often result in the death of innocent persons and lifelong grief to others with whom the knight has no personal quarrel, nor even grounds for quarreling" (ib., p. 65).

B. Walther von der Vogelweide

"[Into the *Minnesang*, Walther] introduced a new combative note and a strain of live criticism. The Troubadour ideal is challenged. '*Minne ist zweier herzen wunne*,' — 'Love is the joy of two hearts' — not melancholy longing but gay fulfillment of a relation in which man and woman share equally, and there is no need for the man's one-sided homage. Sincerity of affection is made the touchstone for both. Judged by this, the lady of high rank is too often found wanting. *The poet turns from her to the girl of low degree in whom he discovers his own ideal of unspoiled womanly goodness and charm*" (Richey, 1943/ 1969, p. 5; my italics).

The trouble with this approach is that it eliminated the revolutionary Provençal triangle of troubadour-lady-lord, and replaced it with the old and universal couple man-woman, be it boy-girl (Romeo and Juliet), or man-girl (Faust-Margareta, Don Juan-Tisbea, DonJuan-Aminta,[230] Don Fernando-Dorotea).

Also, while the triangle troubadour-lady-lord was public for the most part (the troubadour singing poems in front of the lady and the court, and possibly the lord too when he was not away on Crusade), *the couple man-woman* of Walther was private.

In *amour courtois*, troubadour, lady, lord, and the court members exerted a mutual control on what was said and done to be sure that everything was in accordance with known rules, well discussed and formalized. Instead, *the couple*s of Walther love each other in private until either marriage, or unwanted pregnancy, or betrayal and abandonment.

---

[230] The two girls of low social standing (the first a fisher woman, the second a peasant) with unspoiled womanly goodness and charm of Tirso de Molina's *Don Juan*.

Obviously, the German lord did not object to the *Minne B* of his ritter: either *Minne B* of regular love followed by marriage *(Minne B of marriage)*, or the *Minne B* of Faust or Don Juan *(Minne B of seduction and betrayal)*. The lord probably amply practiced *Minne B of seduction and betrayal*, less *Minne B of marriage* because of the necessary practice of arranged marriages. The German lord had no interest in *amour courtois*, in giving power to one or more troubadours, and in empowering his lady.

Wolfram von Eschenbach, Gottfried von Strassburg, and Walther von der Vogelweide wrote as they did because they were part of the *ritter system* shaped by the partial unity of power of the German lords and the semi-Caesaropapism of the German emperors, as discussed in chapter 25 of volume 1. Within this world, they wrote several masterworks, preferably read within the context of the political environment that fostered them.[231]

For them, as *ritter*, the French *amour courtois* (or the German *Minne A*) was meaningless, worse: it was plain wrong. They lived under the UP of the German lords and not under the several DPs (political, religious, cultural, detour abroad) which prevailed first in Provence, then in Champagne, in post-Legnano Florence, and post-Runnymede London.

---

[231] This is particularly true for the many political, and very *ritter*, songs of Walther von der Vogelweide, first written for Emperor Philip, then for Emperor Otto against the Pope, and finally for Emperor Frederick and Archbishop Engelbert (see Jones, 1968, pp. 92-117). "At first glance, Walther's shifting political loyalty suggests a vacillating character at best or a rank opportunist at worst; and his latterday admirers have spilt much ink in explaining how Philip's ardent champion could turn lukewarm toward him, embrace his opponent Otto, and then desert Otto to serve Frederick II as soon as the latter appeared to be in the ascendancy....[However] the chain-of-command now prevalent in our military service was the basis of all political morality in the Middle Ages. If Walther was in the service of a territorial prince, he owed his loyalty solely to that prince, and it was no concern of his whether the prince was in the right or wrong. His was not to reason why; his was to serve his prince, right or wrong.... Walther's only duty was to whichever prince was retaining him at any given time, not to the emperor or to the Empire or even to the German nation as a whole" (Jones, 1968, pp. 113-115). "In battle a vassal was obliged to serve the lord who had nourished him in time of peace (the word 'lord' is derived from 'loaf-ward'), and medieval literature stressed the resulting obligation. The same relationship held between patron and poet, and no irony was intended in the minstrels' motto "Whose bread I eat, his song I sing" *(Swes brot ich ezz, des liet ich sing)*. Walther's duty was to the patron who supported him, a fact that explains how he could ask Otto to pardon the traitor Dietrich of Meissen even though he usually demanded punishment of traitors who disturbed the peace. To be sure, a minstrel was free to seek out a patron whose policies he endorsed, but this was not demanded" (Jones, 1989, p. 539). Very UP, indeed.

This development in the history of German literature moved the Minnesinger down from the castle to the village: "While absorbing much of Reinmar's formal craftsmanship, Walther von der Vogelweide, the greatest lyric poet of the German middle ages, went far beyond the artificial convention by which the *Minnesang* [type A] had been governed, and introduced an element of practical realism, both in his love poetry and in his political, moral and religious *Sprüche*, or gnomic poems, which in the course of the 13th century was to lead ever further away from the world of courtly ideals. By the time of Neidhart von Reuental, a Bavarian squire (d. c. 1250), the knight had turned his attention from the ladies of the castle to the wenches of the village" (Taylor, 1973, p. 534).[232]

In *Minne B*, there is no way to avoid the limitations imposed by the intrinsic simplicity of the chosen female partners: the wenches of the village, or the girl of low degree in whom [the poet] discovers his own ideal of unspoiled womanly goodness and charm. Poor Margareta, how quickly she fell to the charming words of Faust (as did Zerlina to those of Don Giovanni, and both Tisbea and Aminta did to the words of Don Juan).

Walther von der Vogelweide and Neidhart von Reuental may have had the purest of intentions, however, when the *ritter* decided that the courtly love was ruled by artificial conventions, and had to be overcome with the wenches of the village and no longer with well educated and refined ladies, much of importance was lost. Woman was again "natural woman", soon abused and betrayed because of her simplicity, and man was again "natural man", soon guided by his instints and passion, and love for variety.

Also, worth remembering, with *amour courtois* and *Minne B* we are speaking of two different sets of vertical relationship: the first is *upward* to a lady of superior social standing (the wife of the ruler of the castle and surrounding domain) who has to be courted carefully, half in hiding, with much poetry, music, and well educated references; in the second is *downward* "to the girl of low degree in whom [Walter von der

---

[232] On the other hand, those "artificial conventions which governed courtly love" were not so "artificial" south and west of Germany. In Italy for instance, Dante, Petrarch and Boccaccio went beyond those conventions, not by moving down to the wenches of the village, but up to idealized women like Beatrice and Laura, or up to intelligent and daring women like Pampinea, madonna Filippa, monna Giovanna, and the Marchioness of Monferrato of Boccaccio.

Vogelweide] discovers his own ideal of unspoiled womanly goodness and charm" (Richey, 1969, p. 5), a goodness and charm soon to be spoiled.

As reported in the previous chapter: "The essential characteristics that set aside the courtly conception of sexual love and distinguish it from that of previous ages is the cult of the beloved on the one hand—the elevation of the lady loved far beyond the level of ordinary mortals in physical, moral and intellectual worth—and secondly, the ennobling force of that love whereby the lover grows in virtue, worth and esteem" (Denomy, 1946, p. 133). Similarly, as remarked by C. S. Lewis, "our code of etiquette, with its rule that women always have precedence, is a legacy from courtly love, and is felt to be far from natural in modern Japan or India.... French poets, in the eleventh century, discovered or invented, or were the first to express, that romantic species of passion which English poets were still writing about in the nineteenth. They affected a change which has left no corner of our ethics, our imagination, or our daily life untouched, and they erected impassable barriers between us and the classical past or the Oriental present. Compared with this revolution the Renaissance is a mere ripple on the surface of literature" (1937, pp. 3-4).

Can the same be said for *Minne B*? In this, should we criticize Walther and Neidhart and the other poets of *Minne B*? No, as we should not criticize Ariosto for having been a *cortigiano* (as discussed in chapter 3 of volume 2). The German poets of those times were no less under the pressure of their lords than Ariosto was under that of the Dukes of Este. If there is criticism, it should be levelled at the unity of power of the German lords, not at their poets.

C. The Author of the *Nibelungenlied*

The *Nibelungenlied*, a Middle High German epic poem written by an unknown Austrian from the Danube region (c. 1200), is even more critical of the service rendered to a lady. For him too, service is to be rendered to one's lord, and in the lord's real interest, and not to win the lady of one's dreams: otherwise things will go wrong. Indeed, "there can be little doubt that the model of knight-errant offered by the romances was threatening to those who already held power. The knight-errant represented social chaos, a return to the law of the jungle that had been partially limited by the ideas of stability through established power and hereditary succession. The notion that one's kingship was always on the line was not conducive to stability and order. Siegfried himself is the beneficiary of the established law in his own land, but he refuses to use the mechanisms for bride-winning offered by his position and follows instead the model of the chaotic forces of knight-errantry. Any doubt we might have that Siegfried is a representative of the new-fangled literary chivalry is quickly dispersed by his subsequent adoption of the role of courtly lover....

It is obvious from what follows that the ethical models provided by the chivalric literature of the type represented by Hartmann [von Aue; fl. 1190-1210] and the *Minnesänger* did not meet with the approval of the poet of the *Nibelungenlied*.....

The suffering lover caught up in Reinmar-*minne*[233] has but one outlet for his love: *Minnedienst*, the service of the lady. Service and love are inextricably intertwined in the symbolic language of both the lyric and the romance. From the earliest *Minnesang* on we find language denoting feudal service and vassalage used to describe the relationship between the lover and his lady. The lover is 'untertan', 'subject' to his lady, she his 'vrouwe', 'lady' (in the sense of feudal superiority) and so on.

Siegfried needs an opportunity to render service to Kriemhild and this opportunity presents itself in the form of the Saxon war. Siegfried offers his service and it is gratefully accepted.... He renders the feudal obligation of *consilium* and, although his advice is overruled, he is ready to render *auxilium* if his 'lord' is willing to provide the appropriate payment. The payment for Siegfried's service is, of course, Kriemhild's hand in marriage.

The entire Isenstein episode seems to have been put together to demonstrate Siegfried's new subordinate status. Upon their arrival in Isenstein, *Siegfried performs the highly symbolic act of holding Gunther's stirrup for him to mount. In the year 1077 Emperor Henry IV had demonstrated his subjugation to Pope Gregory VII in Canossa with just this symbolic gesture. In spite of the fact that Henry, like Siegfried, had ulterior motives, the scene has remained through the centuries and up to the present day a symbol of humilia-*

---

233 The following, quoted by Jaeger, is an example of *Reinmar-minne* (Minne A in my classification) by Reinmar of Hagenau himself:

"I wish to be known, my entire life, as a master of one thing and one thing only; I seek the world's praise for this one skill, that no man can bear his suffering as beautifully as I. If a woman causes me pain to such an extent that I cannot remain silent night or day, I have so gentle a spirit that I'll accept her hate as a source of joy. And yet, alas, how deeply that discomforts me!" (1985, p. 150). Beautiful, but definitely not *ritter*!

tion. Siegfried feels that he is merely playing a role to gain a much-desired goal, but Gunther cannot resist feeling pride:

> 'The beautiful ladies saw it all through the windows,
> Gunther felt himself made worthier by all of this.'

Gunther's feeling is especially justified by the presence of the ladies, the ultimate judges of one's chivalric worth" (Haymes, 1986, pp. 50-51, 54-55; my italics).

As discussed above by Haymes, Siegfried refused to use the mechanisms for bride-winning offered by his position and follows instead the French model of the chaotic forces of knight-errantry, and then adopted "the role of courtly lover."

D. Gottfried von Strassburg

The Tristan of Gottfried von Strassburg is definitely not an errand knight doing great deeds in honor of his Lady, but—at first, and for quite a while—a courtier in the service of his king:

"A major link joining the world of court clerics with that of the knight of romance is the romance of *Tristan and Isolde* [ca. 1210] by Gottfried von Strassburg.... In this romance clerical, courtier values provide the basis for the hero's character and destiny and far outweigh in their importance chivalric ones. The talents in the foreground are social, artistic, and intellectual. The hero makes his way in the early parts of the work by his eloquence, his learning, his skill in music, and his knowledge of languages, and hunting customs and the French terms for them....

He puts forward his talents with the calculating modesty that Castiglione was to call *sprezzatura*, defusing envy and producing two-fold amazement in the court. He swiftly rises to become the favorite and, finally, the heir of the king. And all this is accomplished before a single tournament is held, before a single battle is fought. These are accomplishments of courtiership, not of knighthood. Gottfried refers to his young hero as '*ein lieber hoveman*' ('a beloved courtier' - 3487).... We are in the secularized and chivalrified atmosphere of the youth of a courtier bishop. The parallel becomes especially clear if we compare the court service of young Tristan with that of [Bishop] Otto von Bamberg [who lived from ca. 1060 to 1139, and served at the court of Emperor Henry IV]" (Jaeger, 1985, pp. 102-03).

Hereafter is one of the several comparisons made by Jaeger between Tristan and Otto:

> 'For at all times he [Tristan] was at [King Mark's] side, as it behooves a courtier, and offered him his service whenever it was called for. Wherever Mark was, wherever he went, Tristan was there like his second self. (*Tristan*, 3399)'
>
> 'When other chaplains were intent on other things, Otto was always at hand... morning and evening and at all times he stood at the emperor's side, psalter in hand, ready to serve. (Herbord 3.34, p. 201)'

"These parallels are not accounted for by literary influence but by the common sociological ground from which the early careers of Tristan and Otto arose: the ideals, ambitions, and representative experiences of clerical court servants" (Jaeger, 1985, p. 104).

Moving to Isolde, "Gottfried's presentation of his hero and heroine would first of all change the significance of the lady.... Gottfried's conception of the feminine element of the love relationship is entirely different from that found in the courtly romance. He rejects totally the idea of love service and the subordination of the man to the woman and substitutes for it a partnership which is based on sexual attraction and recognition in the other of the fulfillment of a need" (Jackson, 1971, p. 64).

But have we become more civilized by rejecting the idea of love service and substituting for it uncontrolled sexual attraction independent of any past and present social obligations?

The answer by Jaeger—in *Medieval Humanism in Gottfried von Strassburg's Tristan und Isolde*—is a clear "No": "Anyone interested in the representation of the individual in Gottfried's poem will be slightly disappointed in the Tristan who emerges after the love potion. He is no longer complex: the possible forms this protean figure can take on seem suddenly limited; his existence unrolls under the aegis of a single force, *minne*; complexities give way to complications.... *It is regrettable that all of this promise funnels into the wearying and destructive love affair; at least from the point of view of the humanist, who professes a philosophy of the perfectibility of man, it is regrettable*" (1977, p. 38; my italics).

Did Isolde gain in self development, self actualization, and valid personal pride? No.

Moving then to the author, as seen by Jackson (1971): "It is only a slight exaggeration to say that Gottfried begins his study of courtly love where the authors of most romances end it. Sexual fulfillment is as important for woman as for the man in Gottfried's view" (1971, p. 64-65). However is sexual fulfillment the *summum bonum* to which everything else must be sacrificed? Does Gottfried hold so low an opinion of woman that he cannot offer Isolde a richer and more civilized life than only sex?

E. In the End, is Gottfried's *Tristan* Minne?

"Of the two great forms of literature at the beginning of the thirteenth century, *minnesong* had as its basis the love of a man for a woman who was either his superior in life and, for that reason, unapproachable, married to another or, for some other reason, unattainable. In the most extreme case, a man could be in love with a woman he had never seen....

In the *courtly epic*, the other great literary form of the period, there is emphasis on action rather than contemplation, and the knight, by his prowess, wins and marries the lady. The idea of love is here completely different, for the lady is won without wooing and the result is marriage, whereas in *minnesong* she is always wooed but never won and the result is normally rejection, certainly never more than a form of distant encouragement. The story of Tristan might, therefore, be considered as a classic *minnesong* situation: the knight in love with the wife of another, her station also being above his. But Gottfried cuts right through the conventional posturing and emphasizes that their love became complete only after their physical union, while the husband, who is normally ignored in *minnesong*, is portrayed in detail as ignoble and essentially guilty. The relationship which is compounded by the epic situation of Tristan having in fact risked his life to 'win' the lady, leads not to marriage or to courtly spiritual love but to 'courtly adultery'" (Batts, 1971, pp. 84-85).

But there are very important reasons not to consider *Tristan* a *minnesong*, because the world it describes has nothing to do with that of *amour courtois*, nothing to do with the world of *Minne* be it of type A or B. The world of Tristan, Isolde, and King Mark is the classic world of the UP, of *supremos* and their courtiers, and of young beautiful girls, married to old and ugly kings, who will soon find ways to fulfill their natural thirst for love and sexual fulfillment[234] (if not tightly controlled in a seraglio).

Indeed, *Tristan* "is not a typical Arthurian romance. This genre in its classical formulation by Chrétien de Troyes places a civilized knight and lover in Celtic fairy tale landscapes, where by dint of chivalric prowess, by constancy of mind and affection, he overcomes fantastic barriers set in place especially to test him. He then wins his lady love and her lands and lives a life pleasing to God and the world ever after. *Tristan* is quite different. The main theater of action, the showplace and testing ground of human talents, is the court, not chivalric combat. The hero is disposed by nature and destiny to suffering and sorrow. His fate is a tragic love affair which sets him irreconcilably at odds with his lord and uncle, and which spells the destruction of both lovers" (Jaeger, 1985, p. 102).

Michael Batts has similar remarks:

- "[Tristan] uses his accomplishments, first as a huntsman, then as a singer and a linguist, to impress the court and make himself welcome there. He gradually unfolds his talents to Mark, who enthusiastically exclaims: "You can do everything I want—hunting, languages, music".Of Tristan [Gottfried] says: "From that time on, Tristan was a courtier much beloved among them. The King and his household kept good company with him, and he too was so obliging to rich and poor alike, that, had it been possible to have done so, he would gladly have pampered them all".... In Cornwall his life has been that of the courtier, passing the time in the exercise of artistic and intellectual accomplishments—quite a different kind of life from that of his father, whose natural habitat had been the tourneying ground and the battlefield" (1971, pp. 49-50).

Clearly, the answer to the section's title, "In the end, is Gottfried's *Tristan Minne*?", is "no": *Tristan*'s is no *Minne* (neither A, B, or

---

[234] In every discussion of the poem (but not while reading it) it is advisable to forget the role played by the magic potion that by mistake linked Isolde to Tristan: "Gottfried does not use it to excuse this relationship which he so strongly affirms, nor is he concerned to demonstrate the unfortunate consequences of the blind choice of love. His affirmation of love does not make the potion superfluous; it is a symbol rather than a necessary adjunct of the plot" (Batts, 1971, p. 55).

Omega), it is a return to the great stories which are created, from time to time, at the fringes of the *insular* world where *supremos* rule and courtiers try to prosper.

4. The Retrograde Courtly Love of Castiglione's Courtier

What differentiates the *ritter* from the courtier of Baldesar Castiglione, is that the *ritter* is first of all a knight, a military man capable of fighting for his lord. If he can also sing, compose poems, and elegantly flatter his master, the better; but only to a point, because often these additional capabilities are those of a poor warrior. With the courtier, instead, these secondary talents take priority.

In turn, both *ritter* and courtier differ substantially from the errand knight and troubadour because, while *ritter* and courtier are creatures of the UP of their lords, the errand knight and the troubadour operate either alone or under the public DP of the troubadour/lady/lord triangle.

As to Castiglione, it is true that most of the third part of his *The Book of the Courtier* (1507-16) is devoted to the ladies–alternately praised and criticized in classical antifeminist fashion; however, the lover who accepted the sovereignty of his mistress and humbly attempted to render himself worthy of her by acting bravely and honorably in society and by performing whatever daring exploit she might command, is forgotten. The terminology of feudalism in which the poet declares himself the vassal and servant of the lady and addresses her as *midon* (my lord) is similarly forgotten. And the notion that the lover's position of servitude to his lady was a privilege: the source of all true virtue and nobility is also forgotten.

In Castiglione's maelstrom of comments on the virtues and sins of women, we encounter comments such as:

- "That women are imperfect creatures and therefore of less dignity than men and incapable of practicing the virtues practiced by men, I would certainly not claim now, for the worthiness of these ladies here would be enough to give me the lie; however, I do say that very learned men have written that since Nature always plans and aims at absolute perfection she would, if possible, constantly bring forth men; and when a woman is born this is a mistake or defect, and contrary to Nature's wishes" (1976, p. 217).

- "I shall say only that, as you know, it is the opinion of very learned men that man is as the form and woman as the matter, and therefore just as form is more perfect than matter, and indeed it gives it its being, so man is far more perfect than woman" (ib., p. 220).

- "I don't see how on this point you can deny that man's natural qualities make him more perfect than woman, since women are cold in temperament and men are hot. For warmth is far nobler and more perfect than cold, since it is active and productive; and as you know, the heavens shed warmth on the earth rather than coldness, which plays no part in the work of Nature. And so I believe that the coldness of women is the reason why they are coward and timid" (ib., p. 222).

Then, on the next page—as was to be expected—we get a new rendition of the *grand canard*: "These deeds had their beginning when the first woman, through her transgression, led man to sin, against God and left to the human race a heritage of death, travails and sorrows and all the miseries and calamities suffered in the world today" (ib., p. 223).

Not all of Castiglione's Part 3 is made of this stuff, but enough to cover with mud any high vision we might have had of the lady and of our duty to behave accordingly.

Once outside of the respectful *amour courtois* for the lady, we fall either into the troubled world of Isolde, Kriemhild and Brünhild, or more often into the down-to-earth world of the girl of low degree with her unspoiled—for a few days or months—womanly goodness and charm, or, worse, down into the ancient world of the woman who is intrinsically "coward and timid."

5. In the Opposite Direction: The DP *Minne Omega* in Dante's *Vita Nuova*

As noted by Charles Singleton, "it was inevitable that between two such divergent conceptions of love as troubadour love of woman and Christian love of God conflict should arise: inevitable because in his 'philosophy' the troubadour found no place for Christian love as such. Neither did Christian love make any room at all for troubadour love" (1958, p. 63).

The first attempt to find a solution was the *dolce stil nuovo* of the school of poets led by Guido Guinizelli of Bologna. Yet, "the so-called *donna angelicata,* the lady-made-angel, of the lyric poetry being written before and in Dante's time in Italy did not bring in herself the solution to the conflict between troubadour love and Christian love" (ib., p. 69).

In that long conflict between the two conceptions of love (troubadour and Christian), Dante's *Vita Nuova* of circa 1293 (nine years before his exile) "begins with troubadour love.... [But, in the end] the revelation is this: that love of Beatrice is something too large to be contained within the ideology of troubadour love. Love of Beatrice reaches far beyond the powers of a God of Love....

For it is charity that bursts the narrow confines of troubadour love. It is the presence of charity, hidden in the beginning, which demands at a midpoint on the way of progression from love to charity that the God of Love be abolished. There is no room and no authority for a troubadour God of Love in the universe of Christian love. That universe is ruled by charity [i.e., by a love of the highest summit, God, and yet socially directed to our neighbors], and the only God who may inhabit it is a jealous God who is Himself charity. *Deus caritas est.*"[235] (ib., pp. 74-75).

"This scheme by which Beatrice holds a medial position, not only in the downward path of love from God to a man but in the upward and returning path of love from man to God as well, is the solution which Dante has brought to the conflict between love of woman and love of

---

[235] The direct translation of this *Deus caritas est* is "God is love." The original Latin of "*Charissimi, diligamus nos invicem: quia charitas ex Deo est. Et omnis qui diligit, ex Deo natus est, et cognoscit Deum. Qui non diligit, non novit Deum: quoniam Deus charitas est*" (I John, 4:7-8) became, in the *Authorized (King James) Version* of the *Biblia Vulgata* (and I now report more text): "Beloved, let us love one another: for love is of God; and every one that loveth is born of God, and knoweth God. He that loveth not knoweth not God; for God is love.... Beloved, if God so loved us, we ought also to love one another. No man hath seen God at any time. If we love one another, God dwelleth in us, and his love is perfected in us.... He that feareth is not made perfect in love. We love him, because he first loved us. *If a man say, I love God, and hateth his brother, he is a liar: for he that loveth not his brother whom he hath seen, how can he love God whom he hath not seen? And this commandment have we from him, That he who loveth God love his brother also*" (4:7-21; my italics to highlight the social nature of this love).

Moving to a modern formulation: "Charity is a theological virtue (along with faith and hope), the highest form of Christian love, whose originating source and ultimate end is God. Although charity in its principal sense is the love we direct towards God—a whole-hearted love awakened, sustained, and fortified by God's own prevenient love for us—it stands as well for the love of neighbor as ourselves, continuously informed and nourished by that love" (Wojda, 1995. p. 300).

God. Love of Beatrice is charity, which means that love of her begins in God and ends in God [by redirecting itself to our neighbors].[236] This love's upward way does not come to a stop at *domina* as did the love of the troubadours. It goes beyond to God, whence it came" (ib., p. 77).

I have entitled this section: "In the opposite direction: The DP *Minne* Omega in Dante's *Vita Nuova*," in which I placed a "DP" in front of "*Minne* Omega," to alert the reader that we are not only moving in the opposite direction of the ritter poetic interest in "the girl of low degree" or "the wenches of the village", but that we are moving away from the partial UP of the German lords toward the DP of communal Italy.

As highlighted by Marwin Becker in *Medieval Italy: Constraints and Creativity*, "The willingness to take chances, both materially and spiritually, was fundamental to the commercial revolution of towns in north and central Italy" (1981, p. 86), and part of those "spiritual chances" were the highest Minne of St. Francis and Dante. Both started as troubadours, but then created the highest form of Minne—Minne Omega, the first for Lady Poverty, the second for Beatrice.

6. *Minne/Kultur* versus *Amour Courtois/Civilisation*

As already quoted: "The literature of Germany in the High Middle Ages is largely a reaction to contemporary literature in France" (Jackson, 1968 p. 55), in turn the reaction against *amour courtois* by Wolfram von Eschenbach, Gottfried von Strassburg, Walther von der Vogelweide, and the author of the *Nibelungenlied* is the first major instance of the many rounds in the fight between the German *Kultur* and the French (and Italian and English) *Civilisation*. In this optic, *Minne B* was *Kultur*: unsophisticated, deeply felt, intimate, rooted in ancestral traditions; while *Amour courtois* was *civilisation:* formalized, over-sophisticated, artificial, superficial, prisoner of exaggerated conventions.

Later-on "the crucial period of divergence between France and Germany was the eighteenth century. At this time, the German intelligentsia and bourgeoisie rejected the aristocratic-courtly emphasis on exter-

[236] My addition to stress the Biblical message of "Thou shalt love thy neighbour as thyself" (*Exodus* 19:18, *Matthew* 19:19 and 22:39, *Mark* 12:31 and 33, *Romans* 13:9, *Galatians* 5:14, *James* 2:8).

nal refinement and emphasized the 'inner enrichment' of the personality" (Gordon, 1994, pp. 90-91).[237]

In his *Reflections of A Nonpolitical Man* of 1915-17, Thomas Mann took up the old distinction, praising the German *Kultur* so different from the French *civilisation*.[238] As summarized by Gordon: "Mann praised Germany's musical, metaphysical, and subjective 'culture.' contrasting it with the more technical, political, and external 'civilization' of France" (1994, p. 89).

And yet—referring to the studies of Norbert Elias and Pierre Bourdieu—Daniel Gordon correctly complained about this idea of a basic "antithesis between culture and civilization treated as if it 'sums up' the differences between the two nations, when it should actually be

---

237 Specifically in the words of Norbert Elias, the primary aspect of this divergence was at first a German internal social contrast "between the [German] courtly nobility, predominantly French-speaking and 'civilized' on the French model, and a German-speaking, middle-class stratum of intelligentsia recruited chiefly from the bourgeois 'servers of princes' or officials in the broadest sense, and occasionally also from the landed nobility. This latter is a stratum [to which, years after, also Thomas Mann belonged] far removed from political activity, scarcely thinking in political terms and only tentatively in national ones, whose legitimation consists primarily in its intellectual, scientific, or artistic *accomplishments.* Counterposed to it is an upper class which 'accomplishes' nothing in the sense in which the others do, but for which the shaping of its distinguished and distinctive *behavior* is central to its self-image and self-justification. And this is the class which Kant has in mind when he speaks of being 'civilized to the point where we are overburdened,' of mere 'social propriety and decency,' of 'the similitude of morality in the love of honor.' It is in the polemic of the stratum of German middle-class intelligentsia against the etiquette of the ruling courtly upper class that the conceptual contraposition of *Kultur* and *Zivilisation* originates in Germany" (1939/1978, pp. 8-9; italics in the original)

238 Here, we must remember that in German, for a long time, *zivilisiert* and *Zivilisation* referred to good manners, and *zivil* was the person who behaved correctly according to the best norms: "***zivil** Adj. 'nach den gesellschaftlichen Normen handeln und auftreten, gesittet, zivilisiert'(17. Jh.)* [civil. adjective, behaving according to the social norms, well-mannered, civilized" (*dtv, Etymologisches Wörterbuch des Deutschen, 2000,* p. 1620). Instead, for the French people, *civilisation*, is "*action de civiliser; résultat de cette action*," and "*civiliser*" is "*Faire sortir de l'état primitif; améliorer sous le raport moral, intellectuel*; *Civil, du latin civilis; de civis, citoyen*. [Civilization is the action and result of civilizing; and to civilize means overcoming the primitive condition; to improve in the moral and intellectual sense; civil, from the Latin *civilis*; from *civis*, citizen]" (*Grand Larousse Encyclopédique*, 1960, *3*, pp. 163-65). So there was often a certain element of confusion—at times, deliberate—in the way some German authors defined and criticized the French *civilisation* by playing on semantic unclarity, as if the French *civilisation* was at best nothing more than *Zivilisation*.

treated as a construct in German intellectual history" (ib., p. 91). However, more than a construct, I would speak of a syndrome which manifests itself after each major period of rapid assimilation of foreign influences.

Also Stephen Jaeger, at the end of *The Origins of Courtliness: Civilizing Trends and the Formation of Courtly Ideal 939-1210*, felt the need to comment on the debate civilization versus culture:

- "To express it pointedly: a good many Frenchmen are persuaded or can be persuaded that whatever is elegant, urbane, courtly, and refined in Germany must have been imported from France. The corresponding prejudice in Germany has it that whatever is effeminate, overrefined, and unmanly in Germany derives from French influence.... The modern perception of the cultural relations between Germany and France is strongly colored by the dichotomy, German '*Kultur*' versus French '*civilisation*'. Art, music, philosophy, the abstract and the mystical, are the province of the Germans; the elegant civil life, to which the arts stand in a subservient position, that of the French. This dichotomy has played a dubious role in the intellectual history of these two lands. It came to its sharpest point during the first world war and in the tortured pages of Thomas Mann's *Betrachtungen eines Unpolitischen*. Mann represented Germany as the land of Luther, of Protestantism; the task of the Germans is 'protest', rebellion against a shallow and corrupting French civilization.... The role of the artist in romanticism generally is not a civilizing one; just the opposite. It is to emancipate the Promethean, and the subversion of a tyrannical or perverse or outdated political order can appear as a devout goal. But to the Germans the isolation of the aesthetic realm meant more than to the French and the English. And it had more prominent spokesmen. It was Nietzsche who so admired Wagner's claim that before his art civilization would dissipate 'like fog before the sun'" (1985, pp. 269-271).

It is important to mention that a similar reactionary and nationalistic syndrome also developed in Russia in the years after the Napoleonic invasion of 1812, when the Slavophiles proclaimed the superiority of Orthodox Russian civilization over the West: Russia was an organic community alien to the European spirit of violence and competition, of rationalism, extreme individualism, competition, strife, conquest and anarchy: "Western civilization was based on false premises, on rationalism, on a split in the consciousness of man of the complete and integral comprehension characteristic of Orthodoxy" (Riasanovsky, 1952, p. 98).

Before this summary comment, Riasanovsky had presented at length the ideas of Aleksey Stepanovich Khomyakov (1804-1860)[239]:

- "[For Khomyakov, in his *History* of 1838-1860], the Slavs had played a leading role in the formation of Hellenic civilization as well as of the modern Greek people. Troy had been a Slavic city, and through Troy the Slavs had been largely responsible for the foundation of Rome and of the Roman state. The Etruscans, the Huns, the Bulgars, the Khazars had been Slavs. Slavic heroes had included Attila, Siegfried, and Parsifal.... The Latin language was derived from the Slavonic.... The Slavs were the best singers and the most musical people on the face of the earth, and therefore one could expect Orpheus to be a Slav.... Siegfried had to be a Slav for several reasons; one of them was that he possessed all the attributes of Slavic mythology, such as the maiden-ruler, the dragon, the cap of invisibility, the dwarfs, the magicians, and the priceless treasures" (Riasanovsky, 1952, pp. 72-73).

Not even Fichte, in his *Addresses to the German Nation* of 1807-8 (see chapter 25 of volume 1) went so far.

Each approach—if properly managed (i.e., kept away from the hand of the *supremo*)—can be a healthy development. Sadly, most often the *Kultur syndrome* falls victim to the propaganda machinery of the unity of power, with dire consequences.

7. Conclusion

As much as the previous chapter on *amour courtois* was a DP chapter, this one on *Minne* is a UP chapter. In both cases we have a dialogue between results and their causes, between some fascinating phenomena and the forces which shaped them.

Clearly every artist is free to write what he wants, yet as seen in the case of Walter von der Vogelweide, he is part of a specific world and its scripts and he is shaped by that world, and shapes it in turn. Once more it is fascinating to study in parallel the evolution of what should be similar in two bordering worlds, and yet evolved differently because of the different power structures and related scripts.

---

[239] "Russian philosopher and lay theologian, the most original, powerful and many-sided of the Russian Slavophil writers" (Schmemann, 1973, p. 332).

8. Neidhart von Reuental's Courtyardly Love

To avoid breaking the continuity of the main line of this chapter, I have opted to put here, at the end, a few additional remarks on Neidhart von Reuental, including a sample of his poetry in both a translation into modern German and into English (my translations), and coupled with a brief comparison with the *ritter Minne* in the second part of Goethe's *Faust I.*

As presented by J. G. Robertson in his *A History of German Literature* of 1970: "The most gifted poet of the later time was the Bavarian nobleman Neidhart von Reuental (Nithart von Riuwental [ca. 1190-ca. 1240]).... He will have nothing to do with noble ladies: he goes out among the peasants, joins in their dance under the village linden, or, if it be winter, in the great 'Bauernstube'. With an often childish pride, he describes his various conquests of village beauties. Here, for instance, is what a daughter tells her mother in one of Neidhart's earlier poems:

> *Der, den ich euch will nennen, den könnet ihr erkennen. Zu dem will ich (also) eilen; er ist genannt von Riuwental; den will ich umfangen. Es grünt an den Asten, dass alle die Bäume davon zur Erde niederbrechen könnten. Nun wisset, liebe Mutter mein, ich folge dem teuren Jüngling. Liebe, hohe Mutter, nach mir klagt er so sehr. Soll ich ihm nicht dafür danken? Er sagt, dass ich die schönste sei von Baiern bis nach Franken"* (pp. 112-113).
>
> [He, that I will name, you know him. To him I want to rush. His name is von Riuwental; him I will embrace. The branches are greening, so much that all the trees are bending earthward to the point of breaking. Now know, sweet Mother mine, that I will follow the dear young man. Beloved and esteemed Mother, he is yearning for me so much. Should I not thank him for that? He says, that I am the most beautiful from Bavaria to Franconia.]

We seem to hear Margareta back home from he first encounter with Faust:

> He bore himself so gallantly
> I am sure he is a nobleman;
> I read it in his bearing
> Else he would not have been so daring.

Similarly, behind all those compliments by Neidhart we can suspect the same final intention that Faust had when he asked the help of Mephistopheles:

> "Faust: Listen, you must get me the girl!
> Meph.: Which girl?
> Faust: The one who just passed.
> Meph.: She? She just came back from her confessor ...Over her I have no power.
> Faust: Unless I have her sweet young blood tonight in my embrace, our pact is broken by midnight."

And Mephistopheles duly helped![240]

Certainly, the interest and feelings are strong, and for a short while sincere and genuine, far from the artificial conventions of *amour courtois*, and the girls are not demanding strange heroic deeds, not even in well organized tournaments. However, these girls will soon be left behind, often pregnant, like poor Margareta, while our noble man moves to the next innocent flower.

---

[240] *A' tout seigneur tout honneur* (Honor to whom honor is due), Goethe did not only write, with *Faust I*, a magnificent *Tragödie* on the *Minne* of *ritter* Faust for a young wench named Margareta, but when he (Goethe) was young—only 25 years old—he had written an equally magnificent novel on the *amour courtois* of troubadour/knight Werther for Lotte:

"[*The Sorrows of Young Werther* (1774)] is the story, in Goethe's words, of an artistically inclined young man 'gifted with deep, pure sentiment and penetrating intelligence who loses himself in fantastic dreams and undermines himself with speculative thoughts until finally, torn by hopeless passions, especially by infinite love, he shoots himself in the head'" (Benét, 1987, p. 919).

In this short novel, there is no sexual conquest, not even a sexual thought on the part of Werther, as it came immediately to Faust when he saw Margareta. Very *amour courtois* is the high—unrealistic—vision that Werther has of Lotte, something that Lotte felt the need to tell him: "Don't you see that you are deceiving yourself, that you are seeking your own destruction? Why must you love me, me only, who belongs to another? I fear, I fear, that it is only the impossibility of possessing me that makes your desire so strong" ("December 20"; 1774/1983, p. 72). Indeed, in the best tradition of *amour courtois*, Werther—conscious of his merits—had already fallen in love with Lotte before having seen her, simply because she had been described as beautiful and unconquerable: "My companion told me, as we drove along through the park to the hunting lodge, that I should make the acquaintance of a very beautiful girl. 'Take care,' added the

Humanity is what it is, and one way—one of the best—was to invent the *amour courtois*, in which "the beloved held a place of superiority over the lover: his desire, anxious and fearful spurred him on to rise in worth to her.... These two elements, the cult of the beloved and the ennobling power of love, are inextricably linked one with the other. The lover grows more virtuous through his love of the beloved and it is love which raises him in the scale of worth and virtue" (Denomy, 1944, pp. 175-76).

---

cousin, 'that you don't fall in love.' 'What do you mean?' I said. 'She is already engaged to a very worthy man,' she replied, 'who is gone to settle his affairs upon the death of his father and apply for a lucrative position.' All this seemed of little interest to me" ("June 16"; 1774/1983, p. 14), and remained of little interest until his tragic end.

In all this, we are reminded of a Flemish parade shield of the 15th century—reproduced in many books, e.g., in Owen's *Noble lovers* of 1975, opposite p. 33—which shows a young armoured knight kneeling in front of an equally young lady. A scroll reports his words: "You or death", while death, as a skeleton, stands closely behind him. Werther could not have Lotte, and chose death instead.

As said, *A tout seigneur tout honneur,* and Goethe was great at both genres: *amour courtois* and *Minne*.

20

Anticipating More Changes: a Chinese *Insular Virtue* from Necessity

1. The Story of the Old Chinese Farmer

Li-jun Ji, Richard E. Nisbett, and Yanjie Su[241] began their 2001 article—on how Chinese and Americans perceive change differently—with a parable:

> There is an ancient Chinese story about an old Chinese farmer. One day his horse ran away. His neighbors came to comfort him, but he said, 'How can you know it isn't a good thing?' A few days later, his horse came back, bringing a wild horse with it. His neighbors came to congratulate the old man, who said, 'How can you know it isn't a bad thing?' A few weeks later, the old man's son tried to ride on the new horse and fell off, breaking his leg. Again, the neighbors came to comfort the old man, who said, 'How can you know it isn't a good thing?' Some months later, a war broke out, and all the young men in the region were recruited for the war. The old man's son was spared because of his broken leg. The story goes on in the same way, as long as the audience's patience hold up! This famous Chinese story suggests that things are changing all the time and people

[241] The first two from the University of Michigan, the third from Beijing University.

> should not be misled by a local state of events. Indeed, an extreme state of events suggests that its opposite is about to occur. (p. 450)

The way the story is told—and used by the three authors—suggests that the old Chinese farmer gave evidence of intelligence and wisdom: he knew that things are changing all the time, and acted accordingly.

However, my reaction to the story was different, and so was the lesson I drew from the experiments by the three authors which showed that the Chinese (63 students from the University of Beijing) anticipate more changes from an initial state than the Americans (46 students from the University of Michigan), and that the Chinese are more likely than the Americans to predict change in the direction of change.

Starting with the story of the old farmer, I was surprised that he had not organized a party to search for the lost horse. He had friendly neighbors who could have helped him, and usually a domesticated horse does not wander far and can be found without too much difficulty. But, no, nothing of the sort was done. At no point does the story say why or how the horse ran away. Did somebody forget to close the door of the barn? Did somebody mistreat it? Similarly, no mention is made of steps taken to prevent a similar loss in the future.

We are then told that the domesticated horse came back with a companion: an event more miraculous than real. Be that as it may, let's accept the arrival of the wild horse. Now, as most people know, the breaking (i.e., domestication) of a wild horse is a very difficult and dangerous enterprise.[242] Yet, the farmer's son tried to ride that wild horse, fell and broke his legs. It could have been worse. I have seen experienced cowboys on crutches with broken bones for having tried to tame a wild horse. Now, what's the use of an old and wise father if he cannot prevent that kind of unfortunate but predictable experience? When the neighbors came to comfort him, at no moment did he say that he was a fool for not preventing the accident. At no moment did he say that he had learned a lesson, and that the next time he would act better.

All changes came to the farmer, none were initiated by him. He was a passive recipient of ununderstood and uncontrolled events.

---

242 For those readers who have never watched a rodeo, I recommend the book by Monty Robert *The Man who Listens to Horses* (1997), especially its pictures of Monty's father's method of breaking a horse. It is indeed a difficult and very dangerous operation, even when done by experts.

Automatically we are miles away from any trace of creativity which demands a personal solution to contrasting demands, to contrasting alternatives. If the solution is always the same "Waiting for Godot", there can be nothing but perpetual waiting; Godot may come, probably he will not.

This is not all. Whoever said that after every bad event comes a good one, and that after every good event comes a bad one? Oh, no! Bad events are known to come in blocks: three, four, five bad events, with not a good one in between. The reverse is also true: several good events may come one after the other, especially to those who act energetically to deserve them. As reported by Koestler, in his *The Act of Creation*, Pasteur seemed "to regard chance as a kind of legitimate reward, casually related to the effort—an almost mystic conception" (1964, p. 145).

So, what are we supposed to learn from the initial story of the old Chinese farmer? Are we to derive a belief in the inevitability of changes—without any effort to understand the cause(s), and no corresponding action? This is no wisdom at all; at best it is the illusion of understanding.

Said differently, I feel that Ji, Nisbett and Su have accepted as true the propaganda of the *Celestial Emperor* which says: In China we live in the best of all possible worlds, where after a bad event comes a good one, and where people enjoy 'free luncheons' as happened to the old farmer who got a second horse for free. However—be careful—all these 'wonders' happen only to those who stay put, do not take initiatives, do not organize themselves in associations as the Americans do [see Tocqueville's praise of this in his *Democracy in America*]. These 'free luncheons' are definitely not bestowed on those who follow such Western proverbs as 'Nothing ventured, nothing gained,' 'Fortune favors the bold,' and 'Faint heart never won fair lady.' No! All initiatives must come from the Emperor and his Mandarins: all other initiatives will be crushed, and the perpetrators punished.

2. Problems with the Ji, Nisbett and Su Study

It is clear, from the way that the study was set up and from the discussion of the results at the end of the article, that the three authors are certain that it is good to be talented at predicting change from an initial state. However, the only true and valid talent is to predict correctly, and if no changes are coming, to predict no change. So, I feel that the tests that the three authors gave, were biased in favor of those predicting change.

More importantly, from the beginning, the authors failed to differentiate between the two worlds in which Chinese and Americans live, and correspondingly to note that "What is good for the goose may not be good for the gander." In other words, in a free world it is better to believe in permanence: the laws will not change from one day to the other, taxes will no triple suddenly, one will not be put to prison for being a capitalist, or an intellectual, or having said a wrong word. In the free world, we may dare to take initiatives, to invest in a new project because of a valid belief in the stability of the system, because we expect few changes (mainly those we initiated). We marry because we believe in the permanence of that union. We have children because we have faith that our partner will not change drastically from one day to the other.

Basically, a lower level of expertise in expecting changes—e.g. in not expecting the secret police to knock at our door, imprison and torture us for some vague suspicion of ideological heresy—may be one of the best compliments we can make to the West. On the contrary, a high level of expertise in change may tell us much about the power system in that country.

Simply praising an expertise in predicting change without major distinctions—as our three authors did—may be a friendly compensation to all those Chinese who unwillingly have been trained in not taking initiative and expecting them from the government, but not more. In China for long centuries, and still now, to be an expert in *enlightened optimistic fatalism*, like the farmer of the story, was better than *pure pessimistic fatalism*: it did not change matters, but it helped in accepting the many changes—and emperors or party secretaries were always introducing changes (e.g., "some months later, a war broke out, and all the young men in the region were recruited for the war"). In such conditions, there will never be sufficient expertise in change: China, especially in the last hundred years, produced more changes than the best experts could foresee.

3. Anticipating More Changes as an *Insular Virtue*

In the title of this chapter I said that anticipating more changes is an *insular virtue* from necessity. Why so? Because the *supremo* is unpredictable. As reported in chapter 23 of this volume, "out of a clear sky" Harun al-Rashid had Ja'far executed. His severed head was exhibited on a bridge and the two halves of his bisected body on two other bridges there to remain for a year.

And "out of a clear sky" Toyotomo Hideyoshi had his adopted son Hidetsugu, his wife, consorts and children, a total of thirty people, all killed, when he changed his mind: no more adopted son, only his two-year-old son Hideyori would be his unrivaled heir.

In China, "in 1433, and again in 1449 and 1452, imperial edicts banning overseas trade and travel, with savage penalties, were issued. Any merchant caught attempting to engage in foreign trade was defined as pirate and executed. For a time even learning a foreign language was prohibited as was the teaching of Chinese to foreigners" (Merson, 1990, p. 77). The many who had been previously involved in overseas trade and travel had better believe in change, because they were immersed in change up to their neck.

Back to Europe:

> Matteo Bandello [1484-1561] issues frequent warnings in his *Novelle* on the precariousness of the courtier's position. He is a valuable witness, for a while an adopted fellow-citizen of Castiglione's and, according to his own account, a personal friend of the latter's family. His tale of Artaxerxes and the seneschal Ariabarzanes allows him to give a long aside on the capricious favours of princes, which better than any data sums up the situation: "Similarly you will see in the courts of kings and princes a courtier who enjoys his lord's full favour. It may be that the prince can do or say nothing without him, and yet, when the said courtier strives his hardest and tries his utmost to maintain or increase his lord's favour, what happens? The prince's mind changes and is turned towards another,[243] and the person who was previously the first man at court finds himself, in a twinkling, the most abject." (Woodhouse,1978, p. 193)

The only solution, under a *supremo*, be he a Caliph in Baghdad, an Italian *signore*, a Japanese generalissimo, a Chinese emperor or Party General Secretary, is for everybody to be an expert in changes, and expect changes from one moment to the other.

---

[243] Bandello, 1554/1910, *1*, p. 38.

4. Views on Change: East versus West

Ji, Nisbett, and Su wrote that "the East and West have different views on change. The idea of change and transformation between two opposite states is the main theme of the *I Ching*, or *Book of Changes*. ..." (ib., p. 450).

However we must distinguish the old West from the modern West, because the old West was as fascinated by changes as China. For instance, for long centuries the West was enthralled by the unpredictable workings of the *Wheel of Fortune*:

> The social and material uncertainty besetting most members of feudal society, an uncertainty which deepened as the internal contradictions of this society grew more acute, was bound to encourage a belief in fate or destiny. Countless illustrations show the wheel of Fortune. She herself sits in the centre, crowned as 'sovereign of the world', and keeps the wheel turning; clinging to the wheel, a youth rises upwards, hopeful and expectant; on top of the wheel, a ruler has enthroned himself in solemn state; further on, we see a man dragged precipitously downwards by the spinning wheel, while under it lies the prostrate figure of a victim of inconstant Fate. The concept of Fate was borrowed by the Middle Ages from antiquity. The Christianization process made Fate subject to God, who has given into her keeping "worldly glitter", to use Dante's expression. (Gurevich, 1985, p. 142)

At about the same time of the belief in the ceaseless rotation of the *Wheel of Fortune*, there was a belief—identical to that of Confucius (discussed by our three authors, p. 451)—that the golden time was in the past, and we need to return to it. For instance, in Western feudal times, there could be no such thing as innovation in law. If a new law was needed it was said to be a rediscovered old law, and its old age vouched for its essential goodness. "Ancient law was good and just" (ib., 1985, p. 166).

However, these attitudes changed gradually in the West under the impact of increasing levels of DP. Already in the 13th century in Italy "the very engine of communal life was impelled by the urge to elaborate collective norms" (Becker, 1981, p. 166). "One presupposition was constant intervention by the commune in every aspect of social and economic activity. This was connected with the confident belief that problems were soluble. *Discussion in council* could find solutions which could then be enforced by legislation. If experience showed this was not the correct solution then another would be tested" (Waley, 1988, p. 59; my italics).

In the above, I have italicized "discussion in council" to stress a meliorist effort by equals, in a city-republic, and not by fearful councillors in front of a mighty king or emperor. One single imprudent word, "merit", said in front of the imperial council about the merit of general Li Ling brought Ssu-ma Ch'ien's death penalty, later graciously converted by the Emperor into castration (see chapter 7 of volume 2 and chapter 25 of this volume). In China, everybody who knew of Ssu-ma Ch'ien and of his fate had every reason to believe in change, in the rotations of the *Wheel of Fortune.*

There may not have been more castrations in China in later times, yet there were plenty of major revolutions, and constant change. In the last one hundred years, especially from 1949 to the present, China has been afflicted by a tornado of change in every field, involving everybody. For instance, "although the Cultural Revolution [of 1966-1976] largely bypassed the vast majority of the people, who lived in the rural areas, it had very serious consequences for the Chinese system as a whole. In the short run, of course, the political instability and zigzags in economic policy produced slower economic growth and a decline in the capacity of the government to deliver goods and services. Officials at all levels of the political system had learned that future shifts in policy would jeopardize those who had aggressively implemented previous policies. The result was bureaucratic timidity. In addition, with the death of Mao and the end of the Cultural Revolution, nearly 3,000,000 CCP [Chinese Communist Party] members and other citizens awaited reinstatement after having been wrongfully purged" (Lieberthal, 1993, p. 152).

The 63 Chinese students who participated in the study in question, and their parents, grandparents, and great-grandparents probably must have seen (even suffered) so many radical changes to make them experts in anticipating change.

In contrast, the Americans have enjoyed governmental tranquility—relatively speaking—with small changes coming one at a time. There was indeed in the United States only one Revolution, that of 1775-1783 (in some respects more a war of independence than a revolution); yes, there was the Civil War of 1861-65, but it brought radical change only to the Southern states.

Americans know that for them changes come gradually, with the participation of all functions of government (legislative, executive and judiciary), and with the open involvement of all interested parties be they trade unions, business, churches, and with a major contribution by a very vocal free press. Every new piece of legislation is discussed at length by the House of Representatives and in the Senate; it must be approved by both Chambers, and finally signed by the President, who has the power of veto. In addition, no law is valid that contradicts the Constitution of 1787 and its few subsequent amendments. This lengthy and complex DP process is a major guarantee of stability (together with a gradual meliorism). Such a situation—combined with the relatively small difference between the two great parties (Democratic and Republican)—gives people a strong confidence in the continuity of their lives, with no major changes ahead.[244] All this explains why the Americans tested by our three authors anticipated less change than their counterparts in Beijing.

Back to China, where science and technology, industry and commerce suffered from two types of change: 1. from the involvement of the state bureaucracy in thousands of details, and 2. from major changes in state policy. The first type of change is illustrated by Etienne Balazs (1968), as reported by David Landes:

> Without exception, Chinese society was highly totalitarian.... No private initiative, no expression of public life that can escape official control. There is to begin with a whole array of state monopolies, which comprise the great consumption staples: salt, iron, tea, alcohol, foreign trade. There is a monopoly of education, jealously guarded. There is practically a monopoly of letters (I was about to say, of the press): anything written unofficially, that escapes the censorship, has little hope of reaching the public. But the reach of the Moloch-State, the omnipotence of the bureaucracy, goes much farther. There are clothing regulations, a regulation of public and private construction (dimensions of houses); the colors one wears, the music one hears, the festi-

[244] To this positive confidence, I attribute, in good part, the fact "that fewer voters show up at the polls in America than in most other democracies: our turnout rate ranks us just above the cellar—narrowly besting Switzerland, but below all twenty-two other established democracies" (Putnam, 2000, p. 31). This comparison is significant: Switzerland—a republic older than the USA, which went through fewer changes, shows even a lower voter participation than the USA. This points to a sense of confidence, by Swiss and Americans, that nobody is attempting to win the election to install a dictatorship.

> vals—all are regulated. There are rules for birth and rules for death; the providential State watches minutely over every step of its subjects, from cradle to grave. It is a regime of paper work and harassment *[paperasseries et tracasseries*], endless paper work and endless harassment.The ingenuity and inventiveness of the Chinese, which have given so much to mankind—silk, tea, porcelain, paper, printing, and more—would no doubt have enriched China further and probably brought it to the threshold of modern industry, had it not been for this stifling state control. (Landes, 1998, p. 57)

Inevitably, a regime of paper work and harassment, endless paper work and endless harassment, forced change, but mostly negative, counterproductive change. Business owners or managers, instead of following business in its normal evolution, had to be constantly vigilant of the rules and whims of bureaucrats.

The second type of change was discussed by Joel Mokyr:

> Most important, however, is the fact that technological change that is generated in large part by public officials and a central government has the nasty weakness of depending on the government's approval. As long as the regime supports progress, progress can proceed. But the government can flip the switch off, so to speak, and private enterprise is then unlikely to step in. Innovation that is run largely by bureaucrats is thus not impossible, but it depends on their goodwill. Because most entrenched bureaucracies tend to develop a strong aversion to changing the status quo, state-run technological progress is not likely to be sustained over long periods. The Chinese miracle is indeed that it lasted so long. It ended when the state lost interest in promoting technological change.
>
> Why the Chinese state changed its attitudes toward technological change is hard to determine. The Ming and Qing emperors were more absolute and autocratic than their predecessors. Before them, coups d'état and regicides occurred frequently, thus introducing an element of "competition" into the Chinese political market. Rigid etiquette and complete obedience and conformism became the hallmark of the Chinese government under the Ming emperors. (1990, pp. 236-37)

Paradoxically, rigid etiquette and complete obedience and conformism are more demanding, and imply more unexpected changes, than free enterprise as shown by the tragedy that fell on Lord Enya (Asano) and his 47 ronin, as narrated in the play *Chûshingura* (see volumes 1 and 3). Complete adherence to a rigid etiquette demands knowing exactly its rules, rules that are not written and that can change at the whim of the *supremo*. Under these conditions, being an expert in change, and expecting it at any moment, is a great virtue, but a virtue acquired out of the harshest necessity, the kind of virtue we hope to spare ourselves and our children, possibly everybody.

In my chapter on the Chinese, in volume 2, I reported the following observation by Wang Liu (1959): "The tremendous emphasis upon conformity also helped to produce a type of personality more introspective and circumspect, and less energetic and active, than what the Confucian theories originally visualized" (p. 94). Usually, one is circumspect for fear of the unexpected, and one good way to be circumspect is to see the possibility of change, probably bad changes. We are miles apart from the belief that *Audaces fortuna juvat* (Fortune favors the bold). This was not the case of the farmer in our story who stayed put, and let things come as they wanted.

In volume 1, in discussing the contrast between *visitors* and *insulars*, I reported the following comments by Yang Kuo-shu on the tripartite division of all social relations among the Chinese: "Demonstrating strong feelings, including the menaced and aggrieved affects of suffering, is dangerous, because it gives others power over relationships and restricts one's flexibility to respond effectively. Ultimately, uncontrolled emotional displays threaten one's position in a world of power" (reported by Kleinman & Kleinman, 1991, p. 288). According to this rule, our old farmer behaved perfectly, with no demonstration of strong feelings: he took everything very calmly.

5. Wanted and Unwanted Change

It may or may not be a good thing to regularly anticipate more change from an initial state, but it is definitely good to obtain the changes one wants, and to avoid the others.

Winston Churchill had much praise for change, but only for the change he wanted:

> Many remedies are suggested for the avoidance of worry and mental overstrain by persons who, over prolonged periods, have to bear exceptional responsibilities and discharge duties upon a very large scale. Some advise exercise, and others, repose. Some counsel travel, and others, retreat. Some praise solitude, and others, gaiety. No doubt all these may play their part according to the individual temperament. *But the element which is constant and common in all of them is Change. Change is the master key*. (1932/1965, p. 7; my italics)

But, did these searched-for changes train Churchill in anticipating more changes from an initial state? Probably not, because the change he wanted—being able to paint well—demanded time, effort and self-control, and the results were a long time in coming.

In the end, we are back to the difference between the *insular personality* (under UP) and the *visitor personality* (under DP) as described in Table 5 of volume 1. There, I listed "fatalism" for the *insulars*, and "free will" for the *visitors*. An obvious fatalist (and *insular personality*) was the old Chinese farmer of our story.

Bernard Shaw wrote: "The reasonable man adapts himself to the world; the unreasonable one persists in trying to adapt the world to himself. Therefore all progress depends on the unreasonable man" (1903/1980, p. 225). Here we have two types of experts in changes, the one in adapting himself to the world, the other in adapting the world to himself. Leaving aside the humor that comes from giving a new twist to the word reasonable and unreasonable, our Chinese farmer was definitely reasonable: he adapted himself to the world, and did not contribute one iota to progress.

Truly, it might have been more instructive if our three authors had asked their Chinese and American respondents to comment on the story of the old Chinese farmer, because, from the article, one might conclude that the American students were less smart—at least on this specific point of expecting and forecasting changes—than their counterparts in China. And it might even be true: a blessing in disguise.

6. "Chinese Patterns of Socialization"

Chapter 1 of *The Psychology of the Chinese People* is entitled "Chinese patterns of socialization: A critical review." What its author said can help in understanding the fatalism of the old Chinese farmer and his expertise in expecting changes:

> A distillate of the literature reviewed may be summarized in the following propositions. At the present state of our knowledge, they represent permissible generalizations which have received relatively strong and consistent empirical support or on which there is overall agreement among observers:
>
> 1. Chinese parents tend to be highly lenient or even indulgent in their attitudes toward the infant and young child, in sharp contrast to the strict discipline they impose on older children. Many observers believe that the change in parental attitudes occurs when the child reaches the age of 'reason' or 'understanding' at around 4 to 6 years.
> 2. Traditionally, great emphasis was placed on obedience, proper conduct, moral training, and the acceptance of social obligations, in contrast to the lack of emphasis placed on independence, assertiveness, and creativity.
> 3. Chinese parents tend to be highly concerned with impulse control. *The child's exploratory, adventurous, or risky activities tend to be discouraged.* ....
> 6. In comparison with Caucasian infants, Chinese infants tend to be less vocal, less active, less likely to smile, *and more apprehensive in social and separation situations* - that is, disposed toward inhibition. (Ho, 1986, pp. 35-36; my italics).

No doubt, our old Chinese farmer lacked independence, assertiveness, and creativity, because of the way he had been socialized. Most probably, as a child his "exploratory, adventurous, or risky activities had been discouraged"; most wisely so, because an *insular* world punishes every form of initiative.

7. More on the Painful Chinese Training in Expecting Change

So harshly have the Chinese, in the last decades, been trained in expecting changes—thereby explaining the results obtained by Ji, Nisbett and Su—that such a "training" deserves to be illustrated. One good way to do this is by referring to Fairbank and Goldman's *China:A New History* of 1998, specifically to chapters 19 and 20. Chapter 19 "The Great Leap Forward, 1958-1960" begins as follows:

In 1958-1960 some 20 to 30 million people lost their lives through malnutrition and famine because of the policies imposed upon them by the Chinese Communist Party [CCP]. Measured by the statistics showing an increase of mortality, this was one of the greatest of human disasters.[245]

In the Great Leap Forward (GLF), we can see several factors at work without being able as yet to assign them their ultimate influence. We begin by noting certain residual aspects of the Chinese inheritance—first, that *the state authorities had unquestioned control over the populace in the villages.* The bifurcation of society into rulers and ruled, the managers and the producers, could now be used by the CCP leaders more intensively than ever before....

All central orders, however, had to be applied by local authorities. Part of China's inheritance was that their state of morale, their loyalty to the center, would be a key determinant of the results achieved. CCP activists had now in a general way succeeded to the local leadership position of the lower gentry of imperial times. They could reassert the old practices of officialism, oriented upward toward seeking their superiors' approval rather than downward to serve the people. (p. 368; my italics)

Chapter 20 "The Cultural Revolution 1966-1976" begins with:

Mao's last decade—from 1966 until his death in 1976—saw a domestic political struggle that convulsed China, constantly amazed the outside world, and achieved appalling destruction. It was a fitting finale to China's twenty lost years begun in 1957 with the Anti-Rightist Campaign that dispensed with so many intellectuals. So pervasive was the cataclysm, directly involving something like 100 million people, that its full history is still far from known or written.

---

245 According to MacFarquhar (1987) "The worst man-made famine in history ... a human catastrophe.... There were 30 million excess deaths between 1958 and 1961" (quoted by Mirsky, 1988, p. 31). "The estimate of China's leading journalist, Liu Binyan, is fifty million" (ib., p. 32).

Also, "[MacFarquhar] shows how in 1958 Mao launched the vainglorious Great Leap Forward, which, by rejecting modern technology and making impossible demands on rural communes, crippled agriculture so thoroughly that there are still parts of China in which children are stunted from lack of food" (ib., p. 31).

> In looking at the Cultural Revolution (CR) in China, we are therefore obliged to imagine *a society that can be run by a Great Leader and a party dictatorship simply because the citizenry are passive in politics and obedient to authority* [*insular personalities*]. *They have no human rights because they have been taught that the assertion of human rights (such as due process of law) would be selfish and antisocial and therefore ignoble. It would also be severely punished.* The problem begins on the ground in the family life of the Chinese village, where the Confucian teaching of social order through dutiful self-subordination has left its mark even today.
>
> One starting point in understanding the CR is to recognize that Mao Zedong had now acquired some of the prerogatives of an emperor.[246] (ib., pp. 383-84; my italics)

> Estimates of the victims of the Cultural Revolution now hover around a million, of whom a considerable number did not survive. To Chinese, so sensitive to peer-group esteem, to be beaten and humiliated in public before a jeering crowd, including colleagues and old friends, was like having one's skin taken off. Generally the victim felt guilty, as anyone may under attack, but especially because they had felt such loyalty and had so venerated Mao and the party. (ib., p. 402)

The above, in turn, can be expanded by the following from Ian Buruma (2000): "The only way [for Mao] to crush all criticism, then, was to destroy all the possible critics by banishing them to remote labor camps, sometimes for life, or ruining their careers, or turning their children against them, or humiliating them in forced public confessions, or simply by murdering them. In a typical example of the Chairman's grim humor he once referred to comparisons made between himself and the Qin emperor. 'Well,' he said, 'what's so special about the Emperor of Qin? He only executed 460 scholars. We killed 46,000!'" (pp. 22-23).

Also "[Philip Short, in *Mao: A Life*, 2000] mentions one instance which reveals the scale of Mao's last horrific campaign. A governor and alternate Politburo member in faraway Inner Mongolia was said to have

---

[246] "Absolute power corrupts absolutely," says the Law of Power of Lord Acton (see volume 1) which bears no exceptions, not even for China.

started a "black party" as a rival to the official Communist Party. It was nonsense of course. But in an effort to "ferret out traitors," 350,000 people were arrested, 80,000 beaten so badly that they were permanently maimed, and more than 16,000 killed" (ib., p. 24)

So, many changes indeed befell an infinite number of Chinese: nearly all bad changes, with few or no good ones in between. By now, that story of the old Chinese farmer is passé, and so is his mild training in foreseeing changes: those 63 students of the University of Beijing knew better.

21

Interpreting Three Recent Studies on the Contrast Between Japanese and American Ethnopsychologies

After completion of chapter 9, "The Strict Ritter Personality of the Japanese", of my third volume, I came across three studies comparing important aspects of the Japanese and American ethnopsychologies. I studied each with much interest, and report hereafter how I interpret them, in the light of my GAM/DP theory.

1. Explaining the Cross-Cultural X

In an article for the *Monitor on Psychology*, Tori DeAngelis (2003) referred to a study by Heine et al. (2001) in which "First, Japanese and American participants performed a task at which they either succeeded or failed. Then they were timed as they worked on another version of the task. 'The results made a symmetrical **X**,' says Heine: Americans worked longer if they succeeded at the first task, while Japanese worked longer if they failed" (p. 62).

These results were what I would have expected based on my prior studies. In countries under Division of Power (DP)—like America—individuals are relatively free to chose which profession they want to pursue in function of their talents/aptitudes/likes and socioeconomic status. A success in an important task points to aptitudes which, if well pursued, will open the door to professional success. Every valid test seems to obey the Delphian command of "Know Thyself", providing knowledge about the self which will then serve as professional and social compass. Accordingly, Americans are culturally primed to try hard at another version of the same task to be sure that they really have an aptitude for it, and that this is a domain worth pursuing and a technique worth learning well.

Instead, failure the first time says: "This is not for me, no sense trying hard a second time. My talents are elsewhere. America rewards a well chosen risk: I must therefore choose correctly based on my true capabilities which, evidently, have nothing to do with the task I have just performed and failed."

On the other hand, in countries under Unity of Power (UP)—like Japan, with its close social structure, strict hierarchy, powerful rulers—what counts is first of all loyalty to the group/ruler, then willingness and ability to do what the ruler asks.

Failure at the first task led the Japanese to conclude something like: "I am not applying myself as I should, and therefore of little use to my group/ruler. This will bring me trouble. I will therefore try harder the second time, be successful and so prove my dedication and capability."

Instead, success the first time says: "I am capable. If the ruler commands, I can do it; no need therefore to prove myself a second time. Anyway, it is the ruler (or group) which will decide what I will do, and be responsible for the group (and my) professional and social success."

These four conditions, two for the Americans, and two for the Japanese, create the Heine symmetrical **X**. The Heine et al. (2001) article confirmed my interpretation:

> One alternative account to these findings warrants discussion. That Japanese worked harder after failure than after success is consistent with a motivation to avoid "sticking out." When Japanese fall short of other's performance, they work harder than when they are leading the pack. This tendency is nicely captured in a word that is sometimes used to characterize a Japanese style of competition: *yokonarabi* (Kumon, 1982). *Yokonarabi* literally means to line up sideways, reflecting that it is more important to keep up with the competition than to break ahead of it. (pp. 612-13)

These comments remind me of the youth of Yamamoto Tsunetomo (ca., 1700) described in chapter 9 of volume 3:

> By the time he was twenty he was still without an official position. It is related that at this juncture someone pointed out to him that his face was 'too intelligent' and warned him that [his lord] Mitsushige disliked such an expression. Tsunetomo states that he spent the next year in front of a mirror trying to correct this fault. (Scott Wilson, 1985, p. 13)

Indeed, the one who breaks ahead, insults his lord who is the only one "capable" of being at the head of the pack! So, to save one's skin it is better not to look intelligent, to wear a smiling mask, and to practice *yokonarabi*.[247]

Another way not to look too intelligent is to avoid looking in the eyes of anybody who could be a superior, as described in the first chapter, of Hiroshi Kagawa's 1997 *The Inscrutable Japanese*, "Why won't Japanese look me in the eyes?" "It has been said that Japanese avoid making direct eye contact during conversations. When compared with other nationalities, especially Americans, this is probably true. Japanese tend to feel that eye contact is rude and signifies an attempt to intimidate or challenge the other person" (p. 19). And, in the past, looking in the eyes of one's lord, or even of one of his samurai could cost one's head, literally.

In chapter 11 of volume 1, I reported the following by Octavio Paz on certain similarities between Mexicans and Orientals:

> Our hermeticism is baffling or even offensive to strangers, and it has created the legend of the Mexican as an inscrutable

---

[247] It is important to realize that Heine et al. and I differ substantially in our understanding of the tested differences between Japanese and North Americans, mainly, because of our different historical frames of reference: short-term for Heine et al., long term for me (going as far back as Canossa 1077 for an understanding of the *visitor personality* of the North Americans, and nearly as far back in the study of the history of Japan for an understanding of the *strict ritter personality* of the Japanese; see my previous volumes 1 and 3).

Indicative of their short-time perspective, for both Japan and North America, Heine et al. said: "Many cultural practices in contemporary North America, such as corporations basing promotions and salaries on individuals' achievements, schools emphasizing the building of self-esteem,... Many practices in contemporary East Asia cultures, including seniority-based systems of promotion and salary, an educational focus on group learning, child-rearing styles that emphasize self-discipline and cooperation with others, and conversational scripts emphasizing constructive criticism, empathy, and sympathy,..." (2001, p. 600).

> being.... The impression we create is much like that created by Orientals. They too—the Chinese, the Hindu, the Arabs—are hermetic and indecipherable. (1950/1985, p. 65)

Further in his book, Paz wrote: "Slaves, servants and submerged races always wear a mask, whether smiling or sullen. Only when they are alone, during the great moments of life, do they dare to show themselves as they really are" (ib., p. 70). And in section 11 "Behind the Smiling Mask", of my chapter 9 of volume 3, on the Japanese, I reported the following from Lafcadio Hearn:

> My cook wears a smiling, healthy, rather pleasing face. He is a good-looking young man. Whenever I used to think of him I thought of the smile.... One day I looked through a little hole in the shoji and saw him alone. The face was not the same face. It was thin and drawn and showed queer lines worn by old hardship. I thought: "He will look just like that when he is dead." I went in and the man was all changed—young and happy again—nor have I ever seen that look of trouble in his face since. But I knew when he is alone he wears it. He never shows his real face to me; he wears the mask of happiness as an etiquette. (quoted by Richie, 1997, p. 145)[248]

All this confirms my interpretation of the Heine **X**: In countries under UP (close social structure, strict hierarchy, powerful ruler; e.g. Japan) what counts first is neither the discovery of one's talents, nor the discovery of a personal compass, but the loyalty to the group/ruler, then the willingness to do what the ruler asks, and finally the ability to do what has been asked. Accordingly, one must succeed either at the first or at the second round of the test.

Under DP, ability comes first, then the willingness to do, and lastly loyalty. If the company/boss does not appreciate my ability, I will quit and search for a more understanding company. So, let's be sure now of my talents by trying a second time at that test in which I did well the first time.

---

[248] See also, in that same chapter 9 of volume 3, similar comments by Alessandro Valignano and by Patrick Smith.

2. Japanese Are Strong in Relative Tasks That Demand Attention to Contextual Information, as Expected from *Strict Ritters*; Americans, instead, Are Strong in Absolute Tasks That Demand Ignoring Contextual Information, as Expected from *Visitor Personalities*.

Kitayama, Duffy, Kawamura, and Larsen reported in their recent article "Perceiving an object and its context in different cultures" that

> in two studies, a newly devised test (framed-line test) was used to examine the hypothesis that individuals engaging in Asian cultures are more capable of incorporating contextual information and those engaging in North American cultures are more capable of ignoring contextual information. On each trial, participants were presented with a square frame, within which was printed a vertical line. Participants were then shown another square frame of the same or different size and asked to draw a line that was identical to the first line in either absolute length (absolute task) or proportion to the height of the surrounding frame (relative task). The results supported the hypothesis: whereas Japanese were more accurate in the relative task, Americans were more accurate in the absolute task. (2003, p. 201)
>
> The current work is consistent with recent theorizing on cultural variation in cognition (Nisbett et al., 2001) in showing that Japanese are more capable of incorporating contextual information in making a judgment on a focal object, but North Americans are more capable of ignoring contextual information. We demonstrated this cross-cultural variation in a nonsocial task that is specifically designed to simultaneously assess the two cognitive competencies of ignoring or incorporating context. (ib., pp. 204-05)

I completely agree, and it makes perfect sense in terms of the DP/UP part of my GAM/DP Theory of Personality and Creativity. The *visitor personality*—most North Americans are *visitors* (see Table 5 in chapter 11 of volume 1)— is "independent, self-directed, self-sufficient and self confident"; the *insular personality*—most Japanese are a blend of *insular* and *strict ritter personalities*—is "dependent, conformist, knowing one's place, intolerant of nonconformity." The *visitor* trusts his mind and pays a lower level of attention to the context, to the opinion of the others; the *insular*, instead, wants to be sure that he knows the opinion of the others, that he knows the party line, so as not to risk his head.

From youth, *visitors* are taught to use their own mind, trust their judgement and stick to it until proven wrong. From youth, *insulars* are prodded to learn quickly what others say, and to adjust accordingly. This training, of both *visitors* and *insulars*, becomes so ingrained that it will reflect itself also in purely abstract tasks as the one devised by Kitayama, Duffy, Kawamura, and Larsen.

A different story is an evaluation of the merits and demerits of each approach. Ideally one should be a relativist for a relative task, and an absolutist for an absolute task. Generally, however, for those born in Japan, it will be safe to be a relativist, an expert in taking into account the context; and indeed so are the majority of the Japanese. For those born in America a higher level of "stick to your guns" is appropriate: it will draw respect from others, and contribute to success.

3. Let's Not Exaggerate: It Is Not True That the Japanese Have no Need for Positive Self-Regard

In their 1999 article "Is there a Universal Need for Positive Self-Regard?" Heine, Kitayama, Lehman, Markus concluded by saying "No": such need is not universal because, for instance, the Japanese do not have it.

> We argue that in North America a key component of constructing the self involves the continual self-affirmation of the individual as an autonomous agent who has functioned, is functioning, and will continue to do so effectively in future, daily social life. In Japan, a key component of constructing the self involves the continual affirmation of the relationships of which the individual is part and thus an affirmation of the self as an active, mutually validating, and validated cultural agent. We elaborate how this suspension of the self in a network of relationships renders the need for positive self-regard necessarily weak and functionally disconnected for many social and psychological pro-

> cesses of the Japanese person. Moreover, we suggest that affirmation of the self as an interdependent agent is achieved not by seeking positive self-regard but rather by maintaining a chronic self-critical view. (p. 767)
>
> Understanding contemporary Japanese-style self-esteem requires a comprehensive grasp of another set of core cultural concepts, some of which are also known and can be experienced in North American contexts but typically are not emphasized or given the same pervasive societal expression as they are in Japan. These concepts include self-criticism, self-discipline, effort, perseverance, the importance of others, shame and apologies, and balance and emotional restraint. (p. 769)
>
> As we have maintained, having and expressing high self-esteem is one significant element of a North American way of being, whereas holding and elaborating self-critical attitudes is an integral part of a Japanese selfway. (p. 788)

With all due respect, I disagree because these comments present a distorted picture of the Japanese ethnopsychology. The Japanese, like every other people, search for positive self regard. This becomes evident, for instance, in the following observation by the eminent Japanese film director Akira Kurosawa:

> Human beings are unable to be honest with themselves about themselves. They cannot talk about themselves without embellishing. This script [of the film *Rashomon]* portrays such human beings—the kind who cannot survive without lies to make them feel they are better people than they really are.... People indeed have immense difficulty in talking about themselves as they really are. I was reminded once again that the human animal suffers from the trait of instinctive self-aggrandizement. (1982, pp. 183, 188)

Donald Richie said similar things about the characters in *Rashomon,* in his book on the films of Kurosawa:

> One can now assign various reasons for the five [key characters] having seen and heard the things that they thought they saw and heard. All the stories have in common one single element—pride. Tajomaru is proud to have raped and fought and

> killed; the wife is proud to have (perhaps) killed; the husband (for now there is every reason to believe that the dead talk) is proud to have killed himself; and the woodcutter is proud to have seen and robbed. They are proud of these actions and we know because they insist upon them. One confesses only what one is openly or secretly proud of, which is the reason that contrition is rarely sincere. But the reasons for the pride, as Parker Tyler has indicated in his fine analysis of the film are not those commonly encountered.
>
> Each is proud of what he did because, as he might tell you: "It is just the sort of thing that I would do." Each thinks of his character as being fully formed, of being a *thing*, like the rape or the dagger is a thing, and of his therefore (during an emergency such as this) being capable of only a certain number of (consistent) reactions. (pp. 74-75)

Practically every sentence by Kurosawa and Richie speaks of people thirsting for positive self-regard. It is equally significant that there is no mention, in this and other books on Kurosawa and his *Rashomon*, of Japanese viewers and critics finding it objectionable that the characters in *Rashomon* were proud individuals, with a strong self-esteem, certainly with none of the self-critical attitudes described by Heine et al.

The point is that the Japanese too have a thirst for positive self-regard as soon as their overlord (emperor, shogun, daimyo, samurai) is not watching them directly or indirectly, as soon as they can be individuals and not cogs in a social machine.

If so, are Heine, Kitayama, Lehman, Markus (1999) completely wrong? No. In fairness to them, things are complicated by the co-existence of more than one ethnopsychology, in the case of Japan *insular* and *strict ritter*.

In chapter 9 of volume 3, I stressed the *strict ritter* side of the Japanese personality because this is what strongly differentiates the Japanese from other people (with exception of the Germans). With the *insular* side of their ethnopsychology the Japanese are similar to many other people, as discussed above in relation to the comments of Paz on the Mexicans and the Orientals.

Some situations bring forth the *insular*, and others the *strict ritter* (or even a blend of the two) from the same person. In this, the Japanese are no exception; every person around the world can be different in different situations[249]; however the two main components of the Japanese personality differ much from each other. Things become simpler when the same reaction can be expected from both *strict ritter* and *insular*, as was the case with the Heine test in which Japanese worked harder the second time if they had failed the first time. Indeed, both the *strict ritter* and the *insular* are aware of their boss, in a way unknown to the Americans. And the same is true for relative versus absolute tests: both the *strict ritter* and the *insular* want to be sure not to do anything that will upset their superior or group. So, here, things are predictable.

When instead, we come to totally unforeseen situations, as those in Kurosawa's *Rashomon*, nothing can be predicted: the samurai may become a coward, but later claim to have been a hero; and the bandit can behave like an heroic samurai. As reported above: "One can now assign various reasons for the five [key characters of *Rashomon*] having seen and heard the things that they thought they saw and heard. All the stories have one single element in common —pride (Richie, 1996, p. 75). It will then be up to a Kurosawa to make a masterwork out of this pride: with characters behaving like Japanese behave, and along his Japanese understanding that "human beings are unable to be honest with themselves about themselves. They cannot talk about themselves without embellishing" (1982, p. 183).

In the end, talking about ourselves, with constant embellishments, is probably a universal trait, provided it can be done without running into trouble with one's superior. When, instead, troubles are certain, we are automatically in the world described by Heine, Kitayama, Lehman, Markus (1999) in which: "[the] affirmation of the self as an interdepen-

---

[249] "Over the past decade a growing body of research has begun to recognize that people's behavior varies stably and predictably across situations (e.g., if situation A then he does X, but if situation B then he does Y), and that such variability reflects the core of personality and provides a window through which to glimpse the underlying goals and motivations of the person.

To illustrate, consider the example of Jack, who is a generally calm, sociable person. Every time his girlfriend attempts to do something with others that does not involve him, however, he becomes hostile and lashes out. Now imagine James, who is also a calm and sociable person. He does not mind so much if his girlfriend wants to do things without him. Instead, he gets very angry and spiteful when his girlfriend tells him what to do. Whereas both Jack and James are calm and friendly in general, each gets predictably angry and hostile in certain situations: Jack when his significant others try to establish independence from him, and James when they try to control him. Thus, being both gentle and hostile—like being smart and stupid—may be equally genuine facets that can coexist in the same, 'real' person" (Ayduk & Mischel, 2002, p. 100).

dent agent is achieved not by seeking positive self-regard but rather by maintaining a chronic self-critical view" (p. 767).

This chronic self-critical view is mandatory, even for samurai as prescribed by Yamamoto:

> Moreover, if at the time that you are asked to perform some function you have deep happiness or great pride, it will show exactly as that on your face. This has been seen in many people and is rather unbecoming. But another type of person knows his own defects and thinks, "I'm a clumsy person but I've been asked to do this thing anyway. Now how am I going to go about it? I can see that this is going to be much trouble and cause for concern." (ca. 1700/1983, p. 36)

So, if even a show of happiness, or great pride for having been asked to perform some function for one's lord, is unbecoming, if the rule is to say "I am a clumsy person", how unbecoming will it be to speak of oneself, and embellish on top!

We are reminded of another rule that says: "Being a retainer is nothing other than being a supporter of one's lord, entrusting matters of good and evil to him, and renouncing self-interest. If there are but two or three men of this type, the fief will be secure" (ib., p. 20).

This last point, of the *fief being secure*, summarizes the main cause of the *ritter* and *strict ritter* personalities: the fief (its land and people, and above them its lord and his heirs, and their name even)[250] is in constant danger of being attacked by a neighboring lord, or by the shogun himself. Everybody and everything must be organized for an efficient defense, or better for an attack against a less secure nearby fief. This, in turn, demanded those *strict ritter* whose training is described in Yamamoto's *Hagakure: The Book of the Samurai,*[251] and in my chapter 9 of volume 3.

---

250 "And on the house of Asano severe punishments, con-comittant to that capital one of *seppuku*, were inflicted one after the other—confiscation of property and extinction of the family name" (Shioya, 1956, p. 18).

251 See also *Yukio Mishima on Hagakure: The Samurai Ethic and Modern Japan.*

22

The Making of a *Special Insular Personality*: Canetti

1. Introduction

Elias Canetti (1905-1994)

Novelist and playwright whose works explore the emotions of crowds, the psychopathology, and the position of the individual at odds with the society around him; he was awarded the Nobel Prize for literature in 1981. Born in Ruschuk (now Ruse), Bulgaria, he moved with the family to Manchester at the age of six. After the sudden death of his father, one year after, he moved first to Lausanne, then to Vienna.

Thanks to the writings of Elias Canetti (the three volumes of his autobiography, his novels, and his massive analysis of crowd psychology) we can study the making of an individual case of *insular personality* (albeit a special one) from childhood onward, and discover that the workings of the unity of power (UP) on a single youth reflect the same results that the unity of power has on large populations, along the centuries.

So far, my studies of *insular* and *visitor* personalities have been group studies—ethnopsychological studies—in which individuals were seen to be endowed, more or less, with the philosophy of life of their society, shaped by centuries of UP or DP. For instance, in chapter 1 of volume 2, I compared the *insular* Burgundians with the *visitor* Tuscans, and related their ethnopsychology to the UP of the Dukes of Burgundy, and the DP of the Tuscan Communes. With Canetti we can instead study the making of an *insular* personality per se, independently of the ethnopsychology of the people around him.

However, we must be careful and not expect too much: Canetti is not a *standard insular,* as described in Table 5 in chapter 11 of volume 1, but only an *insular* (and therefore special) for his praise of the crowd(s)—a praise unique among great thinkers—and for his warm participation in them in some instances.

2. Canetti's Fascination with Crowds, and His Disdain for Individuality

For Canetti, in *Crowds and Power*:

> There is nothing that man fears more than the touch of the unknown. He wants to see what is reaching towards him, and to be able to recognize or at least classify it. Man always tends to avoid physical contact with anything strange. In the dark, the fear of an unexpected touch can mount to panic. Even clothes give insufficient security: it is easy to tear them and pierce through to the naked, smooth, defenseless flesh of the victim.... It is only in a crowd that man can become free of this fear of being touched. That is the only situation in which the fear changes into its opposite. The crowd he needs is the dense crowd, in which body is pressed to body; a crowd, too, whose psychical constitution is also dense, or compact, so that he no longer notices who it is that presses against him. As soon as a man has surrendered himself to the crowd, he ceases to fear its touch. Ideally, all are equal there; no distinctions count, not even that of sex. The man pressed against him is the same as himself. He feels him as he feels himself. Suddenly it is as though everything were happening in one and the same body. This is perhaps one of the reasons why a crowd seeks to close in on itself: it wants to rid each individual as completely as possible of the fear of being touched. The more fiercely people press together, the more certain they feel that they do not fear each other. This reversal of the fear of being touched belongs to the nature of crowds. The feeling of relief is most striking where the density of the crowd is greatest.... Only together can men free themselves from their burden of distance; and this, precisely, is what happens in a crowd. During the discharge distinctions are thrown off and all feel equal. In that density, where there is scarcely any space between, and body presses against body, each man is as near the other as he is to himself; and an immense feeling of relief ensues. It is for the sake of this

> blessed moment, when no-one is greater or better than another, that people become a crowd. (1960/1973, pp. 15-16, 18)

Then, in the second volume of his autobiography, *The Torch in My Ear,* he gave two examples of his fascination with the crowd. The first happened in June 1922, when he was seventeen:

> Early on, about one year after arriving in Frankfurt, I had watched a workers' demonstration on the *Zeil.* They were protesting the murder of Rathenau. I stood on the sidewalk; other people must have been standing near me, watching too, but I don't remember them. I can still see the large, powerful figures marching behind the Adler Works sign.... I sensed a powerful conviction emanating from them; it grew more and more powerful. I would have liked to be part of them; I wasn't a worker, but I took their shouts personally as though I *were* one.
>
> The memory of this first demonstration that I consciously witnessed was powerful. It was the physical attraction that I couldn't forget. I was so anxious to belong to the march, but it wasn't deliberation or reflection and certainly not skepticism that kept me from taking the final leap. Later on, when I gave in and did find myself in a crowd, I felt as if this were what is known in physics as gravitation. (1980/1982, pp. 79-80)

The second time happened in Vienna in 1927 and was much stronger:

> From all districts of the city, the workers marched in tight formations to the Palace of Justice, whose sheer name embodied the unjust verdict for them. It was a totally spontaneous reaction: I could tell how spontaneous it was just by my own conduct. I quickly biked into the center of town and joined one of these processions....I became a part of the crowd, I fully dissolved in it, I did not feel the slightest resistance to what the crowd was doing.... My feelings on that day were all focused in *one direction*. It was the most *unambiguous* day that I can remember, unambiguous only because one's feelings could not be diverted from the day as it went by. (ib., pp. 245, 247)

In the summary words of Dagmar Barnouw: "[For Canetti] only in the densest crowd does the reciprocal fear disappear and is replaced by the alleviation of our limitations. The heavier and denser the crowd, the lighter the weight [role] of the individuality" (1979, p. 51).

Very instructive, in this respect, is a famous passage, in Canetti's major novel *Auto-da-Fé*, "celebrated by other critics as Georg's[252] eloquent disquisition on the futility of individuality and the inevitability of the crowd—a passage that, as we noted above, has consistently been seen as an expression of Canetti's own views on the crowd,... Basking in his ability to mediate the multitudinous roles imagined for him by his psychotic patients, and despairing at his assistants' constitutional incapacity to do so, this preeminent psychiatrist is inspired to explain what distinguishes him from these mundane colleagues" (Donahue, 2001, p. 158):

> Of that far deeper and most special motive force of history, the desire of men to rise into a higher type of animal, in to the mass, and to lose themselves in it so completely as to forget that *one* man ever existed, they [his assistants] had no idea. For they were educated men, and education is in itself a *cordon sanitaire* for the individual against the mass in his own soul.
>
> We wage the so-called war of existence for the destruction of the mass-soul in ourselves, no less than for hunger and love. In certain circumstances it can become so strong as to force the individual to selfless acts or even acts contrary to their own interests.... We know nothing of it; we live still, supposedly as individuals. Sometimes the masses pour over us, one single flood, one ocean, in which each drop is alive, and each drop wants the same thing. But it soon scatters again, and leaves us once more to be ourselves, poor solitary devils....again, possibly in a single country at first, eating its way out from there, until no one can doubt any more, for there will be no I, you, he, but only it, the mass.
>
> For one discovery alone Georg flattered himself, and it was precisely this: the effects of the mass on history in general and on the life of individuals; its influence on certain changes in the human mind. (1935/1946, p. 411)

---

[252] Georg Kien, gynecologist turned psychiatrist.

3. The Making of a *Special Insular*

A. Introduction, on the Making of a *Standard Insular*

This time, instead of referring the reader to chapter 11 of volume 1 of *The GAM/DP Theory*, I report those sections on the making of *insular* and *visitor* personalities which can throw light on Canetti's youth as well as on his fascination with the crowd:

> The *insular personality* is the result of a life, youth in particular, in a society ruled by one domineering power or one dominating hierarchy, (e.g., the emperor, king, local lord, or, in modern times, the government, political party, industry, or professional boss) and where this power is not restricted by any major division of power—in other words, when there is no possibility of maneuver, for the individual or the group, by switching allegiance. Insular people know that they have no recourse, no escape, and that they must stay in the good graces of the power holder or that their only alternative is to fight and seize power: something not open to many.
>
> Visitors, instead, know that they can always resort to Power Holder A, who will surely welcome new supporters in the fight against Power Holder B. In modern times this will mean having recourse to the legislative power if things go wrong with the executive or appeal for help to the judiciary, one of the political parties, the clergy or the unions. Visitors know that they can never be harmed too badly, never be crushed because somebody else will come to their aid....Without recourse against the power above, insular persons must protect themselves through the strongest ties at their own level and at the hierarchical levels under them. This means maximum emphasis on family ties. Only the closest relatives (parents, spouse, children) will help them in case of trouble. The family must be united; the family must be the secure haven where each can take off the hated mask of smiling obedience. The children must know that they can count on their parents regardless of whatever happens to them. (2001, pp. 142-43)

Turning to the "Summary Table of *Visitor* versus *Insular* Personality Characteristics", its very first set lists, for *visitor*: "Free will and individualism", and for *insular*: "Determinism, fatalism, and group thinking" (p. 141).

B. The First Episode

From Falk (1993):

> [As narrated in volume one of his autobiography, *The Tongue Set Free (TF)*], Canetti's earliest memory takes him back to the summer of 1907 when he was only two years old. His parents were vacationing in Carlsbad that year and they had taken a young Bulgarian girl with them to serve as a nanny. Canetti remembers a maid carrying him from his room into the hallway, where they meet a man who approaches them and tells the child to stick out his tongue. The man takes a knife from his pocket, advances the knife toward the child and says, "Now we'll cut off his tongue" (*TF*,3). The frightened child is unable to withdraw his tongue to escape the advancing knife, but at the last moment the man retracts the blade, returns the knife to his pocket, and tells the child, "Not today, tomorrow" (*TF*,3). Apparently, each morning began with this terrifying ritual.
>
> Ten years later Canetti learned from his mother the details surrounding that early memory. Presumably the nanny had met the young man in town and would visit his room at night. The purpose of the morning episodes was to make sure the child did not speak about these encounters. The threat was seemingly successful and "the child quite literally held his tongue for ten years" (*TF*,4)....The young man had ordered the child to be silent and the child was obedient not only at the time, but for 10 years. The command was issued by a source of greater strength, there was "no hope of fighting" it, and it was not "open to doubt" at the time (*Crowds and Power*, 305).

C. Additional Fears, and the Discovery of the Crowd

As narrated by Canetti:

> Wolves were the first wild animals I heard about. In the fairy tales that the Bulgarian peasant girls told me, there were werewolves, and one night, my father terrorized me with a wolf mask on his face.
>
> It would be hard to give a full picture of the colorful time of those early years in Ruschuk, the passions and the terrors. Anything I subsequently experienced had already happened in Ruschuk. (1977/1983, p. 4)

> Five or six of them [Bulgarian peasant girls] always lived in our home. They were quite young, perhaps ten or twelve years old, and had been brought by their families from the villages to the city, where they were hired out as serving maids in middle-class homes.... In the evening, when my parents went out, I stayed at home with the girls. Low Turkish divans ran all the way along the walls of the huge living room. Aside from the carpets everywhere and a few small tables, they were the only constant furnishing that I can remember in that room. When it grew dark, the girls got scared. We all huddled together on one of the divans, right by the window; they took me into their midst, and now they began their stories about werewolves and vampires. No sooner was one story finished than they began the next; it was scary, and yet, squeezing against the girls on all sides, I felt good. We were so frightened that no one dared to stand up, and when my parents came home, they found us all wobbling in a heap. Of the fairy tales I heard, only the ones about werewolves and vampires have lodged in my memory. Perhaps no other kinds were told. (ib., pp. 9-10)

Here, finally, the little boy has found a warm environment that combines the excitement, needed by a "surpassingly impressionable"[253] youth, with the protection afforded by squeezing against friends on all sides, in something which will later be found in the crowd: "In that density, where there is scarcely any space between, and body presses against body, each man is as near the others as he is to himself; and an immense feeling of relief ensues" (Canetti, 1960/1973, p. 18).

D. The Mother: First her Absence in his Life in Carlsbad, then her UP in Lausanne

There is much missing in Canetti's story of his great fear at age two in Carlsbad, when the lover of his nanny scared him with a jack-knife. Especially missing is why it took him ten years before he had the courage to ask his mother about what had happened. One possible explanation is that the boy felt, somehow, that his mother was at fault, and that it was better not to bring up a story in which she had not played the role of the protector, a story in which her little boy had not seen in

---

[253] Leonard, 1982, p. C21.

her the obvious protector against the man who told him he would cut out his tongue if he would speak. The man was Power A, and the mother should have been the logical Power B which opposed A. But the child did not go to her, it simply did not come to his mind to do so. He was in the charge of the nanny, and if the nanny did nothing to protect him, that was the end of the story: he had to suffer, period.

There is something that Canetti told, on how his mother was before the death of her husband, which throws light on why the boy did not dare to go to her for help, on why he could not see her as Power B: "Earlier I had known her as impatient and autocratic, overbearing, impulsive; the movement I most clearly remembered was her ringing for the governess to get rid of us children" (1977/1983, p. 61).

One year later, in 1913, we find mother and son in Lausanne where the mother taught him German during the summer in preparation for his schooling in Vienna. She invented a strange way to teach the new language by withholding the English-German grammar from him:

> She read a German sentence to me and had me repeat it. Disliking my accent, she made me repeat the sentence several times, until it struck her as tolerable. But this didn't occur often, for she derided me for my accent, and since I couldn't stand her derision for anything in the world, I made an effort and soon pronounced the sentence correctly. Only then did she tell me what the sentence meant in English. But this she never repeated, I had to note it instantly and for all time. (ib., p. 67)

Soon she derided him for his slow progress:

> When she became particularly impatient, she threw her hands together over her head and shouted: "My son's an idiot! I didn't realize that my son's an idiot!" ...I now lived in terror of her derision, and during the day, wherever I was, I kept repeating the sentences...After all, I had no book to check myself in; she stubbornly and mercilessly refused to me have it [the English-German grammar], though knowing what friendship I felt for books and how much easier it would all have been for me with a book. But she had the notion that one shouldn't make things easy for oneself; that books are bad for learning languages; that one must learn them orally, and that a book is harmless only when one knows something of the language. She didn't notice

> that I ate little because of my distress. She regarded the terror I lived in as pedagogical. (pp. 68-69)

Later his governess tried to put in a good word for him, but to no avail:

> Of course, I listened when Miss Bray said to Mother: "The boy is unhappy. He says Madame considers him an idiot!" "But he *is* one!" she was told. "Otherwise I wouldn't say so!" That was very bitter, it was the word on which everything hinged for me. I thought of my cousin Elsie in Palatine Road, she was retarded and couldn't speak properly. The adults had said pityingly: "She's going to remain an idiot." (p. 69)

This, indeed, is not the kind of mother to whom a child will go for help when he is scared by a jackknife. This is the kind of mother with whom it is better to wait until the age of 12—when she was affected by the death of her husband and in need of the love of her son—before asking her to explain what happened with the man who told him he would cut out his tongue. This is the mother who, indeed, may discourage individuality—by insisting on her teaching method larded with the worst of all insults "idiot!"—and foster a love for the crowd: in the crowd, a least, the mother [or mother substitute, or substitutes of the man with the jackknife, or the werewolves and vampires] cannot reach him; in the crowd he is no idiot, all are equal, and equally clever.

With her teaching methods, the Mother in Lausanne is Power A over the poor boy (with the governess unable to play Power B), as much as she failed to be Power B in Carlsbad when the child needed that power to balance the Power A of the man with the jackknife. In both instances a strong *insular* element was forced on Canetti's personality, both stronger than the fears from the stories of werewolves and vampires, which were conquered finally by wobbling in a tight heap with the young Bulgarian girls.

Back to the title of the first volume of Canetti's autobiography, *The Tongue Set Free*, "It would be wrong," wrote Falk (1993), "to believe that Canetti 'held his tongue' and remained completely silent until he was 12 years old. Virtually every episode that follows this first memory is filled with stories about speaking and the power of language" (p. 4). Yet that conversation with his mother must have had a major liberating effect—beyond his "disobeying the order not to talk"—if he chose this event as the title of his autobiography.

> In his study [*Crowds and Power*], Canetti points out that "every command consists of momentum and sting. The momentum forces the recipient to act, and to act in accordance with the content of the command" (*CP*, 305). The two-year-old child, however, did not carry out the command to the fullest because he finally—after 10 years—disobeyed the dictate by speaking to his mother about the morning episodes. Consequently, the sting, that element of the command that is retained in the mind of the victim, has been evaded after all, making the force of the command ultimately ineffective. (Falk, 1993, p. 4)

However, with all due respect, I disagree. Once the family had returned to Bulgaria, there was no way in which the young man could avenge himself on the boy for talking. In no way could an intelligent boy, like him, think that there could still be a residue of the command not to talk: some sting that could hurt him. The reason for his waiting ten years before asking what happened rests, far more, in knowing how much his mother was "impatient and autocratic, overbearing, impulsive; the movement I most clearly remembered was her ringing for the governess to get rid of us children." Asking a question, too early, would only have made her ring for the governess to get rid of him.

The way I read Canetti, the "tongue set free" of the title of the book is the tongue which became free when he dared to talk to his mother about the man with the jackknife. Certainly, at the end of the joint explanation of the story, it is the mother who is at fault because she was not there to defend her son against the jackknife. But, was she ever there for him? Only after that conversation is the boy free, free of his mother,[254] but not free of the need for the warmth of the crowd, or free of the fascination for the nature of the crowd. In the absence of a balancing Power B in his youth, her Power A made him an *insular* (albeit partial or *special*).

---

[254] Only to a point, as can be found in reading all three volumes of Canetti's autobiography; yet free, at least, tȯ talk .

E. The UP of Grandfather Canetti

> When Elias was six years old, his parents decided to leave the oppressive dominance of the grandfather's home and business [in Ruschuk] and join the mother's brothers in their business in Manchester, England. The grandfather was outraged at their decision and made every effort to thwart their plans. The most terrible event, however, took place prior to their departure, when the grandfather "cursed his son solemnly" in the presence of a horrified family (*TF*, 33). Canetti recalls that he "heard them speaking about it; Nothing, they said, was more dreadful than a father cursing his son" (*TF*, 33). What made this childhood experience especially poignant was that the father died very unexpectedly a year later. Both the boy and the grandfather attributed the death to the curse. The seven-year-old boy, then and later, perceived the death as an example of the power of words. (Falk, 1993, pp. 7-8)

As it had been the case for Canetti himself, the father in Ruschuk also had no Power B which could oppose the Power A of the grandfather: "No curse is more dangerous and no curse more feared. My father had stuck to his guns, and not much more than a year after arriving in England he was dead" (1977/1983, p. 62).

Everybody was powerless to prevent the curse, as the boy had been powerless against the jackknife first, and then against his mother's teaching methods and insults. In each case, man alone—or in a small group, as in Ruschuk—is powerless. The only safe alternative is the crowd, a big crowd, like the one in Frankfurt in 1922: "They marched in serried ranks and cast defiant glances around. Their shouts struck me as though addressed to me personally. More and more of them came. There was something consistent about them, not so much in their appearance as in their conduct. There was no end of them... it grew more and more powerful. I would have liked to be part of them.... It was the physical attraction that I couldn't forget. I was so anxious to belong to the march" (1980/1982, pp. 79-80).

Back to Ruschuk, in July 1911, it was tragic that nobody had the courage—better yet, the power—to stop grandfather Canetti from cursing his son: nobody to shout at him "Stop you fool! Since when is it a crime for a 29-year-old man, married, with three children, to leave his father, and go to work elsewhere? What is this stupid story of cursing one's son for a most normal step in life? Stop it!" The only thing we read is that all relatives, present in the courtyard, listened in horror, which is just what the grandfather wanted. With the years, he had acquired sufficient power over them to prevent even a shade of a rebellion.

F. Back to the Mother, and a Comparison of Miss Bray with Ishiguro's Miss Kenton

Two years after the curse of the father, Miss Bray, Canetti's governess, had tried—in vain—to stop a similar curse, this time on young Canetti, when the Mother called her son 'idiot!'.

Poor Miss Bray, she had so little power against the mother: she could not insist on defending the boy, as Miss Kenton could not insist, as described by Ishiguro in his novel *The Remains of the Day*:

> "Mr. Stevens [said Miss Kenton], I am outraged that you can sit there and utter what you have just done as though you were discussing orders for the larder. I simply cannot believe it. You are saying Ruth and Sarah are to be dismissed on the ground that they are Jewish?"
>
> "Miss Kenton, I have just this moment explained the situation to you fully. His lordship has made the decision and there is nothing for you and I to debate over."
>
> "Does it occur to you, Mr. Stevens, that to dismiss Ruth and Sarah on these grounds would be simply - *wrong*? I will not stand for such things. I will not work in a house in which such things can occur. ... I am warning you, Mr Stevens, I will not continue to work in such a house. If my girls are dismissed, I will leave also. ..."
>
> "Miss Kenton, let me suggest to you that you are hardly well placed to be passing judgments of such a high and mighty nature. The fact is, the world of today is a very complicated and treacherous place. There are many things you and I are simply not in a position to understand concerning, say, the nature of Jewry. Whereas his lordship, I might venture, is somewhat better placed to judge what is for the best". (1989, pp. 148-149)

In the end the two maids were fired, and Miss Kenton did not resign. A year after, the subject came up again between the two:

> "You probably have no idea, Mr. Stevens, how seriously I really thought of leaving this house. I felt so strongly about what happened. Had I been anyone worthy of any respect at all, I dare say I would have left Darlington Hall long ago.... It was cowardice, Mr. Stevens. Simply cowardice. Where could I have gone? I have no family. Only my aunt.... Whenever I thought of leaving, I just saw myself going out there and finding nobody who knew or cared about me. There, that's all my high principles amount to. I feel ashamed of myself. But I just couldn't leave, Mr. Stevens. I just couldn't bring myself to leave." (ib., pp. 152-53)

Miss Bray could have said the same. She had been fired already once for teaching Christian songs to the boys, and finally she had been kept only because George, the youngest, was very attached to her. Now, she was in Lausanne, possibly without the money to pay her fare back to England. Like Miss Kenton, she could not risk leaving her present job without a letter of recommendation for a new position back home. She could not do more for Canetti, like the relatives in Ruschuk for his father.

3. Additional Considerations

By paying close attention to the *insular* in Canetti, by looking at his youth as an example of the working of the unity of power which led him to his protective passion for the intimacy of the crowd, we have neglected the *universalist-miner-leadsman* in him: *universalist* because of the early death of his father, *miner* because of the strange love-hate relationship with his mother, and *leadsman* because of his Jewish origins but life from age six outside of a Jewish community or regular Jewish family.[255]

---

[255] For instance, as already intimated, at age seven, after the death of his father, his governess, Miss Bray, taught him Christian songs: "Many of the songs mentioned the Lord Jesus, she told us his story, I wanted to hear about him, I couldn't get enough, and I couldn't understand why the Jews had crucified him.... [The mother discovers this, and Miss Bray tries to explain what she did.] She spoke to her about Jesus, saying he had died for us too, I [Canetti] butted in, fully won over by her, Mother got into a dreadful temper and menacingly asked Miss Bray whether she didn't know that we were Jews, and how could she dare lead her children astray behind her back?" (1977/1983, p. 64).

On the other hand, it is Canetti himself who in all his books is forcing us to pay close attention to his *insular* fascination with crowds. The early death of his father certainly contributed to the *universalist* component of his personality and creativity: his desire to come to an all encompassing understanding of crowds and power, in all countries and all times. However, that death was immediately overshadowed by the pains and problems of the mother,[256], and later by the efforts of the son to understand the causes, of the death of the father, which converged on the conclusion that the mother was at fault because—as discussed in detail in Canetti's third volume of his biography, *The Play of the Eyes* (1985/1986; pp. 221-229)—of the pains the father suffered from the discovery of her literary/language infidelity with her doctor in Bad Reichenhall, Germany, where she went for an extended period of cures.

Certainly much more could be said on Canetti, *universalist-miner-leadsman*, possibly also *reformer* from the resented maternal domination. Yet, I see already much—and fascinating—that can be learned from his life and works on the impact of the unity of power on a single individual, independently of the action of the large UP or DP affecting a whole nation or civilization.

---

[256] "During the early months, the disconsolate widow relied heavily on the strength and support of her oldest son [Elias], perhaps more than should have been expected of a boy of his age. He was her constant companion; they took their meals together; he even slept in the same room with her, preventing her from committing suicide on numerous occasions" (Falk, 1993, p. 12).

23

Against UP, Therefore Against Castiglione

1. Introduction

Peter Burke told his readers, at the beginning of his 1996 *The Fortune of the* ***Courtier*** *- The European Reception of Castiglione's* ***Cortegiano***: "Since I first read it in the late 1950s, Castiglione's *Cortegiano*, or *Courtier*, has been one of my favourite books.... I have written primarily for people who already know and love the *Courtier* and would like to known more about its context" (pp. viii-ix).

With such a "caveat", I should not have read Burke, but having bought his book, I read it anyway, curious to understand why anybody in the second half of the 20th centuries could love the *Courtier*: how anybody could find pleasure in a book instructing people on how to please their masters at the cost of their *liberté, egalité, et fraternité.*

Given that *De gustibus non est disputandum*, I can only present my reason for disliking Castiglione: behind his work I see the corruption induced by the Law of Power of Lord Acton—*Power tends to corrupt and absolute power corrupts absolutely*—and among those corrupted—and then corrupting others—was Count Castiglione himself. This said, the best way to explain my dislike for Castiglione is to start by reporting a number of criticisms of him, his book, and the world of the courtier.

2. Criticisms

George Bull (1976) said, in the introduction to his translation of *The Book of the Courtier*:

> "Today the book still shocks, not of course because of its mild indelicacies but because of contemporary impatience with the fundamental values which it enshrines. It is hard, indeed, to think of any work more opposed to the spirit of the modern age.[257] At an obvious level, its preoccupation with social distinction and outward forms of polite behaviour creates an intense atmosphere of artificiality and insincerity. (When James Joyce first read *The Courtier* his brother told him he had become more polite but less sincere.)
>
> The great virtues it proposes for a gentleman are discretion and decorum, nonchalance and gracefulness. As Luigi Barzini comments in his satirical book on the modern Italian, quoting Castiglione as the model, the 'show' is all. The courtier must watch his dress, his speech, his gestures chiefly because of their effect on his reputation. If he fights well in battle, he must make sure his commander sees him do so....The combination of intense and selfish individualism with appalling snobbery naturally repels the modern reader, and also the modern historian, who is suspicious of the assumption that history consists in the exploits of the ruling class and the favoured few....
>
> More seriously, *The Courtier* offends modern susceptibilities because it is a flight from the truth. Throughout the book, to be sure, there runs a vein of natural melancholy, partly attributable to Castiglione's constant awareness of the fickleness of Fortune and the inevitability of death. But by its very nature it is a book which turns aside from the realities of life to its idealization. War—which Castiglione experienced and disliked—is glorified; the criminal behaviour of some of the gay companions he knew at Urbino is glossed over or ignored; the crudeness of Court life in sixteenth century Italy—an earlier version of *The Courtier* was far nearer in this respect to the historical reality—is refined away. The political discussions—using the language and concepts of the ancient world with regard to the rule of the one, the best or the many—are totally unrealistic" (pp. 15-17).

Then in the words of Sydney Anglo (1987): "Castiglione has no evil motive: yet there are two dominant features of his work which are susceptible to sinister interpretation. The first is his stress on seeming

[257] And the spirit of the previous *communal age* of Dante and Boccaccio.

rather than being, upon appearances rather than reality.... The other striking feature of *Il Cortegiano* is its stress on the art of pleasing. Castiglione's courtier is both dissimulator and consummate toady. He is agreeable to great men, gentlemen, and ladies; and every day alters his manner according to the company he keeps. At one point, it's suggested that the courtier is to serve the prince with devotion and, 'in his will, manners and fashions to be altogether pliable to please him'" (p. 15).

Also Vincent Cronin (1969) in his *The Flowering of the Renaissance* called a spade a spade: "The courtier's aim in life should be to keep on good terms with his prince, not least by telling amusing stories—such jests, however, should never be anti-religious—in order to instruct him in virtue, check any wilful use of his power and incline him to peace. The courtier cheerfully accepts one-man rule, because this most resembles God's government of the earth, and because in the battlefield, on board ship and in constructing a building responsibility is given to a single man.... Ideally, the prince should act on the advice of a council of nobles and a council of the commons—Castiglione had been impressed by the English constitution—but human nature and society being what they are, the courtier would have to influence his prince indirectly, by purveying 'honest and honourable pleasures'. His political role, in short, was rather humble: to 'sweet the potion, gild the pill'" (p. 99).

But probably the best description and critique of the courtier is that by La Bruyère (1645-1696), the French essayist and moralist who knew the courtier from direct experience. In his "De la cour," La Bruyère wrote:

> # 2. "*Un homme qui sait la cour est maître de son geste, de ses yeux et de son visage; il est profond, impénétrable; il dissimule les mauvais offices, sourit à ses ennemis, contraint son humeur, déguise ses passions, dément son coeur, parle, agit contre ses sentiments. Tout ce grand raffinement n'est qu'un vice, que l'on appelle fausseté, quelquefois aussi inutile au courtisan pour sa fortune, que la franchise, la sincérité et la vertu*"[258] (ca. 1690/1951, p. 215).

---

[258] (The capable courtier keeps his gestures, his eyes, and his countenance under control; he is consummate, impenetrable; he conceals the bad assignments, smiles to his enemies, masters his temper, disguises his passions, belies his heart, speaks, behaves against his feelings. All this big refinement is nothing but a vice, which we call falsity, at times as useless to the courtier for his fortune, as frankness, sincerity and virtue.)

And even more telling of the power of the *supremo* and of the abject condition of the courtier: # 13. *"Il n'y a rien qui enlaidisse certains courtisans comme la présence du prince: à peine les puis-je reconnaître à leurs visages; leurs traits sont altérés, et leur contenance est avilie"* (ib., p. 217).[259]

3. The Court

Moving from the courtier to the court, the following, from Ralph Albanese Jr.'s discussion of life at court in France in the times of Henry II[260] (r. 1547-1559), confirms and expands on what said up to now:

> The court is, in short, a deleterious presence, just as much a locus of vulnerability as one of sociability: its ordered or controlled agitation—*'une sorte d'agitation sans désordre'*—encompasses the intrigues inherent in the courting ritual. Dissemblance was thus an integral part of the maintenance of appearances, since life at court entailed a constant observation of the actions of others as well as one's own.... The perpetual watchfulness of society assumes the dimensions of a spy system at court. Given the radical subordination of the private self to the imperatives of public scrutiny, one witnesses the systematic crushing of intimacy, the virtual impossibility of private relationships.... The constraining influence of the court can best be measured by the fact that all courtiers, regardless of rank, are subjects of the king. Ceremonies serve to exalt royal greatness and elaborate patterns of social ritual exert a powerful pressure on the courtier, since *bienséance* requires his presence at courtly activities at all times....This public obligation to remain on permanent display is a cornerstone of the courtier's existence. Representation of the self precludes the possibility of solitude. The courtier's highly problematic status stems from his intensive infighting with others to win royal favor: his raison d'être depends upon his

---

[259] (There is nothing which makes some courtiers uglier than the presence of the prince. I could hardly recognize their faces: their traits are perverted, and their bearing debased.)

[260] As described in Mme. de La Fayette's novel *The Princesse de Clèves* of 1678. As an additionial piece of information, Henry II was the son of that Francis I who in 1516, at Bologna had warmly urged Castiglione to publish *The Courtier* (Hacket, 1937, p. 171).

> ability to reside (i.e., succeed) at court.... If the fundamental idleness of the courtiers helps to account for the collective amorality, artificiality, and emptiness of individual lives, an irreducible link exists between the enslavement of courtly existence and the enslavement of passion. This notion of undifferentiated sameness subverts an aristocratic code governed by the profound desire to be "distinguished," that is, to be perceived as radically different from others. (pp. 91-93)

4. A Substantial Literary Achievement?

For George Bull, "as a handbook for gentlemen, *The Courtier* conceals the most shameless opportunism under the cloak of a tiresome refinement; as a memoir of life at the Court of Urbino, it touches up history to the point of distortion. Nonetheless, even though many of Castiglione's values are questionable (though for their time they shone out in an uncouth world), *The Courtier* cannot be so easily dismissed. It is historically significant and instructive. It is, at the very least, an entertaining book. Most of all, it is a work of substantial literary achievement" (1976, pp. 15-17).

Contrary to George Bull, I do not find *The Courtier* entertaining, while in no way negating its substantial literary achievement, provided that we remember that it is a work at the service of princes, fostering an *insular mentality*. To be clear, let me briefly discuss another substantial UP literary achievement, *Thousand and One Nights* (also known as *Tales from the Arabian Nights*).

Thousand and One Nights begins with what king Shahryar did when he discovered the stark infidelity of his wife. He had her killed, and he in person slew all people involved in the misdeed, concubines and attendants:

> He also sware himself by a binding oath that whatever wife he married he would abate her maidenhead at night and slay her next morning to make sure of his honour, "For," said he, "there never was nor is there one chaste woman upon the face of the earth." Then [his brother] Shah Zaman prayed for permission to fare homewards; and he went forth equipped and escorted and travelled till he reached his own country. Meanwhile Shahryar commanded his Wazir to bring him the bride of the night that he might go in to her; so he produced a most beautiful girl, the daughter of one of the Emirs, and the King went in unto her at even-

> tide and when morning dawned he bade his Minister strike off her head; and the Wazir did accordingly for fear of the Sultan. On this wise he continued for the space of three years; marrying a maiden every night and killing her the next morning, till folk raised an outcry against him and cursed him, praying Allah utterly to destroy him and his rule; and women made an uproar and mothers wept and parents fled with their daughters till there remained not in the city a young person fit for carnal copulation. Presently the King ordered his Chief Wazir, the same who was charged with the executions, to bring him a virgin as was his wont; and the Minister went forth and searched and found none; so he returned home in sorrow and anxiety fearing for his life from the King. Now he had two daughters, Shahrazad and Dunyazad. (Burton transl. 1885/1978, p. 14)

At this point a *visitor reader* may ask with what right the Sultan raped and killed a thousand girls? Indeed, under a minimum of DP, the Vizir would instead have refused the Sultan's initial request, and have sharply reminded him of his duties as ruler.

However under the UP described in *Thousand and One Nights*, the Vizir was terror-stricken. Obviously, there was no other power to help him against the Sultan, no other major script which told the Sultan to think and behave differently. Under that UP any attempt by the Vizir to act in defense of the people would have resulted in his being sliced to death, and all the members of his family killed in the most horrible fashion. The Sultan had soldiers, paid by him, who obeyed him completely. So nobody would have escaped, starting with the Vizir and his daughters.

Moving from literature to history, the following reported by Philip Hitti on the famous Caliph Harun al-Rashid (764-809)[261] explains why the Vizir of Sultan Shahryar could not even envisage opposing minimally the UP of his master:

> In January 803 the caliph, as if from a clear sky, had the thirty-seven-year-old Ja'far executed. His severed head was exhibited on a bridge and the two halves of his bisected body on two other bridges, there to remain for a year. His aged father

[261] This is the ruler who exchanged diplomatic missions with Charlemagne "in the course of which the Caliph sent, among other gifts, and elephant to the Frankish Emperor, and granted special facilities to Frankish pilgrims visiting the Holy Places in Palestine" (Saunders, 1965, p. 115).

> and his brother died in prison. The Barmakid property in the amount of 30,676,000 dinars was confiscated. Ja'far's palace became the residence of Caliph al-Ma'mun. The vizirial family was wiped out of existence, but the vizirate continued and its incumbents under al-Ma'mun belonged to its school. Likewise the Barmakid name survived to the present as a synonym of unbounded generosity. In the absence of an official explanation of the reasons for the tragic ending of such a prominent family, several reasons—such as impiety, pro-'Alid or pro-Persian proclivities—have been offered with no supporting evidence. One plausible explanation is that the caliph at last could no more tolerate the vizirial tutelage which had become hereditary and assumed the aspects of being a state within the state. In Harun's caliphal firmament there was no room for more than one sun. But why mutilate poor Ja'far's body remains a mystery. (1973, pp. 90-91)

Also the courtiers of Castiglione knew well not to oppose their lord:

> "What you must do," answered Federico, "is to obey your lord in everything that redounds to his profit and honour, but not as regards things that bring him loss and shame. Therefore, if he were to order you to commit some treacherous deed not only are you not obliged to do it but you are obliged not to do it, both for your own sake and to avoid ministering to your master's shame. It is true that many things which seem good at first sight are evil, and many things which seem evil are good. Thus it is sometimes allowable, in the service of one's master, to kill not just one man but ten thousand, and to do many other things which on a superficial view would appear evil although they are not." (1507-28/1976, p. 131)

5. Castiglione: *The Perfect Courtier*

"Baldesar Castiglione, courtier and diplomat, poet, scholar and soldier, is generally thought of as himself typifying the gentlemanly virtues expounded and extolled in *The Book of the Courtier*. The comment of Emperor Charles V, when he heard of his death, has become famous: '*Yo vos digo que es muerto uno de los mejores caballeros del mundo*'—I tell you, one of the finest gentlemen in the world is dead'" (Bull, 1976, p. 9). Accordingly, Julia Cartwright entitled her biography of Castiglione: *The Perfect Courtier*.

The perfect courtier pleases his master(s)—one after the other (as did Walther von der Vogelweide discussed in chapter 19): "Two drafts of the text [*The Courtier*] have survived, and a close study of the process of revision has revealed a number of significant changes. For example, scholars have caught Castiglione the author in the act of deleting references to France and inserting references to Spain at the time when Castiglione the diplomat was advocating an alliance between the pope and Charles V. The author also gave increasing space to noblemen at the expense of the ladies and also of commoners, such as the sculptor Giovan Cristoforo Romano, whose roles shrank in successive versions of the text" (Burke, 1996, p. 36).[262]

Cecil Clough (1978) had similar comments: "Indeed, as occasion demanded so Castiglione revised his text. Between 1521 and 1524 Castiglione incorporated a flowery reference to Charles V, for instance, and another subsequently (probably after 23 May 1524) to the emperor's grandparents, Ferdinand of Aragon and Isabella of Castile. During the period 1521-1524 Castiglione changed the section dealing with love (newly introduced in the second version of 1518-20) so that the emphasis was on its spiritual rather than its carnal aspects. One can suppose that the former additions arose from Castiglione's having an eye to the emperor as a potential patron, and that the latter change was made with an eye to showing his suitability for papal service. By analogy one can speculate that Castiglione probably added the interpolation concerning the King of France early in 1516 because he was then thinking in terms of that king as a potential patron" (p. 26).

The flowery reference to Charles V, which drew Clough's attention, is the following:

> Then Bernardo Bibbiena remarked: "Very great promise is also shown by Don Carlos, Prince of Spain [the future Emperor Charles V], who although he is not yet ten years old displays already such intelligence and such convincing signs of goodness, prudence, modesty, magnanimity and every kind of virtue that if, as many think, he is to be the Emperor of Christendom it is to be

---

[262] That same Charles V, in 1529, shortly before the sudden death of Count Castiglione—at that time Papal Nuncio (of Clement VII Medici) in Madrid—made Castiglione Bishop [sic] of Avila (Cartwright, 1927, *2*, pp. 414-415).

> believed that he must eclipse the name of many ancient emperors and equal the fame of the most famous men who have ever lived." (1507-28/1976, pp. 312-13)

6. The Cynicism of the Courtiers

Cecil Clough, in his 1978 article, "Francis and the courtiers of Castiglione's *Courtier*" studied not only Castiglione but his friends and colleagues Pietro Bembo, Lodovico Canossa, Bernardo Dovizi, Federico Fregoso, and Michel de Silva:

> "The careers of the principal male courtiers of Urbino strongly suggest that as a group the courtiers had no deep loyalty to a prince or a ruling dynasty; their concern appears to have been to get to the top as courtiers, and therefore self-interest was the primary motive. Castiglione abandoned the Gonzaga of Mantua in 1504 when he believed there was greater opportunity for him in the service of Duke Guidobaldo da Montefeltro. Canossa bettered his career prospects by leaving Duke Francesco Maria's service for that of the pope, and subsequently left the latter for Francis I. The general attitude of self-interest appears common to Italian courtiers and was not limited to the Urbino circle; the Sanseverino family, for instance, is sufficient example to make this point" (p. 39).

However the political situation changed with the intervention in the Italian affairs of the armies of Francis I and of Charles V. This, among others, reduced the number of independent principalities and the wealth of their princes, while the church still had some money:

> "The problem for Italian courtiers of the old order, therefore, was that with the number of courts decreasing, and with less opportunity of advancing themselves by switching from one court to another, particularly as princes were turning to the services of bureaucrats, their courtly career was blocked.... It is striking testimony to the situation indicated above that of the courtiers of *The Courtier*, Bembo, Canossa, Federigo Fregoso, Dovizi, Castiglione himself, and Michel de Silva, to whom Castiglione dedicated his *Courtier*, all chose to advance their career by way of the Church. All save Castiglione reached the rank of cardinal....

Finally, it should be noticed that the years after about 1530 witnessed developments which changed the situation outlined above. The shock of Luther's stand and the quickening of reforming zeal, which the self-interest of Italian clerics of all ranks had done much to stimulate meant that only spiritually dedicated Italian ecclesiastics were likely to advance their careers in the Church. The last years of the courtiers of Urbino that were alive after about 1530 show how they, characteristically, were able to trim to the winds of change. Both Federigo Fregoso and Pietro Bembo spent the last ten years of their lives devoting themselves to spiritual matters. In so doing one might cynically suggest that to the end they remained true to self-interest." (Clough, 1978, pp. 46-50)

7. In Reality, against the Unity of Power

Castiglione, like the others–as noted by Woodhouse, 1978–was a courtier because he had no alternative: "The many trivial accomplishments detailed by Castiglione could have been, and often had been, ignored with impunity by the universal men of the earlier Renaissance. Castiglione's contemporaries, by contrast, simply could not afford to ignore any possible preparation which might ingratiate them with a patron and so ensure for themselves a relatively safe haven.[263] It was a period when, outside the protective confines of the court, for the sophisticated individual, there existed only chaos and danger. The development is even more striking by the time Tasso composed his dialogue on the court, *Il Malpiglio* (c. 1586)[264]. There, adulation and sycophancy are

[263] That was already the case with Petrarch, who in 1353—as discussed in volume 2—accepted the invitation of the ruler of Milan feeling that his individual liberty and his dignity as a man of culture would be better assured by a strong aristocratic regime. "This choice was seen as a betrayal by his Florentine friends who were stongly against a 'tyrannical' regime as that of the Viscontis" (Ferroni, 1991, p. 239). A similar displeasure for the Perfect Courtier can be read in the following by Clough: "Castiglione consistently had his eye on the main chance, and one can suppose he calculated concerning patronage of his *Courtier* as carefully as he did all else. He ultimately decided against Francis I, and not surprisingly he received nothing from that king" (1978, p. 38).

[264] Specifically: *Il Malpiglio: A Dialogue on the Court* (1586/1973) conducted between Vincenzo Malpiglio, his son Giovanlorenzo Malpiglio, and a Neapolitan Stranger.

Already, the title is intriguing because it may simply refer to father or son Malpiglio, but it may also refer to the common Italian word *malpiglio* which means

taken for granted as obvious requirements, almost, indeed, as professional qualifications in themselves, and Tasso, aware of this, often comments on the contrast between the tone of his own precepts and those of Castiglione. Just fifty years after Tasso's dialogue, in Niccolò Strozzi's private admonitions, ingratiation and deceptive adulation are essential requirements, while cultural attributes and qualifications are completely ignored in favour of practical and often less scrupulous means of obtaining and maintaining a secure position at court....

Yet the most splendid minds of the period had, since the early Renaissance, been forced to submit to the pettiness of their masters' whims. From the biographies of scholars and statesmen such as Pierpaolo Vergerio, Lorenzo Valla, and Giovanni Pontano, to name but three, the toughness of their endurance may be judged. Among Castiglione's distinguished contemporaries, Michelangelo, Ariosto, Machiavelli, Matteo Bandello, Pietro Aretino and Benvenuto Cellini complain, at times bitterly, about the caprices of their patrons, the wickedness of envious fellow-courtiers or diplomats, the physical discomfort of working conditions or living accommodation, the dangers of insanitary and plague-ridden cities, the perils of campaigning alongside untrustworthy allies in wars for which they had no enthusiasm. Yet all were willing to tolerate personal hardships in order to guarantee their own and their family's safety and avoid the uncivilized insecurity of life outside the court" (Woodhouse, 1978, pp. 189-191).

In the end therefore—as was the case with the mandarins (see chapter 7 of volume 2), or with Ariosto (see chapter 3 of volume 2)—the fault was less with Castiglione than with the UP of the *signori*.

8. An Eminent Pupil of Castiglione: John Donne

John Donne (1572-1631) was the leading poet of the 17th century English "Metaphysical school," and also an eminent pupil of Baldesar Castiglione as discussed by Peter DeSa Wiggins in his (2000) *Donne, Castiglione, and the Poetry of Courtliness*. In the words of Matthew Woodcock, in his review of Wiggins' book:

"angry look, frown, scowl," thereby conveying Tasso's angry attitude toward the court. At the next level, the word *malpiglio* can be seen as the combination of the *mal (male)* [bad/evil] and *piglio* [I take, from the verb *pigliare*, to take]. In this last interpretation *Il Malpiglio* could refer to the gifts and honors the courtiers accept from the lord, or take from the simple people. Also, in the words of Dain Trafton, "Hints that Tasso puts in the Neapolitan Stranger's mouth lead to harsh conclusions about the court: it is infected by envy, founded on the exploitation of courtiers in whom it encourages a base servility, and dominated by a prince whose power is absolute, even over men who are equals or superiors" (1973, p. 10).

> [Wiggins] emphasizes a performative sense of identity and demonstrates the ways in which *The Book of the Courtier* provided Donne with a model for how to write poetry of political ambition and personal advancement. First published in Italian in 1528, Baldesar Castiglione's *The Courtier* was translated into English in 1561 and into Latin ten years later; its repeated reprintings throughout the sixteenth century testified to its enormous popularity. It was, for those of Donne's generation, the quintessential grammar for self-construction and self-presentation, offering in dialogue form an extended guide to manners, conduct and the rhetoric of social advancement. At the opening of his fifth Satyre, Donne admits that Castiglione's "rules" may make good courtiers, but jestingly asks who makes courtiers good[265]....Wiggins concentrates on four main features or "codes" that link Donne and Castiglione: a general tolerance of dissonant views and voices, and attendant resistance to final judgments; a preference for provocative play over serious debate; a characteristic pose of effortless artistry (*sprezzatura*) and concealment of one's labours; and a conscious acknowledgement of the (potentially dangerous) power available to those skilled in casuistry. (Woodcock, 2002, p. 24)[266]

---

[265] Thou shalt not laugh in this leafe, Muse, nor they
Whom any pitty warmes; He which did lay
Rules to make Courtiers, (hee being understood
May make good Courtiers, but who Courtiers good?)
(quoted by Wiggins, 2000, p. 30).

[266] Specifically in the words of Wiggins on the first and fourth code:

1. "Distinguishing Castiglione's dialogue is its participants' tolerance of the dissonant in crucial areas of their experience, to the point where their tolerance becomes the mark of a disabused mentality. Castiglione's courtiers have the discretion to recognize that their exchanges sometimes reach a point of impasse where to attempt even provisory resolutions would be to engage in self-deception.... 4. At risk of being taken for Iago's or Bosola's[*], Castiglione's courtiers make casuistry their business, dedicating much time (especially in Book III) to demonstrations that artifice in their case, whether in cultivating provocative repartee or a nonchalant demeanor, is the exception to the rule that dissimulation is evil. They (along with Donne and his speakers) demonstrate their sincerity by openly conceding that others might be at risk in dealing with them and that their talent for prevarication is dangerous" (pp. 18-19).

[*] Bosola is Duke Ferdinand's evil tool and spy in John Webster's *The Duchess of Malfi* (c. 1613).

Donne's question on "who makes courtiers good" is a jewel of understatement. He knows well that nobody makes him good, and that the whole process is one of increasing corruption. In turn, while "a general tolerance of dissonant views and voices, and attendant resistance to final judgements" can be equated with Isaiah Berlin's praise for pluralism, it may also be equated to the courtier's acceptance of the evil deeds of his master notwithstanding some personal reservations.

24

More on Lord Acton's Law of Power

1. Introduction

The corruption induced by power—so vividly decried by Lord Acton—on both the power holder and those under him, is so pervasive and manifold that it must be studied in ever deeper circles, like Dante studied the nine circles of *Hell*. I have already discussed several of these circles of power in my previous volumes; I continue here my travel with a visit to the circles of the royal counsellors of the Ottoman interpreters and of the royal-aristocratic *honra* described by Calderon de la Barca in *El médico de su honra* (The surgeon of his honor).

2. The Royal Counselors

For centuries, some of the best thinkers around the world thought that the solution to the inevitable corruption of kings and emperors consisted in finding *an uncorrupted and incorruptible counselor* whom the ruler could trust knowing that "the Incorruptible" would never seek personal advantages.

In his 1995 essay on *King Lear*, Ben Schneider, Jr. considered various explanations for the King's mistakes, chiefly among them the lack of an honest advisor, an absence decried, long before, by Seneca, who had written:

> I will show you what the highest in the land stand in need of, what the man who possesses everything lacks: someone, assuredly who will tell him the truth, who will deliver him from the

> constant cant and falsehood that so bewilder him with lies that the very habit of listening to flatteries instead of facts has brought him to the point of not knowing what truth really is. Do you not see how such persons are driven to destruction by the absence of frankness and the substitution of cringing obsequiousness for loyalty? No one is sincere in expressing approval or disapproval, but one person vies with another with flattery, and, while all the man's friends have only one object, a common aim to see who can deceive him most charmingly, he himself remains ignorant of his own powers, and, believing himself to be as great as he hears he is, brings on wars that are useless and will imperil the world. (1995, pp. 4-5)

Having reported this, Schneider concluded that "the evidence strongly suggests that [the Earl of] Kent diagnosed Lear's case at the outset of the play correctly as an example of 'power' seduced by 'flattery' (1.1.148)" (ib., p. 6). In other words, the King—not knowing Seneca—was cursed with the absence of an incorruptible counselor, and had around him only self-interested flatterers.

Not only Seneca, but also Baldesar Castiglione in *The Book of the Courtier* urged in no uncertain terms that a courtier must tell his Lord "the truth about all he needs to know, without fear or risk to displease him" (1507-16/1976, p. 284), lest he fall prey to the flatterers that surround him.[267]

The trouble with this theory is that, in truth, the flatterers are there because the flattered wants them, and he brutally eliminates those who tell him the truth. Already poor Seneca—who thought he could be the incorruptible counselor to an uncorrupted emperor, in his case Nero—became increasingly irksome to his master who, in A.D. 65, ordered him to commit suicide.

Already, in A.D. 59, the incorruptible had been corrupted by his master, and Seneca had been compelled to be privy to Nero's plan to murder his (Nero's) mother Agrippina the Younger, a great-granddaughter of Augustus. So, even for Seneca there had been no escape from the rule that he who touches pitch will be defiled: "What had commenced as

---

[267] However, earlier in his book, Castiglione had given contradictory instructions by saying that "Our courtier will avoid foolish arrogance; *he will not be the bearer of bad news*;... [He will] engage in conversation which will be pleasing and agreeable to his master, in order not to disturb his peace and quiet" (ib., pp. 127-7; my italics).

a reign devoted to clemency and regulated by philosophy declined into vicious intrigue and matricide. And upon every politic occasion, Seneca must compose a letter, or devise a speech. That Seneca was considerably enmeshed in the toils of Imperial vice there can be no doubt. And there can likewise be no doubt that Seneca suffered acutely for his role in much of this business. He himself speaks feelingly of the suffering one must endure from one's conscience after being involved in wrongdoing" (Motto, 1973, p. 32).

In Japan, when things went wrong, the fault could only be attributed to the counselors, never to the great man. So, for those unhappy with political developments, the solution was always the same: kill the "bad" counselor so that a "good" one could take his place. The sad counterpart was that the great man felt that this was as things had to be: his own mistakes did not count, the counselors, even if they had opposed him, had to pay. He—like pharaoh belonged to a superior world.

In the life of Tokugawa Keiki/Yoshinobu (1837-1913), the last shogun, we encounter a succession of chief counselors: Nakane, Hiraoka, and Hara:

> Just as Nakane's death had paved the way for Hiraoka, so now that of Hiraoka would do the same to Hara. Written purposely in a woman's hand by an anonymous family retainer, the poem was a pointed reminder that nothing guaranteed Hara himself would not meet the same fate. "Fate, is it?" murmured Keiki to himself, gazing at the message in his garden and wondering whether it was meant as a slander of himself. He then dismissed it from his mind. For a liege lord, there could be neither weakness nor defeat. *All blame for weakness and defeat went, rather, to his advisers. This was a defining principle of the relationship between lord and retainer*, comprising the chief glory of the feudal age. Having been born and raised a feudal lord, even Keiki, a man of unparalleled sensitivity, was dense in this one matter. That his acrobatics in the political arena cost lives, leaving the road littered with the bodies of his loyal strategists, was something he never understood.[268] (Shiba, 1967/1998, p. 138; my italics)

---

[268] Three years after, "On September 11, 1867 [when already officially shogun], his most trusted aide, Hara Ichinoshin, was cut down by antiforeign fanatics. Certain hatamoto in the shogunate had long despised Hara, blaming him for what they saw as

In China, "while the Son of Heaven [i.e., the Emperor], who held the ruling power through the Mandate of Heaven, could do no evil and was not to be held liable for any mistakes he made, his eunuchs were eminently culpable, becoming tailor-made scapegoats when and if something went wrong with the despotic institution" (Tsai, 1996, p. 221).

Similar things are true for most countries, the result of the universal efforts by the UP to place the *supremo* above suspicion. The *supremo*, by definition, was endowed with all possible virtues. This, for instance, is the classic image we encounter in many oriental stories, for instance in the *Seven Princesses* (or *Seven Beauties*) by Nezami (1141-1204) the greatest romantic epic poet in Persian literature.

Part of the story is a description of the great king—who will marry the seven princesses—who was endowed with all possible virtues and a troubled kingdom. A troubled kingdom? Sure, said Italo Calvino: "The ideal sovereign—so we think—should have a prosperous kingdom with happy subjects. Not at all! These are prejudices of our down to earth mentality. That the king be a wonder of all perfections does not exclude that his kingdom be oppressed by the most cruel injustices perpetrated at the hand of perfidious and greedy ministers. But, given that the king enjoys the celestial blessing, the moment will come when the sad condition of his kingdom will be revealed to him. Then, he will punish the vile Vizier and give satisfaction to everybody who will come and tell him of injustices they suffered" (1991, p. 67).

In other words, it is only thanks to the existence of perfidious and greedy ministers that there can be an ideal sovereign endowed with all possible virtues: all the faults are theirs, with none left for the king.

Back to Europe, also Gottfried of Strassbourg, in *Tristan and Isolde*, narrated how King Mark suffered from the absence of incorruptible counselors, in other words how King Mark loved only 'yes men':

Yoshinobu's bizarre and inscrutable behavior. Being a samurai of the loyalist house of Mito, he must be attempting to persuade Yoshinobu to sell out, they reasoned. Two men participated in the assassination, both of them shogunal vassals. They stopped by Hara's residence in the morning, announcing themselves as 'men of Mito,' then forced their way inside. Coming upon Hara as he was dressing his hair, they raised their swords, killed him, and fled with his head" (ib., pp. 201-202). Once more, the blame and punishment went to the innocent counselor, and not where it belonged.

> [The King, having exiled the two lovers from his court, felt lonely and changed his mind: He] at once summoned his councilors and kinsmen, at court and in the country, to ask them for their advice. He told them how he found the lovers and declared himself no longer willing to believe that Tristan and Isolde had misconducted themselves. His councilors realized immediately which way his wishes were tending, and that he had made his declaration with a mind to having Tristan and Isolde back. *Like the wise men they were*, they advised him according to his own desire and inclinations, namely, that since he knew of nothing contrary to his honor, he should recall his wife and his nephew and take no further notice of slanderous talk concerning them. (quoted by Cantor, 1995, p. 155; my italics)

That "like the wise men they were...." is both a narrative and psychological jewel. King Mark, like King Lear and Tokugawa Yoshinobu, wanted counsellors who would rubber stamp his decisions, and then take the blame if things went wrong. Obviously, in all these cases, what was missing was not the incorruptible counsellor, but the incorruptible king who could have proven Lord Acton wrong.

3. The Ottoman Interpreters

There is another way, to see how illusory Seneca's solution was. Interpreters, by their humble station, should have no interest in distorting the information they convert from one language to another, yet the corrupted powers-that-be are unable to refrain from punishing the messenger for unwelcome news. Thus they corrupt the interpreter who soon learns to doctor the message so that it says what the *supremo* wants to hear. Already, Count Castiglione had advised his courtier *not to be the bearer of bad news*. And we know from volume 1 the tragic end of the ten survivors of the 100,000 horsemen sent by Muhammad Tughlak in 1337 to conquer China: "These ten returned to Delhi with the news of the disaster and there, at the command of the Sultan, were all executed" (Canetti, 1973, p. 242).

As discussed by Bernard Lewis, the Ottoman interpreters were often criticized for their disloyalty,

> certainly justified by the evidence, [however] they were too frightened to do their job properly..., and the [foreign] embassies

> almost all agreed that they [the Levantine dragomans[269]] were far too scared of the Turkish authorities to deliver any unpalatable message honestly. Thus, for example, when the British or the French or the Austrian ambassador wanted to deliver a severe message, the severity disappeared entirely. The severe message, as transmitted by the dragoman to the *reis efendi* or whatever other Ottoman official he dealt with, became a humble supplication.[270] (1999, p. 14)

To better explain the constant fear under which the dragomans lived, Lewis reported from a letter that one of them sent to the Aga of the janissaries, the highest military officer in Turkey. One passage will suffice to give the flavour of such documents:

> "Having bowed my head in submission, and rubbed my slavish brow in utter humility and complete abjection and supplication to the beneficent dust beneath the feet of my mighty, gracious, condescending, compassionate, merciful benefactor, my most generous and openhanded master. I pray that the peerless and almighty provider of remedies may bless your lofty person, the extremity of benefit, protect my benefactor from the vicissitudes and afflictions of time, prolong the days of his life, his might and his splendour and perpetuate the shadow of his pity and mercy upon this slave." It goes on like this at some length. If this was how a dragoman addressed a high Ottoman functionary, one can understand a certain concern on the part of European diplomats about the form in which their words—written or spoke—were transmitted to their Ottoman addresses. (ib., p. 14)

As said, the lowly position of the interpreters should have excluded the possibility of their searching for personal gain in making a distorted

---

[269] Dragoman: In the Near East, an interpreter or professional guide for travellers.

[270] "From the Ottoman point of view, the Ottoman Sultan was the ruler of the world; outside there were enemies or vassals, and Ottoman protocol was not willing to use the full titles which these outside rulers claimed for themselves. Thus, in letters, addressed to Queen Elizabeth—polite, friendly letters—she is addressed as ... 'Queen of the vilayet of England'. The Holy Roman Emperor himself, in Vienna, is called 'the king of Vienna'.... The letters themselves reveal the same sort of approach, so that when the Sultan writes a friendly letter to the Queen of England, the purpose of what he says is that he is happy to add her to the vassals of his imperial throne, and hopes, in the formal phrase, that she will 'continue to be firm-footed on the path of devotion and fidelity'" (Lewis, 1999, p. 14).

translation; yet they would have been the first to pay for conveying bad news. Accordingly, they became fearful: as corrupted in Ottoman Constantinople as were those rhetors in Byzantine Constantinople who sang the panegyric of the emperor (see volume 1).

Back to King Lear, his problem was not the absence of a truly honest counsellor but the King's own power. Furthermore, having Shakespeare set his tragedy in pre-Christian times, there were no cardinals or bishops to oppose the king's power, or to remind him of the virtues extolled in the *Sermon of the Mount* (Matthew, 5:3-12). In other words the king's power was not limited by a power different from his, independent of his favors.

Something similar is true for *Hamlet* which, while set in Christian times, does not bring forth, at any moment, an independent voice—of a priest or monk or bishop—who could have mitigated the rottenness and cruelty at Elsinore. Strictly speaking, we encounter a priest in the last act of *Hamlet*: he leads the procession with the corpse of Ophelia; the King, the Queen, Laertes and mourners following. In answer to a question by Laertes, the priest says:

> Her obsequies have been as far enlarged
> As we have warranty. Her death was doubtful;
> And, but that great command o'ersways the order,
> She should in ground unsanctified have lodged
> Till the last trumpet; for charitable prayers,
> Shards, flints, and pebbles should be thrown on her;
> Yet here she is allowed her virgin crants,
> Her maiden strewments, and the bringing home
> Of bell and burial. (Act V, Scene 1)

Here, the Church is in no way opposing or even mildly criticizing royalty. It is not demanding honesty, fairness and generosity. No, here, the Church is nothing but a petty servant of the state, bureaucratically blind, unhappy for having to "enlarge the obsequies." Here the Church is not acting as DP moderator, brake, or opponent of the State, she is part of the UP.

4. The Royal-Aristocratic *Honra* as described in Calderon de la Barca's *El Medico de su Honra*

Around 1598, Lope de Vega had written:

> ***Honra*** *es aquella que* ***consiste en otro****.*
> *Ningún hombre es honrado por si mismo,*
> *que del otro recibe la honra un hombre...*
> *Ser virtuoso un hombre y tener méritos,*
> *no es ser honrado ... De donde es cierto,*
> *que* ***la honra está en otro y no en él mismo****.*
> (Lope de Vega: *Los comendadores de Córdoba*; quoted by Castro, 1972, p. 77; Castro's italics here rendered in bold font)

In my translation.[271]

> *Honra* is based on the others.
> Nobody is *honrado* by himself,
> because a man receives *honra* from others.
> To be virtuous and well-deserving,
> does not imply being *honrado*... From this it is certain
> that *honra* stems from the others, not from oneself.

Forty years later, similar observations were made by Calderón de la Barca in *El medico de su honra* (The surgeon of his *honra*):

> For now the public voice is so much feared
> And so great is one's terror of opinion
> That most would sooner lose their fame in secret
> Than keep it pure in face of public scandal.[272]
> (I, 14; 1635/1960, p. 20)

---

[271] As discussed in detail at pp. 175-77 of my volume 2, the Spaniards, besides the word *honor* (meaning the same as the English word "honor") have the word *honra*, close to "honor" yet different, as can also be seen from the reported verses by Lope de Vega. Accordingly, every time it is a question of *honra* and *honrados*, I have kept them unchanged in my translation.

[272] Pierre Guenoun—in the introduction to his translation of *El burlador [seducer] de Sevilla [Don Juan] y convidado de piedra*, of 1620, by Tirso de Molina—said something similar when he noted that the women in the play are "*nullement soucieuses de leur corps, mais hautement attachées a leur réputation, hautement sensibles à leur établissement*" [not at all concerned with their bodies, but highly attached to their reputation, highly sensitive to their position in society] (1968, p. 11).

El medico de su honra is important because it narrates of so much evil done in the name of the *law of honor/honra* (while, in reality, that evil had been done in the name of the *law of power*).

As commented by the eminent literary critic Menéndez y Pelayo:

> *Don Gutierre Alonso de Solís no es un celoso, sino un esclavo, un súbdito dócil de la ley del honor.... El Médico de su honra no se venga de agravio alguno; su esposa es inocente, pero él quiere borrar hasta la más leve ombra, el más imperceptible átomo de deshonor, y lo hace por el sangriento medio de la incisión en las venas de su mujer.*[273] (1944, iii, p. 241)

For Salvador de Madariaga, in his *Englishmen, Frenchmen, Spaniards* (1928/1969), the driving force behind the "fair play" of the English is *action*, behind the "raison" of the French is *thought*, and behind the "honor" of the Spaniard is *passion*. However, in the case of Don Gutierre, and many other Spaniards, it is cold passion: not the passion of the man of strong emotions, but the cold passion of the slave of the UP of public opinion, and behind that of the UP of the king. So, everything is kept secret and done on the sly. In the ruminations of Don Gutierre: "but it must not be known to anyone. Secrecy can achieve great victories. When an affront is secret, it demands a secret vengeance. Let Mencía die in such a way as no one would imagine" (III, *3*; 1635/1960, p. 64).

In the end, Don Gutierre—thanks to the prudent way in which he killed his wife, and covered his actions—is not punished, but rewarded by the King who knows the whole story.

The King: "Don Gutierre, you'll need some consolation for so great a grief, an equal gain to compensate your loss, so give your hand to Leonor—At the same time her honor to restore and keep your worth and great deserts" (III, 19; ib. p. 80).

---

[273] [Don Gutierre is not jealous, but a slave, a docile subject of the law of honor.... The surgeon of his *honra* does not avenge himself of any offense; his spouse is innocent, however he wants to erase even the faintest shade, the most imperceptible atom of dishonor, and he does through the bloody incision in the veins of his wife (Mencía)]

"The King, the epitome of justice, unwittingly and indirectly supports injustice in order to protect his royal honor[274] from being sullied" (Hesse, 1960, pp. xxiii-xxiv). But already earlier, in the first scene of Act III, the hearing in front of the King "reveals that Gutierre needs no tangible reason to confirm his suspicion of his wife's infidelity:

> ...men like me do not require to see: it is enough
> That to imagine, to suspect, to fancy,
> To have an inkling, entertain a doubt,...

The irony is compelling as the King, who is extolled as the symbol of justice, accepts Gutierre's explanation of his conduct, and is willing to allow a figment of his subject's imagination to stand for the truth" (Hesse, 1960, pp. xvi-xvii). In the final comments by Everett Hesse:

> The conviction of an innocent woman on the false charge of infidelity is shocking and the inhuman and barbarous cruelty of having her bled to death is offensive to the modern reader, especially of the Anglo-Saxon race. And it is made doubly so when the guilty ones go unpunished, and when Gutierre, the worst offender, is allowed to marry Leonor, the girl he originally loved. Mencía, the innocent victim, is sacrificed on the altar of "honor". Menéndez y Pelayo points out that the public would never have tolerated a "Christian" ending in which the husband pardoned the offense of the wife in matters of jealousy and adultery ....The ending has perplexed critics through the centuries. Is it immoral as Menéndez y Pelayo seems to believe? He points out that it is difficult to reconcile this opinion with Calderón's tirades against the harsh laws of honor, the church's approbation of his plays, and the usual moral tone of his theater in general.... It is only within recent times that a few critics have begun to consider this play a forceful condemnation of an honor code which, because of the nature of its principles, had to operate in secrecy and encouraged all manner of deceit in order to protect one's reputation in the eyes of a suspicious society more inclined to believe

---

[274] It had been the King's brother, Prince Henry who through his loving attentions for Mencía had caused her to take a number of steps to protect her honor which roused the suspicion of her husband. In order to protect her honor, Mencía did things she would never have done otherwise but which led her husband to kill her to protect his honor, a crime which, in turn, is not punished to protect the king's honor.

> the worst rather than the best about people. Calderón's criticism is achieved so unobtrusively that it has gone almost unnoticed through the years. His singular contribution is the presentation of a situation in which the audience or reader is left to form his own opinion about what the play really means. (pp. xix-xxiii)

However, from this perspective, and probably the correct one, what is condemned by Calderón is not only the exaggerated and distorted code of honor, but the king himself whose "Don Gutierre, you'll need some consolation for so great a grief, an equal gain to compensate your loss, so give your hand to Leonor" (III; ib., p. 80) is a payment by the *supremo* to one of his key officers: in essence, "stay faithful to me, and I will be faithful to you."

In the last scene, the King addresses Gutierre a last time, which is followed by this brief, scary dialogue between Gutierre and Leonor:

> King: Well, give that hand to Leonor. Your worth deserves her.
>
> Gutierre: Yes, I give it. But mark well, Leonor, that this hand is stained with blood.
>
> Leonor: That does not matter. It means nothing to me. And neither can surprise nor frighten me.
>
> Gutierre: You see I am the Surgeon of my Honour. It is a science I do not forget.
>
> Leonor: Then cure my own when it's in need of it.
>
> Gutierre: On that condition, then, Leonor, I give this right hand. (ib., p. 82)

So, not only does the King know the whole story, but also Leonor who is willing to marry the blood stained Surgeon and to be similarly cured in case of need. The message to an *insular audience* could not be clearer: this is the way for a man to defend his honor, and for women to accept the cure in case of infraction. The play, therefore, is "immoral as Menéndez y Pelayo seems to believe" as long as an *insular audience* is confirmed in its prejudices, but that same play is moral when a *visitor audience* is confirmed in the truth of the Law of Lord Acton that "Power tends to corrupt and absolute power corrupts absolutely". It corrupts not only those who hold power, but the whole society under them: that society which "would never have tolerated a 'Christian' ending in which the husband pardoned the offense of a wife in matters of jealousy and adultery," even when the wife had done absolutely nothing wrong.

25

Castration—As Punishment or for Eunuchism—Is a Specific UP Horror

1. Introduction

> Castration of the male is referred to in Chinese medical history as early as 1100 B.C. and was at first imposed as punishment for crime. Subsequently castration was sought voluntarily as a means of obtaining the lucrative and influential position of eunuchs in the imperial court....There are likewise references to castration in pre-Christian times in other oriental kingdoms such as Egypt, Babylon, and Persia. (O'Donnell, 1967, p. 194)

Instead:

> The Jews, Greeks, and Romans disapproved of castration and emasculation. The practice was considered inhumane and forbidden by God (Deuteronomy 23:1).... Castration of human beings and animals was just as despised by the prophet Muhammed (ca. 570-632) as it was by the Jews. Emasculation was a punishable offense. This prohibition also prevailed during the Umayyad dynasty (661-750). (Scholz, 2001, pp. 28, 197-198)

Under the relatively high DP in which Jews, Greeks, the Romans of the Republic, and the Arabs of the first centuries of Islam, lived, castration was shunned. The same is true of the Japanese who lived under the partial UP of their daimyo/shogung/emperor,[275] and even more so of all the people who lived/live under the high DP of the West. Instead castration—as punishment or for eunuchism—was widely practiced in the strong UP empires of China and Byzantium, and eunuchs in large numbers were bought for service in the harems of the UP post-Umayyad Arab dynasties and in the harems of the UP Turkish empire. Therefore—leaving aside castration as practiced by primitive people—it would seem that castration for punishment and eunuchism is a specific horror of large countries living under UP.

Sadly, all this makes sense. Under UP, only the will of the great ruler counts: his likes and dislikes, his habit of punishing severely (by castration) for even the smallest infractions, and his paranoiac fear of infidelities by his secluded and oppressed wives and concubines which he tried to prevent by having only eunuchs serving in his palaces.

However, already under partial DP and partial UP, the ruler(s) needs the drive, initiative and courage of his officers and best soldiers (e.g., samurai), and this is possible only if these men have a reasonably high sense of autonomy, self-confidence, and self-drive: all virtues that disappear the moment that castration is widely practiced.

2. Castration as Punishment in China

Already in volume 2, I referred to the castration of the great Chinese historian Ssu-Ma Ch'en (145-87 B.C.) ordered by Emperor Wu Ti as punishment for the historian's imprudent word "merit", said in front of the imperial council: the merit of General Li Ling (see Boorstin, 1983, p. 561). Here, I will first mention another case of similar punishment by that same emperor, then expand on the tragedy of Ssu-Ma Ch'en through his own words.

A. The Punishment of the Courtier Hsü Kuang-han

> Owing to a mistake or accident he had had the misfortune to suffer the punishment of castration. The incident reveals the

---

[275] As seen in volume 3, in the past, the Japanese could be gruesome when torturing or punishing(*), yet, nowhere have I found mention of castration. The absence of this ultimate form of degradation, is probably the result of the partial UP which fostered the generalized ethos of the samurai.

(*) e.g., the Tokugawa Bakufu's propensity, in the XVIIth century, for *hanging people in the pit*, and before, in the XVth century, for Saito Toshimasa's fondness "for torture and boiling people" (Turnbull, 1977, p. 145).

starkly cruel nature of the times, for he was subjected to such treatment as a matter of mercy. Hsü Kuang-han had once been attending Wu Ti on an expedition to the shrines and summer retreat of Kan-ch'üan, and on this occasion he had saddled his horse with a fellow courtier's saddle. For this all too understandable error he was brought up on a charge of carelessness and robbery, and it was decided that the crime merited the death penalty. It was only due to the intervention of an imperial edict that such punishment was commuted to that of castration. (Loewe, 1974, p. 124)

B. The Punishment of the Grand Historian Ssu-Ma Ch'ien

In the words of Pan Ku who, around 60-80 A.D., wrote a biography of Ssu-ma Ch'ien: "For the next seven years the Grand Historian devoted himself to writing and arranging his book. Then he encountered the misfortune of the Li Ling affair and was plunged into the dark, in bonds. He sighed bitterly and said, 'Such is my fault, such is my fault, that I have brought mutilation to my body and may never again serve my lord!'" (reported by Watson, 1958, p. 54).

Then from a letter that Ssu-ma Ch'ien had written to a friend, and reported by Pan Ku:

> But I remember that I am no more than a mutilated being who dwells in degradation. Anything I might try to do would only meet with censure; should I try to help others I should only succeed in doing them injury. Therefore I am "in sadness and despair with no one to speak to".... But one like myself, whose very substance is marred and mutilated—though I might possess the worth of the jewels of Sui and Ho, though my conduct might be as pure as that of Yu and I, in the end I could never achieve glory, but on the contrary would only succeed in arousing laughter and bringing shame upon myself.... No deed is more hideous than bringing shame to one's ancestors, and no disgrace greater than the palace punishment (castration). That a man who has undergone such punishment is fit no longer to be associated with is the opinion not of one age alone but has been held since ancient times. So from old times men have been ashamed to associate with eunuchs. If even ordinary men are loath to have dealings with eunuchs, how much more so in the case of gentlemen of virtue and feeling? Although our court today may be in

> need of good men, what business would I, a mere "remnant of the knife and saw," have in trying to help and recommend the finest and most worthy men of the world? ... From this it is obvious that I am a worthless person who by mere chance has been tolerated at court.... If I could not at that time introduce any great precepts or present any of my ideas, now when I am no more than a slave who sweeps the paths, mutilated and ranked among the low and worthless—now should I try to lift up my head and look lordly and discourse upon right and wrong, Would I not show contempt for the court and bring shame to the gentlemen of my time? Alas, alas! A man like myself—what can he say now? What can he say now? ...
>
> Then it happened that I was summoned into council, and I took the chance to speak of Li Ling's merits, hoping to broaden His Majesty's view and put a stop to the angry words of the other officials. But I could not make myself fully understood. Our enlightened Ruler did not wholly perceive my meaning, but supposed that I was trying to disparage the Erh-shih General and plead a special case for Li Ling. So I was put into prison, and I was never able to make clear my fervent loyalty. Because it was believed that I had tried to defame the Emperor, I was finally forced to submit to the judgment of the law officials. My family was poor and lacked sufficient funds to buy commutation of the sentence.... And then I was thrown into the "silkworm chamber" [where castrations were performed].... I met this misfortune because of the words I spoke. I have brought upon myself the scorn and mockery even of my native village and I have soiled and shamed my father's name. With what face can I again ascend and stand before the grave mound of my father and mother? Though a hundred generations pass, my defilement will only become greater. This is the thought that wrenches my bowels nine times each day. Sitting at home, I am befuddled as though I had lost something. I go out, and then realize that I do not know where I am going. Each time I think of this shame, the sweat pours from my back and soaks my robe. (quoted by Watson, 1958, pp. 58-66)

Back to what I wrote in volume 2, it was not only Hsü Kuang-han and Ssu-ma Ch'ien who were befuddled; not only they had lost their dignity and did not know where they were going; not only they were thinking with shame about what had happened, but the countless num-

> were small boys sent to witness the ghastly scene, but they were made to visit alone the place in the darkness of night and there to leave a mark of their visit on the trunkless head. (Nitobe, 1905/2001, pp. 44-45)

Closer to our times, James Moloney in *Child Training and Japanese Conformity* reported that:

> in the development of *ko* (filial piety), the father is proclaimed the titular and revered head of the family, while the mother, in comparison to the father and to the sons, especially the eldest son, is without status. Even the babe in arms cannot escape obeisance to the father. Filial respect and conformity are foisted upon him while he is still an infant: the mother pushes the head of the suckling into a bow as the father enters the room. Later, if he is not duly obsequious, the baby is chided for his infantilism and unfavorably compared with more dutiful children. If this fails, as a last resort the need for obedience is made clear to him by the burning of dried leaves on his skin. One subjection to this moxa cautery usually suffices, but if it does not, the second searing or branding accomplishes the complete riddance of the child's self-assertiveness and full compliance to the institution of filial respect. (1954/1962, p. 214)

Similar things were said by Arthur Koestler in *The Lotus and the Robot* of 1960:

> A traditional form of correction in Japanese education was moxa - the burning of a cone containing the leaves of a special plant on the child's skin, where it often left a life-long scar. But this was considered as a cure rather than a punishment—.... Thus the apparent cruelty of Japanese education is really a form of conditioning, which does not exclude love and affection—on the contrary, it is based on the axiomatic conviction that the child in its innermost being wants to be properly reconditioned that way, as the flower welcomes the wire rack to display its petals to best advantage. (pp. 199-200)

B. England

For England, it is easier if we move backward in time, starting with the 1962 observations by Philippe Ariès in his *Centuries of Childhood*:

> The relaxation of the old scholastic discipline [in France in 1763-1770] corresponded to something else: to a new orientation of the concept of childhood, which was no longer associated with the idea of the weakness of childhood and no longer recognized the need for its humiliation....In England this concept of education appeared much later. In its educational records, the eighteenth century in England appears as a period of violence and brutality. 'Flogging' became increasingly common. According to the historian of Winchester College, H. Cook, in the seventeenth century pupils were flogged only on Saturday, 'the bloody day'. Later they were flogged every day. George III, meeting a boy from Eton in the course of a walk near Windsor, asked him: 'Well, well, my boy, when were you flogged last, eh, eh?'...A reformation was called for: it came at the end of the first third of the nineteenth century, and it was carried out by Thomas Arnold, the headmaster of Rugby. This reformation was inspired by the same principles as the French reformation of the 1760s: the idea of awakening the man in the child. But the English schools did not adopt a single one of the French methods. The English schools preserved the old discipline — corporal punishment and the monitorial system, which had been abolished in France — but they managed to change its spirit completely. For example, if the birch was retained, it was no longer simply as a punishment but above all as an instrument of education, *an opportunity for the boy being flogged to exercise self-control, the first duty of the English gentleman*. (pp. 264-265; my italics)

In this, Thomas Arnold himself would have completely agreed, he who wrote in his *Education in School* of 1835 that:

> The fault of the old system of flogging at Winchester was not its cruelty, but its inefficiency; the punishment was so frequent and so slight as to inspire very little either of terror or of shame... Punishment, then, inflicted by a parent or a master for the purpose of correction, is in no true sense of the word degrading; nor is it the more degrading for being corporal. To say that corporal punishment is an appeal to personal fear is a mere

> abuse of the terms....The total abandonment, then, of corporal punishment for the faults of young boys appears to me not only uncalled for, but absolutely to be deprecated. It is of course most desirable that all punishment should be superseded by the force of moral motives; and up to a certain point this is practicable.... But it is one thing to get rid of punishment by lessening the amount of faults, and another to say, that even if the faults are committed, the punishment ought not to be inflicted. Now it is folly to expect that faults will never occur; and it is very essential towards impressing on a boy's mind the natural imperfectness and subordination of his condition. (1835/1970. pp. 121-25)

Later in the book, Dr. Arnold explained some of the features of the English system which—and he knew it—seem stranger to people of other countries, including the Scots. In this text Arnold explains the internal logic of fagging[282] within the English boarding system, but without discussing the plusses and minuses of the boarding system.

> It seems to me, then, that the complaints commonly brought against our system of school discipline are wrong either in their principle or as to the truth of the fact. The complaint against all corporal punishment, as degrading and improper, goes, I think, upon a false and mischievous principle: the complaint against governing boys by fear, and mere authority, without any appeal to their moral feelings, is perfectly just in the abstract, but perfectly inapplicable to the actual state of schools in England. By 'the power of fagging', I understand a power given by the supreme authorities of a school to the boys of the highest class or classes in it, to be exercised by them over the lower boys for the sake of securing the advantages of regular government amongst the boys themselves, and avoiding the evils of anarchy,—in other words, of the lawless tyranny of physical strength....
>
> I have said that fagging is necessary for a multitude of boys when living together; for this will show how the system may be required in the public schools of England, and yet be wholly needless in those of Scotland. The great Scotch schools are day-schools—those of England are boarding-schools. Now

282 fag: A boy in an English public school who acts as a servant for another boy in a higher form or class.

> the difference between these two systems is enormous. In the Scotch schools the boys live at their own homes, and are under the government of their own relations; they only meet at school for a certain definite object during a certain portion of the day. But in England the boys, for nearly nine months of the year, live with one another in a distinct society; their school life occupies the whole of their existence; at their studies and at their amusements, by day and by night, they are members of one and the same society, and in closer local neighborhood with one another than is the case with the ordinary society of grown men. At all those times, then, when Scotch boys are living at home in their respective families, English boys are living together amongst themselves alone; and for this their habitual living they require a government. It is idle to say that the masters form, or can form, this government; it is impossible to have a sufficient number of masters for the purpose; for, in order to obtain the advantages of home government, the boys should be as much divided as they are at their respective homes. (1835/1970, pp. 127-29)

So, the English system was different in many respects by including: boarding away from home for a long time, corporal punishment, and fagging. The practice of boarding away was already commented on by an Italian visitor to England[283] in 1497 in his report back home:

> The want of affection in the English is strongly manifested toward their children, for after having kept them at home until they arrive at the age of seven or nine years at the utmost, they put them out, both males and females, to hard service in the houses of other people, binding them generally for another seven or nine years. And these are called apprentices, and during that time they perform all the most menial offices, and few are born who are exempted from this fate, for everyone, rich he may be, sends away his children into the houses of others, whilst he in return receives those of strangers in his own. And on inquiring

[283] Probably a noble Venetian who accompanied the Venetian ambassador Francesco Capello to the court of Henry VII. The manuscript, *A relation or rather a true account of the Island of England* (*about the year 1500*), was formerly in the library of the Abati Canonici at Venice, then in the possession of the Rev. Walter Sneyd. Finally it was translated by Charlotte Augusta Sneyd, and published in 1847, in the original Italian and English translation, by The Camden Society.

> their reason for this severity, they answered that they did it in order that their children might learn better manners. *The best manners in question may well have been hardiness (being unspoiled), and imperturbability, two virtues difficult to learn at home from one's parents* (1497/1847, pp. 24-25; my italics).

C. A. Sneyd, the translator of this text, added the following note:

> The severity of parents at this time was so great, that the children perfectly loathed the sight of them; nor did their discipline relax as they grey older, 'for daughters, grown women, and sons, gentlemen of thirty and forty years old, might not sit in their presence without leave, but stood like mutes bare-headed before them'. (1847, p. 77)

However, Barbara Hanawalt, in her book *Growing up in Medieval London: The experience of childhood in history*, found fault with the Italian visitor:

> I have carefully refrained from mentioning the observations of the late medieval Italian traveler until now because he has caused so much mischief among twentieth-century historians. He has convinced them that the English did not love their children and therefore sent them off to live with other families at a tender age. His English informants, in their turn, claimed that he misunderstood their customs (should we picture him as an early anthropologist). They loved their children so much that they sent them off to other families so that they could learn better and be disciplined. Should they have undertaken such an important trust themselves, they would not have been able to beat the poor youngsters sufficiently to see that they were well raised. Historians, taking the side of the anthropologist rather than the natives, have accepted the traveler's interpretation... It is time that we listen again to the voice of the native. It speaks clearly in the field notes and is corroborated in independent evidence.
>
> London citizens valued the apprenticeship experience for their sons....Few fathers wished to have their sons serve as apprentices in their households. Only four apprentices served with their fathers among the fifteenth-century Mercers; a similar number did so among the Goldsmiths. Most parents preferred that

> their sons expand their social contacts and learn additional trade secrets through another master.
>
> The move of a child from the natal home to that of the master was certainly a major event in the lives of the child and the parents. In the fourteenth century in particular, the parents were sending a child of perhaps no more than fourteen a very long way from home to live with a comparative stranger...Having a child enter into an apprenticeship represented a considerable emotional and monetary investment for the parents. They perceived the arrangement not as getting rid of a child, but as establishing his future and perhaps theirs. (1993, pp. 146-47)

Leaving aside the fact that Sneyd (1847) had concurred with the Italian visitor on the parental severity, there is a way to agree with Hanawalt, while accepting much of the severity of the times, and this by paraphrasing for England what Koestler said about the severity of Japanese parents: 'Thus the apparent cruelty of English education is really a form of conditioning, which does not exclude love and affection—on the contrary.'

Said differently, those English parents were simply following the Biblical instruction of "He who spares the rod hates his son, but he who loves him is diligent to discipline him"[284] (Proverbs 13:24).

According to Koestler, the Japanese severity was "based on the axiomatic conviction that the child in its innermost being wants to be properly reconditioned that way, as the flower welcomes the wire rack to display its petals to best advantage" (pp. 199-200). On the other hand, there is no mention by Koestler of children being sent away from home either to apprenticeship or boarding schools, possibly making the English system stricter than the Japanese (leaving aside the very strict training of the samurai).

This said, why should the English have adopted educational methods stricter than those of the rest of Europe, stricter than those of the Austrians, Dutch, Danes, French, Germans, Italians Norwegians,

---

[284] There may be another explanation for Hanawalt's reaction to the Italian visitor: Hanawalt concentrated her studies on London, as indicated by the title of her book: *Growing up in Medieval London*. Big international harbors, like London, are often more advanced, on many fronts, than the inland, and this certainly was the case with the way parents treated their children. In this respect, the reader will remember my praise for creativity of the international harbors, as discussed in my chapter 6 of volume 1 "Creativity via a Detour Abroad."

Russians, Spaniards and Swedes? One answer would be to look at the impact of the last major sudden and global conquest in Western Europe, that, in 1066-1067, of England by the fierce French-speaking Normans[285] of Guillaume le Bâtard, later upgraded to William the Conqueror.

David Douglas reports in his 1964 study of William the Conqueror much that could have contributed to the making of the English imperturbability:

> On his route he savagely devastated the land through which he passed, sparing no male and leaving nothing behind him which could support life. Just before Christmas [1069] he reached York, and there in a burnt city, surrounded by a desolated countryside, he celebrated the Nativity of Christ.... An eleventh-century campaign was inevitably brutal, but the methods here displayed were widely regarded as exceptional and beyond excuse, even by those who were otherwise fervent admirers of the Norman king. (pp. 220-21)
>
> There is no doubt, for instance, that William shared in the savagery which marred so many of the secular rulers of his age, and in one respect he was held in England to have been exceptional in his wanton disregard of human suffering.... This was a strong and pitiless king. But he was not simply a self-regarding tyrant, nor was he regarded as such by those he ruled. The severity of his administration they had felt for themselves, but having experienced also 'the good security he made in this country', they left it to God to judge his ruthless suppression of disorder.
>
> It was a charitable verdict, for the personal portrait which emerges is undoubtedly repellent.... Assuredly he was a man to fear. 'Earls he had in his fetters—he expelled bishops from their sees, and abbots from their abbacies; he put thegns[286] in prison and finally he did not spare his own brother Odo'. (ib., pp. 371-375)

---

[285] According to the *Encyclopaedia Britannica*: "Normans, originally NORTMANNI, also NORTHMEN, a term used generally in medieval western Europe to denote the barbarian heathen pirates (Vikings) from Scandinavia who between 800 and 900 pillaged or occupied many coastal areas. More particularly, and in modern usage, it refers to those Vikings who settled in what later became the duchy of Normandy in northern France" (1978, *vii*, p. 391).

[286] In pre-conquest England the typical warrior was the thegn; by the time of William's death he was the knight.

This comment on the Norman invasion is so important for our discussion on the origins of the English imperturbability that it deserves to be confirmed by another historian, this time John Bowle in his *The English Experience* of 1971/2000:

> "This was a fatal day for England," wrote William of Malmesbury two generations after the Norman Conquest, "a melancholy havoc for our dear country brought about by its passing under the dominion of new lords." The event had brought most of the Anglo-Saxon aristocracy to ruin and led to ferocious exploitation....The conquerors were French-speaking foreigners....The Normans and Angevins were not, like the heathen Knut, respectful to the insular culture; they regarded the Anglo-Saxons with contempt....The Conqueror was a ferocious character. William, Duke of Normandy since 1035, was the bastard son of Duke Robert I and Herleve, probably the daughter of a tanner of Falaise, where tanneries flourished. Duke since he was eight, William had survived frightful risks and humiliations. (pp. 81-82)
>
> William's sons who succeeded him were both strong men: William II, Rufus, a famous and ferocious soldier in whose reign the Normans mastered Cumberland, Westmorland and part of the Welsh marches and established the frontier with Scotland.... Rufus was as ruthless as his father: when the Count of Eu lost a trial by battle, he had him blinded and his stewards hung. (pp. 88-89)
>
> In Normandy the barons, heirs of Norse pirates, had held their property by right of settlement and waged private wars: in England all men now held their land from William, and his peace and the prerogative power to enforce it pervaded the land. He imposed his feudal right much more stringently and exploited rebellion to confiscate and redistribute fiefs, to pin men down to their obligations to provide the garrison of the new territory. For such were the Normans, Bretons, and Flemings who had come over with the Conqueror, their power clamped on the country by the crude "motte and bailey" strongholds, later elaborated into stone keeps, such as the White Tower built by William himself to overawe London. (p. 93)

If later the education in England was sterner than elsewhere in Europe, it may have been the way to train people to survive in a very harsh world. We are reminded of the tripartite division of all social relations among Chinese, discussed in chapter 11 of volume 1, in particular the third for dealing with strangers in which "demonstrating strong feelings, including the menaced and aggrieved affects of suffering, is dangerous, because it gives others power over relationships and restricts one's flexibility to respond effectively. Ultimately, uncontrolled emotional displays threaten one's position in a world of power. Balance, blandness, control provide greater access to power and protect one from the feared effects of power: loss of resources or status or life" (Kleinman & Kleinman, 1991, p. 288). Imperturbability and impassivity of the Chinese kind, while not the same, are first cousins.

Similarly, for centuries, when faced with the Normans, the Anglo-Saxons had best not show strong feelings, including the menaced and aggrieved affects of suffering.

4. And Presently?

Are the British still imperturbable? Should they still be so? Basically, imperturbability is a virtue when opposing the enemy be he a man, a grizzly, or when climbing for the first time the Matterhorn, as Whymper did in 1865. Instead, imperturbability is a defect when dealing with friends. With friends: *Shared joy is double joy, shared pain, is half pain;*[287] or as narrated by Richard of his friend Tony, in Robin Maugham's *The Servant* of 1949: "Then he looked down so that I should not see how moved he was. I wonder who began the gag that the English are unemotional?" (p. 33).

If we go back to the poem by Kipling, we see immediately that he is painting a world of enemies:

> If you can keep your head when all about you
> Are losing theirs *and blaming it on you*,
> If you can trust yourself when all men doubt you,
> But make allowance for their doubting too;

---

[287] From the original by Christoph August Tiedges (1752-1841): *Geteilte Freud' ist doppelte Freude, geteilter Schmerz ist halber Schmerz*, in turn a variant of Ciceror's *Et secundas res splendidiores facit amicitia et adversas partiens communicansque leviores.*

> If you can wait and not be tired by waiting,
> *Or being lied about*, don't deal in lies,
> *Or being hated*, don't give way to hating,...
> (my italics)

In such a world, yes, imperturbability is badly needed.

So, to be imperturbable or not? The answer, as usual, is that it depends, like when to be androgynous: "On the *Bem Sex Role Inventory*, in order to be androgynous, one needs to be forceful and dominant ('masculine' items) and also shy and soft-spoken ('feminine' items). It is hard to see how a person could be all of those at once. Of course, the ideal androgyn would be expected to display different characteristics in different situations, depending on what was most appropriate; but knowing what is most appropriate is also rather difficult" (Hyde, 1985, p. 101).

Even more difficult is to fine-tune old national scripts and institutions such as the English imperturbability. In the end, as always, *Felix qui potuit rerum cognoscere causas*.[288]

But this is not the whole story. For Barzini, the British are still imperturbable (at least as late as 1982, when he wrote his book):

> Clearly such [stoic] virtues, admirable as they still are, are outdated, unfit (except for rare occasion) for the complex and corrupt times we live in.... They surely have not aided in the construction of an integrated Europe. Take but one of the stoic virtues, imperturbability. In the past it made the British dangerous adversaries in negotiations, on battlefields, or when governing turbulent mutinous colonies. It was this virtue that enraged Napoleon at Waterloo, who cried, "This man Wellington is so stupid he does not know when he is beaten and goes on fighting."
>
> Roberto Ducci, a brilliant man, Italian ambassador to London for years, one of the team who wrote the Treaty of Rome [1957], and a *desenchanté* believer in Europe, thought imperturbability was today one of the main impediments to efficient

[288] (Fortunate is he who has been able to learn the causes of things) (Virgil's *Georgics*).

> collaboration with the British. He discovered it was always difficult, if not impossible, to make the Foreign Office aware of the gravity of an approaching crisis and of the necessity of taking some immediate decisions. The British foreign secretary or high official, who evidently considered him an excitable and emotional Latin to be humored and calmed like a jittery horse, told him there was no crisis, not really, no need to rush into rash decisions, and that the best thing to do was to keep cool, do nothing for the time being, and wait for the murky situation to clear up. This approach, the very best in the days of sailing ships and armies marching on foot, is clearly inadequate in a world in which events follow each other at dazzling speed and opportunities vanish forever within hours. (p. 64)

Here I disagree with Barzini. The political imperturbability he described is passé—not because of the need, now, of quicker decision than in the past (Wellington made quick decisions during the battle of Waterloo[289]; and the negotiations for the Treaty of Europe did not require instant decisions)—but because nowadays relations among countries (European countries or members of the United Nations) must be friendlier than in the past.

At the individual level, this applies, for instance, to Roberto Ducci who, while openly complaining of the lack of prompt decisions by the Foreign Office, was probably more upset for having been treated as "an excitable and emotional Latin".

The Anglo-Saxons of old had no need of being sensitive to the feelings and problems of the Normans, but the British of today (and this is true of everybody) have to be sensitive to the needs of other people and nations.

Indeed, national characteristics are changing quickly all over the world, but how quickly and in depth? As long as things are quiet, many of the old characteristics stay hidden, but may come forth in times of stress, as noted by Barzini: "definitely stoic were the sober newspaper comments on the brilliant, skillful, and daring Falklands war, in which most ancient British virtues surprisingly shone once again" (ib., p. 63). Here, too, *Felix qui potuit rerum cognoscere causas.*

---

[289] For a description of Wellington's imperturbability and quick decisions at the right time, at the Battle of Waterloo, see for instance Parker, 1983, pp. 186-90.

PART III

The Creative Process
Epistemology (Truth)
Positive Psychology
Synopsis

27

The Creative Process

1. Incubation, and Delayed Insight

Anthony Storr (1999), in his review of a recent book on creativity, complained that there was "too little discussion of the phenomenon known as 'incubation'; that is, the mysterious fact that an idea can strike a person, be put on one side, and, even years later, prove to be the essential seed from which a new creation is born. Consideration of unconscious processes and dreams does not appeal to hardheaded experimentalists, but this does not mean that these phenomena are unimportant" (p. 31).

To avoid a similar criticism, I will begin this chapter by discussing the incubation that preceded Fermi's discovery of nuclear reactions affected by slow neutrons, a discovery which (together with his identification of new radioactive elements produced by neutron bombardment) won him the 1938 Nobel prize in physics. In an article titled "The Miracle of the Two Tables", Gerald Holton discussed how Fermi made his discovery:

> [In 1934, at the end of Summer vacation, when Fermi and his collaborators reassembled at their *Instituto*, things began to go astray]. Their readings turned out to depend on the tables they used to support the equipment. One table, on which the spectroscopes of an earlier day had sat, was made of wood; the other of stone. When an experiment on inducing radioactivity in a target made of silver was made on the wooden table, a markedly greater activity resulted than if the same experiment was tried on stone. The group christened it the "miracle of the two tables".

To get to the bottom of the puzzle, Fermi initiated a systematic set of observations. The group reasoned that perhaps the lead housing around the target affected the neutrons which reached the target in those two cases, and they observed that the interposition of a block of lead changed the activation somewhat. Therefore, Fermi decided to make a lead filter in the shape of a wedge of varying thickness, to insert it into the neutron beam....

[In Fermi's own words] "Instead of my usual custom, I took great pains to have the piece of lead precisely machined. I was clearly dissatisfied with something: I tried every 'excuse' to postpone putting the piece of lead in its place. When finally, with some reluctance, I was going to put it into its place, I said to myself: 'No! I do not want this piece of lead here; what I want is a piece of paraffin.' It was just like that: with no advanced warning, no conscious prior reasoning. I immediately took some odd piece of paraffin I could put my hands on and placed it where the piece of lead was to have been."

The result was immediately obvious: a great increase in the radioactivity induced in the target by the neutron beam, even when the experiment was done on the marble shelf....The miracle of the two tables was now unmasked. Fermi realized that in those earlier, puzzling experiments, some of the original neutron beam had first gone into the wooden table. Those neutrons, being of about the same mass as the hydrogen nuclei in the wood, lost some of their energy by collision; and then some of those slowed neutrons had been scattered up to the silver target, producing that unexpected effect on it. The nuclei in the marble in the other table, with nuclei much more massive than neutrons, could do all this only very poorly.... But in that story, there is a haunting puzzle. Fermi was the most rational and least impulsive of scientists; yet, by sudden, determined action, with, as he said, "no advanced warning, no conscious, prior reasoning", he had placed the crucial piece of paraffin before the neutron source. This seems an example of the kind of intuitive intelligence, which, for better or worse, often guides brilliant scientific minds in the earlier phases of their work. Scientists these days tend to shy away from speaking about such an indefinable notion. (2002, p. 13)

2. Enter Incubation, and Delayed Insight

Here is my interpretation of what happened to Fermi: his memory registered what he and others had noticed: "on wood far better than on marble"; shortly afterwards, an active agent in his brain—which, in the absence of a better word, I will name the *head librarian* (HL) of the library of memory—expanded that first piece of information with the following one from the library of memory: "plenty of hydrogen in wood, none in marble." Next, the HL sent a member of his staff in search, through the long corridors of the library of memory, for something tagged "solid" and "very high hydrogen content". Once the HL was told by his staff that paraffin was that solid, his job was to pass that piece of information to the *chief*, (i.e., to Fermi's consciousness).

In the case of Fermi, the problem was less one of incubation but of delayed insight. Everything was ready for a quick insight, but his "group" involuntarily had thrown a red herring: "The group reasoned that perhaps the lead housing around the target affected the neutrons which reached the target in those two cases, and they observed that the interposition of a block of lead changed the activation somewhat". The group should have focused on why the table of wood was doing better than the table of marble, instead of looking at the lead housing. Because of the group discussion, Fermi as *chief* instructed Fermi as *HL* to search the memory for enough information to give the right instructions to the technician for making a lead filter in the shape of a wedge of varying thickness to be inserted into the neutron beam. The *head librarian* was clearly unhappy with the instructions he received, and looked at ways to sabotage the lead wedge until he had the *chief* calmed down, and was able to pass the results of his homework to the *chief*: a piece of paraffin, and forget the lead!

What we must remember is that *chief* and *head librarian* (and his staff) operate in totally different environments. The *chief* is at the receiving end of all outside information, be it only a strange rumor or the opening of a door. The *chief* receives data from all his senses, and must often take immediate action. Among others, the *chief* must pay attention to other people: their ideas, their feelings. Because of this, Fermi as *chief* had to talk to his collaborators, and take their comments seriously.

On the other hand, Fermi's *HL* could more easily concentrate on intellectual problems because he was more isolated than the *chief* from the outside world. Equally important, as discussed further on, a good *HL* is constantly under "Zeigarnik tensions" which keep him at work day and night, more at night for serious tasks when the *chief* is not pestering him with petty assignments.

"Fermi was the most rational and least impulsive of scientists; yet, by sudden, determined action, with, as he said, 'no advanced warning, no conscious, prior reasoning', he had placed the crucial piece of paraffin before the neutron source." Yet, "somebody/something" in him had done the homework based on information in his memory (wood rich in hydrogen, marble not, paraffin more).

Let me put it clearly, the *head librarian* is no genius, because he lacks the contacts with the outside world which the *chief* has; and long-range creativity depends on inputs from the outside world which only the *chief* has. Furthermore, only the *chief* can make use of the internal homework of the *HL* and his staff. Only the *chief* can direct the hand toward a block of paraffin because the senses report to the *chief* and not to the *HL*.[290]

With Fermi we had *incubation through elimination* of internal stumbling blocks, and this process may last months and years until, for instance, some ideas slowly "degrade" from the impact of additional

---

[290] As an aside: moving from the act of creation to the personality of the creator, the following on the youth of Enrico Fermi can be related, within the GAM theory, to the youth of Jean Piaget as discussed in volume 3:

"In the winter of 1915 [when Fermi was not yet fourteen] a dramatic occurrence struck the Fermi family and upset their spiritual balance. Giulio [Enrico's older brother by one year] developed a throat abscess. It grew to the point that breathing became difficult, and the doctor advised surgery. It was to be only a minor operation, and the boy would be allowed to return home right after it.... [However] the boy died before the anesthesia was completed. The blow could not have been heavier, nor the family less prepared to receive it.... [The mother] never recovered from the blow of his death. She had been a lively, witty woman; now she was given to long spells of crying, and she let herself go into moods that were upsetting to her family, as she was too intelligent not to realize.

Mrs. Fermi found an outlet for her grief in her tears. But Enrico's mute sorrow may have been deeper than hers. His brother had been his steady companion and sole friend. There had been no need for others because the two completed each other to form a unit, as two atoms unite to form a molecule. They had no free valency to hook onto others. Now lonely, but shrinking from any show of feelings, he kept his grief to himself.... There was one thing that Enrico could do alone to fill the melancholy hours: study. And study he did, following an avid interest in science" (Fermi, Laura, 1954, pp. 15-17).

Not mentioned (as in the autobiography of Piaget) is the impact on the father from the changed conditions of his wife. He too was not able to assist his son as before, not even to lead a normal life as before. Much had died in the family life alongside the death of Giulio, as if both parents had, in part, died. Accordingly, Fermi, like Piaget, developed into a GAM *universalist* at the search of universally valid truths.

information until they stop blocking a valid insight. Then there is the *incubation by addition*, in which an idea is strengthened through additional information. In the case of Fermi, this could have been reading, at a later time, of the hydrogen content in wood, should he have forgotten what he had learned on wood in his courses of chemistry. Two other variants are the *incubation by bridge*, in which a new piece of information helps to link two previously unrelated elements, and the *incubation by stress* in which a new event lashes the *HL* to work with renewed energy.

3. The Zeigarnik Effect

Expanding on what I said before, the drive behind active incubation is the *Zeigarnik effect*: "A phenomenon named after Lewin's pupil Zeigarnik, who showed in 1927 that interrupted tasks are remembered more proficiently than uninterrupted tasks, depending on involvement in the dynamic field" (*Encyclopedia of Psychology*, 1972, *3*, p. 400). Specifically:

> Among the elegant and precise experiments designed and executed by Lewin and his group of graduate students were those of Bluma Zeigarnik on forces involved in remembering finished and unfinished tasks....The story is sometimes told that Zeigarnik's interest in the problem of memory as a tension system within the individual was sparked by her observations of the fantastic feats of memory performed by waiters in Berlin sidewalk cafés. Though each waiter was responsible for several tables and the orders of several parties at each table, none of the waiters ever wrote down the orders. Nor did any of them forget "who got what" up to the moment the order was served. After each customer's bill was tallied in the waiter's head, and his task was therefore complete, he quickly forgot that customer and his order. This persistency of memory until *completion of the task* suggested to Zeigarnik that a tension system was operating up to the point of carrying out the intention. (Monte, 1977, pp. 433-34)

Clearly, the waiters were not constantly thinking of who had ordered what, yet a number of specific data (customers and their consumptions) were in their minds ready for action (to be paid by the right customer and for the right amount).

In the case of Fermi, the hidden tension, within his memory, was "why wood and not marble?" Until he got an answer the tension remained. As the waiters of Zeigarnik were not conscious of all the data and pertinent tensions in their minds, so was Fermi the *chief* not conscious of the wood/marble problem, but Fermi's *HL* knew of it, and was under tension to find a solution.

I am not sure about others, but I often wake up, in the middle of the night, with the answer to problems that I had been dealing with during the day or on preceding days. These answers come to me in the form of brief suggestions or snatches of a few words. I have learned not to wait long: I grab a piece of paper and jot down what my *HL* has worked out for me. In my case, my *HL* has greatly improved the quantity and quality of his work along the years. I do not remember him as energetic, clever and knowledgeable 10 or 20 years ago. It would seem that the *HL*, like a good muscle, improves through exercise.

Am I speaking of dreams? No. Never have answers or suggestions come to me in the form of a dream (i.e. in form of a story, or episode, or images). No, the answers are always verbal, dry and short, quasi telegraphic. It is not that I do not dream, on the contrary I dream nearly every night, and remember part of my dreams. But never do these dreams bring me anything useful.

Also, while practically all my dreams elicit emotions of joy, or annoyance, or fear, messages from the *HL* are, at first, plain annoyance: they wake me in the middle of the night when I would prefer to sleep, resting for the next day. It is only after a while that I am thankful for messages which are delivered in a fairly dry and emotionless way.

Others, instead, seem to be inspired by dreams as discussed by Owen Flanagan in *Dreaming Souls: Sleep, Dreams and the Evolution of the Conscious Mind* (2000):

> Many people have told me that they are inspired by dreams. This is not surprising in itself, since dreaming is a kind of thinking and certain thoughts, at least, are inspiring. But then it is added that many of the greatest poems, novels, and paintings were first hatched in dreams. This is not entirely false if the point is that we produce or develop certain ideas in our sleep which we then take back into the world and do something with by the light of day.... Indeed, I take it that dreams were [sources of inspiration] in both the Coleridge and Shelley cases. I simply want to deny the strong, and surprisingly widely held view, that dreams are a

> unique and privileged place where creative juices flow. Dreaming is part of living and as such contributes in varied and sundry ways to who we are, what we do, and what we make. But making art takes work, invariably work by the light of day. (pp. 185, 189)

However, there is no doubt in my mind, that the quality of dreams—and the same is true for all incubations, i.e., the work of the *HL*—is in function of the overall quality of the preceding conscious thinking and feeling. In order to dream of Kubla Kahn in Xanadu where Alph the sacred river runs, Coleridge had to have read of these things in his conscious hours.[291] So, paraphrasing Pasteur, I will say that "Quality incubation (or dreams) belongs to minds which are prepared by patient study and persevering efforts."

Back for a moment to the *Head Librarian*, I suspect that his work can be related to recent results by researchers of the Massachusetts Institute of Technology who said that "they have entered the dreams of rats and found them busily working their way through the same lab maze they negotiated during the day"[292] (Associated Press, as reported in the *Houston Chronicle*, Jan. 25, 2001, p. 12A).

4. Immediate Insight Without Incubation, yet Zeigarnikian and Pasteurian

---

[291] "Coleridge's own account of the genesis of 'Kubla Khan' is as follows. It was published in 1816 with the poem:

'.... In consequence of a slight indisposition, an anodyne had been prescribed, from the effect of which he [the author] fell asleep in his chair at the moment that he was reading the following sentence, or words of the same substance, in "Purchas's Pilgrimage": "Here the Kahn Kubla commanded a palace to be built, and a stately garden thereunto. And thus ten miles of fertile ground were included with a wall'" (Lowes, 1927/1964, pp. 324-325). In turn, Xanadu is the *Ciandu* of Marco Polo (and the present Shangdu), the summer capital that Kublai Khan decreed in the grassland of Inner Mongolia, 200 miles northwest of Beijing. As to Alph, this was the river Alpheus of Virgil, "and Coleridge, like every schoolboy, knew it: '... *Alpheum fama es hunc Elidis* ...'" (ib., p. 359).

[292] I am not sure if the MIT scientists are able to distinguish between dream (a complex story which, in humans, can be told to others) and subconscious thinking without a story (the way I feel the *HL* works for me). It is quite possible that in the case of those rats there had to be some vision/dream of the maze in order to solve it (arriving at the goal in the shortest way).

Arthur Koestler, in his *The Act of Creation*, analyzed the famous "Eureka" of Archimedes in terms of a powerful Zeigarnik effect (even if he did not use that term). Based on additional readings on the life and works of Archimedes, I will hereafter elaborate on Koestler's interpretation by showing how Archimedes' past interests were the foundation of his Eureka, while "the stepping in the bath tub" was only a secondary accident, among many possible, which triggered the answer. Seen from the point of view of Pasteur, the Eureka insight was fully deserved, and quasi inevitable. In Koestler's words:

> Hiero, tyrant of Syracuse and protector of Archimedes, had been given a beautiful crown, allegedly of pure gold, but he suspected that it was adulterated with silver. He asked Archimedes's opinion. Archimedes knew, of course, the specific weight of gold,... If he could measure the volume of the crown he would know immediately whether it was pure gold or not;... Blocked situations increase stress.... One day, while getting into his bath, Archimedes watched absentmindedly the familiar sight of water-level rising from one smudge on the basin to the next as a result of the immersion of his body, and it occurred to him in a flash that the volume displaced was equal to the volume of the immersed parts of his own body—which therefore could simply be measured by the pint [Eureka!]....
>
> Neither to Archimedes nor to anybody else before him had it occurred to connect the sensuous and trivial occupation of taking a hot bath with the scholarly pursuit of the measurement of solids....The creative stress resulting from the blocked situation had kept the problem on the agenda [Zeigarnik effect] even while the beam of consciousness was drifting along quite another plane. (1964, pp. 105-07)

Koestler's final explanation was *bisociation*[293] "the perceiving of a situation or idea, L, in two self-consistent but habitually incompatible frames of reference, M1, and M2 .... so that a familiar and unnoticed aspect of a phenomenon—like the rise of the water-level—is suddenly perceived at an unfamiliar and significant angle" (pp. 35, 108).

---

[293] "Bisociation means combining two hitherto unrelated cognitive matrices in such a way that a new level is added to the hierarchy, which contains the previous separate structures as its members" (Koestler, 1967, p. 183).

But, from a Pasteurian Point of View, "Who was Archimedes?"

Archimedes was a Greek mathematician, scientist, and inventor, the founder of theoretical mechanics. "Archimedes himself set no value on the ingenious mechanical contrivances that made him famous, regarding them as beneath the dignity of pure science and even declining to leave any record of them except in the case of the sphere he constructed to imitate the motion of heavenly bodies.... In accordance with the expressed desire of the philosopher, his tomb was marked by a sphere inscribed in a cylinder, *the discovery of the relation between the surface and the volume of a sphere and its circumscribing cylinder being regarded by him as his most valuable achievement*" (Waerden & Heath, 1978, p. 1087-88; my italics).

So, the Eureka discovery was made by somebody curious about volumetric relations, be it of a sphere inscribed in a cylinder or of his leg "inscribed" in a bath tub. The discovery was made by somebody who had trained himself in adding the tag "volume" to many of his observations, and in recalling things tagged "volume".

Hiero's demand was constantly in Archimedes's mind (Zeigarnik) until "chance favored the mind that was prepared" (Pasteur).[294]

However, the way I see it, that specific chance event of paying attention to his leg entering the bath tub played only a secondary role in the Eureka: that same day, or the next, Archimedes might have noticed the change of volume in a cup of milk when dipping a piece of bread into it, or when pouring some gravy on a plate of dumplings and realizing that the larger the quantity of dumplings, the less gravy he could add without overflowing. Basically, his mind was primed to the tag "volume", and it reacted quickly when his memory received two related pieces of information tagged "volume": 1. more of his body entering the bath tub, and 2. more water rising in the tub. The Eureka then consisted in linking the static volume of the crown to the dynamic volumes of his immersed body and corresponding water rising in the tub. The crown was still static, but it could be made dynamic by immersing it in a tub of water as he did with his leg.

Basically, what I have done is to add, to the Zeigarnik effect (tension), the Pasteur effect of long studies and experiments (home-

---

[294] Pasteur had said: "Chance only favours inventions for minds which are prepared for discoveries by patient study and persevering efforts" (quoted by Koestler, 1964, p. 145).

work) whose data were carefully and richly indexed and cross indexed in his memory.

5. Multiple (*divergent*) Indexing in Memory

Without inputs there can be no outputs: *De nihilo nihil*. Accordingly, the foundations of creativity are not to be found in quality *divergent recalling* (aka *divergent thinking*) which is, for instance, coming up with one or two solutions to a given problem, or—in a divergent thinking test—as many ways to use a brick. No, the way I see it, the foundations of creativity are to be found, long before, in quality *multiple (divergent) indexing in memory*, specifically in *multi-purposed indexing* and *cross indexing* of new information.

Nobody will ever mention, in any divergent recalling, the apprehensive young Eve of the *Sistine Chapel* if that person has never visited the Sistine Chapel (or studied books on it) and paid attention to the girl surrounded by the left arm of God while He is creating Adam with His right hand. Does her face show fear of the unknown, or joyful expectation? It is not clear (and Michelangelo has magnificently done so on purpose): her shoulders are turned away as to escape her fate, while her head is courageously turned toward her assigned husband.

Now, useful recollections of the Sistine Chapel will surge vividly at the right moment only if at the time of storing that piece of information in memory it was larded with many *tags* and *cross-tags* (e.g., Michelangelo, Rome, Vatican, Pope Julius II, Renaissance, Pope Julius & Michelangelo & Vatican, Creation of Adam, Pre-creation of Eve).

Obviously, all this indexing and cross-indexing—all this tagging for easy retrieval—must be done immediately at the moment of storage by the *chief*, or by the *head librarian*. If all this tagging is not done, then the subsequent use of that piece of information will be limited.

Coded Under Specific Tensions

Howard Gruber and Doris Wallace (2001) are right in stressing that:

> Networks of enterprises have dynamic properties. They represent, among other things, all the unfinished business of the creator, his or her plans and intentions. The Zeigarnik effect—the tension and drive to complete an unfinished task (Zeigarnik, 1927)—is, in the psychologist's laboratory, not a very robust phenomenon (Prentice, 1944). However, it takes a new force when

> seen in the light of a creator repeatedly returning to the hard work of a beloved but exasperatingly intractable project or a task located somewhere in his or her network of enterprise. (p. 34)

My approach is more inclusive. The genius—already the young challenged personality—goes around with hundreds of interconnected Zeigarnik effects in his/her mind. Any new experience, any interesting passage of a new book, is immediately related to one or more of those Zeigarnik tensions, and accordingly coded and stored in memory. For instance, a given comment by Kant will be rapidly coded and stored under Kant, UP in Germany, *ritter personality*, Enlightenment, Isaiah Berlin, creativity, to name only few.

6. Moving to the GAM/DP Theory

GAM

The same (outward) experience, the same book, will elicit more reactions and a richer encoding in memory in the minds of *challenged personalities* who are constantly under many strong Zeigarnik tensions: their minds are on the *qui vive?* for answers in their *war of the scripts* with society. The many Zeigarnik tensions, the many years of focused homework leading to a huge store of extensively-coded and cross-coded data (in other words a trained *chief*, and a quality *head librarian*) explain the high creativity of *challenged personalities*.

DP

Under a strong division of power, for instance the fights between papacy and empire at the time of the *War of Supremacy* of 1075-1313, the minds of people are under many societal Zeigarnik (dialectical) tensions, surrounded by countless data (e.g., the propaganda, demands and offers by pope or emperor). The continuous homework to sort these data, and discuss them with others, was a Pasteurian effort which inevitably brought rewards in the form of "chance discoveries" like those of Fermi and Archimedes", i.e., of high creativity.

28

Truth

1. Truth?

In *Objective Knowledge* Karl Popper said: "I accept the commonsense theory (defended and refined by Alfred Tarski) that truth is the correspondence with the facts (or with reality); or more precisely, that a theory is true if and only if it corresponds to the facts" (1975, p. 44). This, too, is my position.

But when can we ever fully know Truth? When can we be certain that a test, even if repeated successfully many times, will say something valid about reality? My answer is: when the new discovery fits within a major web of different, well-tested pieces of knowledge, which are theoretically integrated, and lead to fully confirmed original forecasts.

For instance, we can be certain that water is a compound of hydrogen and oxygen, i.e., that its formula is H_2O, and not another formula, say NaCl, because many independent strands of theoretical and practical knowledge confirmed and integrated each other on the basis of the discovery that water is H_2O. For instance, water is formed—together with CO_2—when burning methane CH_4 (the main component of natural gas) with air; water is formed when reacting (burning) hydrogen with oxygen; and both hydrogen and oxygen can be obtained from water by either electrolysis or decomposition at high temperature. The city gas of the past (when no natural gas was available) was a blend of carbon monoxide and hydrogen produced by treating hot coal with water: $C + H_2O = CO + H_2$. Hydrogen can be produced (even if not economically) by treating iron with water at high temperature: $3Fe + 4H_2O = Fe_3O_4 + 4H_2$. Photosynthesis, fundamental for life on this planet, is in its essence $6CO_2$ (carbon dioxide in the air) + $6H_2O$ (drawn from wet earth via the roots) + light falling on a green plant (chlorophyll) = $C_6H_{12}0_6$ (glucose) + $6O_2$ (the fresh oxygen produced by the green plants during the day).

Ergo, there is no way in which water could not be H_2O, and that any water produced with process A will not behave like water coming from process B or C.

It is also important to stress how this knowledge came from totally different fields. For instance, those who produce hydrogen by electrolysis of water may not know or care about the making of ethyl alcohol by reacting ethylene with water: $C_2H_4 + H_2O = CH_3CH_2OH$ (as true and good alcohol as that made by fermentation by old Noah of Gen. 9:21), in a synthesis which would be impossible if water was not H_2O.

Taking the problem from a different angle, scientists understood the chemical basis of water long before the discovery of photosynthesis; in other words we are not dealing with a chain of assumptions, or partial theories, all based on a first link, or nucleus, which if proven wrong makes the whole castle collapse. Here each piece is true by itself, and yet shown "more true" by the other experiments, and industrial methods, which use H_2O in a different way or context.

Here is Truth, period.[295] Here, the statement that water is a compound of hydrogen and oxygen "corresponds with the fact (or reality)" and leads to forecasting the results of thousand new chemical reactions in both the organic and inorganic fields. Furthermore, the success of these forecasts is now so frequent that it is not reported in the literature

---

[295] Henry Cavendish developed his theory that water is a compound of oxygen and hydrogen around 1784, more than two hundred years ago, and William Harvey his theory of the circulation of blood between 1616 and 1628 nearly four hundred years ago, therefore giving the lie to W. H. Newton-Smith of Balliol College, who in *The Rationality of Science* wrote that "there is inductive support for a pessimistic induction: any theory will be discovered to be false within, say 200 years of being propounded" (1981, p. 14).

In the same way, Sir Bernard Katz, at the 1994 meeting in London on creativity and discovery in biomedical science, sponsored by the Royal Society of Medicine, said: "Nowadays it would, in my opinion, be almost ridiculous, and in any case an unjustifiable waste of time, to subject Hermann's local circuit hypothesis to yet another attempt at falsification" (1994, p. 226). A fortiori this can be said for Cavendish's discovery that water is H_2O. Here truth is Truth, not waiting for a Popperian falsification that will tell us that it is something different. In this context it is good to remember that Einstein did not falsify Newton's law of gravity the way Copernicus falsified the geocentric theory of the universe. For instance in the study of the movement of artificial satellites or the travel of humans to the Moon, scientists still use the Newtonian equation.

save when a new useful substance is produced, such as a new important pharmaceutical product.[296]

2. For GAM/DP not the Freeman Dyson's model of science, but that of William Harvey

According to Freeman Dyson, "the way a scientific argument goes is typically as follows: We have a number of theories to explain what we have observed. Most of the theories are probably wrong or irrelevant. Then somebody does a new experiment or a new calculation that proves that theory A is wrong. As a result, theory B now has a better chance of being right" (2002, p. 4).

Instead, according to Pye-Smith and Singer (1973), William Harvey followed a different model in his discovery of the circulation of blood:

> The method by which Harvey arrived at his solution of the most fundamental and difficult problem in physiology is well worthy of attention. He had not only furnished himself with all the knowledge that books and the instructions of the best anatomists of Italy could give, but, *by a long series of dissections*, had gained a far more complete knowledge of the comparative anatomy of the heart and vessels than any contemporary had. Thus equipped, he began his investigations into the movements of the heart and blood by seeing their action in living animals. He minutely describes what he saw in dogs, pigs, serpents, frogs and fishes, and even in slugs, oysters, lobsters and insects, in the transparent shrimp, and lastly in the chick while still in the shell.

---

[296] I insist on what is known as true (and not simply self-evident, and not known elsewhere and before; and interrelated in a large system of explanations and confirmed forecasts), because in many circles it has become a habit to focus on what was wrongly stated at a given time, and therefore needed correction. Those errors are quite normal, and in no way diminish the truth of what was later so amply explained, confirmed and integrated with the rest. The history of both chemistry and medicine is paved with magnificent errors, for instance, for a long time, chemistry was not a science but alchemy. But out of alchemy came chemistry as a science. In essence, the pessimists are entitled to concentrate on past errors and proclaim a continuation forever of errors on top of errors. Yet I see them wrong—including Karl Popper and Bryan Magee (1997)—for not focusing on what has been proven right, solid as granite, fully integrated, fully confirmed, rich of confirmed forecasts, marvelously explained at different levels, as water being a compound of hydrogen and oxygen, or the heart being the "pump" responsible for the circulation of blood in humans and many animals.

> He particularly describes his observations and experiments on the ventricles, the auricles, the arteries and the veins. He shows how the arrangement of the vessels in the fetus supports his theory. He adduces facts observed in disease as well as in health to prove the rapidity of the circulation. He explains how the mechanism of the valves in the veins is adapted to favour the flow of the blood to the heart. He estimates the capacity of each ventricle, and reckons the rate at which the whole mass of blood passes through it. He elaborately and clearly demonstrates the effect of obstruction of the bloodstream in arteries or in veins, by the forceps in the case of a snake, by a ligature on the arm of a man, and illustrates his argument by figures. These results can be explained only by the constant circulation of the same blood. (p. 143; my italics)

Harvey did not simply discredit a previous theory A, giving a better chance to theory B of being right, but he studied carefully not only the human body, but also that of many other species. As the result of all this work, Harvey's theory became a discovery, i.e. the truth on the matter blood circulation.

With the GAM/DP theory I have tried to do the same by studying personality and creativity at both the individual and societal levels, and in so many different ways. At times the readers may have asked if there is a real benefit in studying people and events so far back in history, in such distant civilizations. My answer is the same that Harvey must have given to those who asked him the purpose of so many studies, of beings so different from men.

I may or may not have achieved the same certainty as Harvey, but my aim is the same: truth, full validity, not another theory that chases the previous one, in turn chased by the next in rapid succession.

Back to Harvey:

> In one point only was the demonstration of the circulation incomplete. Harvey did not see the capillary channels by which the blood passes from the arteries to the veins. This gap in the circulation was filled 30 years later by the great anatomist Marcello Malpighi, who described the capillary circulation four years after Harvey's death. But the existence of the channels first seen by Malpighi was already clearly pointed to by Harvey's reasoning. (ib., p. 143)

Here too, I hope to follow the Harvey model of science. There is much in the GAM/DP theory which is incomplete and demands the work of others, possibly for many decades or even centuries. There is still an enormous amount that needs to be understood in the field of personality and creativity, as there was still much to be learned on the blood circulation after Harvey and Malpighi, and there still is much to be learned nowadays.

3. And the GAM/DP Theory?

I see the gradual emergence, for GAM/DP, of an integrated system of explanations, evidence, successful forecasts, of a coherent net of observations, experiments, hypotheses and valid deductions.

For GAM, I see how the studies—by Roe (1952), Eiduson (1962), Albert (1983) on the high incidence of early parental death among eminent people; by Albert and Runco (1986) on the families of creative children; by Csikszentmihalyi (1991) on the *complex creators [dedicated]* and the *pathological creators [challenged]*, by Garmezy (1985), Werner and Smith (1982), and Werner (1989) on *resilient children*; by Weeks and James (1995) on eccentrics; by Adler (1958) on "organ inferiority" and derived inferiority or superiority complex; by Ott (1993) and Safranski (1998) on Heidegger; by Brent (1981) on Darwin—agree with each other and with GAM. Each study stands alone, and yet each stands stronger when integrated with the other, via GAM. For instance, one can start with Werner and Smith, move to Weeks and James, and end with GAM, or vice versa, and the result is the same.

In the same way, the biographies of Heidegger agree with those of Newton and Darwin, if read as examples of the working of GxAxM.

Obviously, all this is tall talk if not "taken with a few cosmic grains of salt," as recommended by Koestler (1968, p. 267) in his study of Kepler. In my case, these cosmic grains of salt will come from that *friendly-hostile co-operation of many scientists*, as discussed by Popper (1971, 2, p. 217).

For **DP**, I see a similarly integrated web of results, and explanations, of the studies by Kohn (1969) on class conformity; by Harwood and Miller (1991) and Harwood and Lucca Irizarry (1992) on what contemporary American and Puerto Rican mothers consider adequate conduct in their children; by Diaz-Guerrero (1990) on the personality differences between American and Mexicans; by Barzini (1965), Bennassar (1979), Kleinman and Kleinman (1991), Le Goff (1988), Madariaga (1969), Marías (1987), Patai (1983), Paz (1985), Pryce-Jones, and Sanchez-Albornoz, on history and ethnopsychology of various people; by Berlin (1967) on *hedgehogs* versus *foxes*; by Véliz (1994) on *gothic foxes* versus *baroque hedgehogs*; by Park (1928) on the *marginal man*; by Morris (1972) on the discovery of the individual 1050-1200; by Geertz (1975) on the Western conception of the individual; by Simonton (1984) on the plusses for creativity of political fragmentation; by Mokyr (1990) on technology and science, China vs. Europe; by Quint (1996) on Ariosto; by Gianni (1967) on Pulci; by McClelland on *The Achieving Society* (1967).

29

Positive Psychology

In the words of Martin Seligman and Mihaly Csikszentmihalyi: "A science of positive subjective experience, positive individual traits, *and positive institutions* promises to improve quality of life and prevent the pathologies that arise when life is barren and meaningless. The exclusive focus on pathology that has dominated our discipline [psychology] results in a model of the human being lacking the positive features that make life worth living. Hope, wisdom, creativity, future mindedness, courage, spirituality, responsibility, and perseverance are ignored or explained as transformations of more authentic negative impulses" (2000, p. 5, my italics).

All along, I have highlighted the fundamental contribution of *positive institutions* because this is the area in which the DP theory can best contribute with its emphasis on division of power, on power sharing, on the assumption of power—and derived advantages and responsibilities—by every human being. And the institutions which the DP theory favors most are the generous, well organized volunteer organizations which give power to each and every volunteer.

I already discussed, in volume 1, the power which volunteers acquire by being able to move from volunteer organization A to volunteer organization B, as soon as they feel that the organization is betraying its proclaimed goals, or top management is getting ossified or self-enriching. Here, I want briefly to mention another degree of freedom or power which consists in being able to move from one part of the organization to another with a degree of flexibility unknown and impossible in business. If running in trouble with Line Division A, a volunteer can switch to Line Division B, or to Administration, Recruiting, or Training, all without the approval of top management, as needed in industry or commerce.

Back to the build-up of a *Positive Psychology*, I see the "insertion of the individual within *positive institutions*" as one of the greatest improvements ever, because in such a case we deal with adding power and dignity to individuals instead of robbing them. In the past, psychologists have too often focused their attention on negative institutions which disempowered people. One classical example is the series of experiments conducted in 1961 by Stanley Milgram on obedience: "Milgram's aim was to find how much pressure to obey would be created by the experimenter in the lab coat when he said 'Teacher, you have no other choice; you must go on!' How far would subjects go under these circumstances?... To everyone's horror, when the experiment was conducted, 62 per cent of the teacher-subjects went all the way to 450 volts!" (Dworetzky, 1985, p. 536). Milgram, here had created a situation in which people were shown incapable of independent thinking, shorn of power.

A similar result was obtained by psychologist Philip Zimbardo, in 1973, when "one group of students were randomly assigned to be guards, another to be prisoners. Before long, the prisoners began to rebel and manifest resentment against the guards, while the guards became increasingly insensitive and even brutal to the prisoners. *The power of the structure of the community to mold behavior of both the inmates and the authorities was dramatically illustrated*" (Roediger, Rushton, Capaldi, & Paris, 1984, p. 553, my italics). Once again, we have people following instructions in a condition of ignorance of goals and methods; once again we have people who have been disempowered, including the "guards" who have been trapped in a role which, in their case, led to insensitivity and brutality.

At one level these studies deserve to be praised for pointing to the force of *negative power situations* which bring people to act unethically. At another level, and a more important one, we suffer from the shortage of equivalent studies on *positive power situations* where *positive structures* of the community bring forth the best in man. In 1943, in Trent, Italy, during the war, a young woman of 23, Chiara Lubich, began a modern lay religious movement which soon took the name of *Focolare* (hearth in Italian), and whose members were called *Focolarine* (the women), and *Focolarini* (the men). From the very beginning, charity was one of their top priorities, even when spending so many hours in air-raid shelters because of the constant bombing by the Allies:

> "The mother with her children was there a few feet away. One of my companions would take a child in her arms and look after it, another would help the old people or get them something to eat. We would hurry home to make an enormous pot of soup, and then we would run to distribute it to the hungry. The poor were our brothers. They also came to our house. And all around the table you would see little groups of two-one of us and someone in need."... "Give and it will be given unto you." People who knew them in Trent saw what the girls from the 'little house' were doing, and soon goods of all sort began arriving at 2 Piazza Cappucini. The little house became a warehouse, or rather a clearing house, as clothes, furnishings, utensils—anything, in fact, that the owners didn't need—arrived there, only to be passed on and distributed to those who did need them.... "You should have seen our little house at that time", she [Chiara] says. "Goods were piled up and then distributed, only for more to arrive shortly. So with the apples. As soon as we received them we went off to distribute them to the poor. On our return there were some more apples! This sequence of events was repeated three times in succession, until, in the evening, a suitcase with apples arrived!"... Within months of the establishment of that first little community at Piazza Cappucini, over 500 other people became associated with the little group. They were scattered in 158 towns and villages in the region of Trent. (Gallagher, 1997, pp. 38, 40-1)

Here we have a completely different situation from the one studied by Milgram and Zimbardo. Here is a situation which promotes friendship and generosity among many people, from Chiara to her first Focolarine, and to the many people in Trent who brought goods for the poor who were being assisted by Chiara and her friends. In this, all involved were empowered: given the possibility to do good, to take initiative, to show the best in their personality. Everybody was a winner.

"When they were called to speak in the outlying villages and towns, Chiara would give her companions written notes, the fruit of her contemplation—but always with the instruction that they were to read them and then destroy them. *She was adamant that she did not want the other girls to be reliant on her. They were to understand Scripture for themselves and then get rid of the intermediary*, as it were" (ib., p. 45, my italics). This is a wonderful example of how Chiara and her Movement were empowering people, who had to rely on their knowledge, understanding, personality, and creativity, who were not bound by the words of the Founder. These people were empowered, not disempowered, within a freely-chosen positive social structure.

I have chosen to discuss the Focolarini because I have known some of them, for a long time, and have admired their friendly and generous initiatives for the common good. Yet there are many similar examples in other religious or non-religious volunteer organizations. The point, in this whole discussion, is to forget the old emphasis on the study of the effects of a *UP negative psychology*, on pathological cases, on how society can disempower people, and to move, instead, toward a *DP positive psychology* as fostered by Seligman and Csikszentmihalyi and the 15 contributors to the January 2000 special issue of *American Psychologist*, on "Happiness, excellence, and optimal human functioning."[297]

[297] The 15 articles discussed "such issues as what enables happiness, the effects of autonomy and self regulation, how optimism and hope affect health, what constitutes wisdom, and how talent and creativity come to fruition" (Seligman & Csikszentmihalyi, 2000, p. 5).

30

Synopsis of the GAM/DP Theory of Personality and Creativity

The **GAM/DP theory** deals with personality and creativity at both the individual and societal levels: what makes an individual creative, and what makes a society creative.

G stands for genetic endowment, **A** for assistances, **M** for misfortunes, **DP** for division of power (**UP** for unity of power).

Major misfortunes of youth (e.g., early parental death, physical infirmity) reduce the acquisition of normal societal scripts[298] from parents, other relatives, school, teachers, and playmates. This condition, when backed by a good genetic endowment (e.g., for intelligence, temperaments, specific talents), and by valid assistances (e.g., from parents, schools, books), stimulates the build-up of new personal scripts which will then conflict with those of society in a life-long *war of the scripts*: the tension and material for creativity. Also, a vacuum of normal scripts leads naturally to ask "Who says so?" and "Why?"—in other words to the development of critical thinking—when confronted with statements that appear self-evident to others. The final potential for creativity can be expressed by the formula **GxAxM**.

Different kinds of misfortune have a different impact on personality and creativity; so much so that eminent creators can be profitably studied in comparison with others affected by the same misfortune of youth. Accordingly the theory has studied fourteen *challenged person-*

[298] Scripts are habits, routines, common ways of thinking, taboos, schemes, traditions, cognitive-structural units, paradigms.

ality families, one for each of fourteen major misfortunes (e.g., early parental death, physical infirmity, parental failure of character, parental domination, suffered lack of parental love).

The second part of the theory says that long periods of division of power, **DP** (e.g. when power is fought or grudgingly shared between emperor and pope, between state and church; when two or more major ideologies fight each other) give origin to the *visitor personality*; and that, instead, long periods of unity of power, **UP** (one single ruler or dominating hierarchy, one single dominating ideology) give origin to the *insular personality*.

DP fosters critical thinking. People begin to mistrust the power holder. They have the courage to speak their mind, to be creative on matters that are not those wanted by the power holder. *Visitors* go places and take risks (they can do so: in case of trouble with power A, they can ask the help of power B which will give it gladly). Automatically the *visitors* take initiative for the common good because they do not trust the *supremo* who is more interested in his personal power and glory. Soon the *visitors* are firm believers in the Law of Power of Lord Acton: "Power tends to corrupt and absolute power corrupts absolutely." *Visitors* therefore are highly creative, and so are the **DP places** which nurtured them (e.g., the Florence of Dante and Boccaccio, Leonardo and Michelangelo). Every civilization was highly creative at first, under a fair measure of DP, then declined when the UP took control, and most people became *insulars*.

In the end, individual personality and creativity potential can be expressed by the formula **GxAxMxDP**, in which **DP** has a positive impact already on assistances and misfortunes of youth; later on all aspects of adult life.

At the individual level, the GAM theory studied such eminent creators like Mozart (e.g., in the chapter "Why Mozart and not Salieri"), Bach, Wagner; Cervantes, Goethe, Tolstoy; Newton, Darwin, Einstein; Toulouse-Lautrec, Klee, Picasso. At the societal level the DP theory studied a long string of civilizations and countries in chapters like "The DP Origins of the Scottish Enlightenment", "The Origins, Decline and Fall of the Egyptian Civilization", "Did the Mandarins Kill the Chinese Civilization?", "The Low Creativity and High Unity of Power of the Byzantine Civilization".

The theory is presented—by the author, William A. Therivel—in four volumes published by Kirk House (2001 to 2004) entitled *The GAM/DP Theory of Personality and Creativity* (volumes 1 to 4).

Picture Credits

References

Abulafia, David (1988). *Frederick II: A medieval emperor*. New York: Oxford University Press.

Adler, Alfred (1958). *What life should mean to you*. New York: Capricorn Books.

Ahmad, Aijaz (1992). Orientalism and after: Ambivalence and metropolitan location in the work of Edward Said. In Ahmad's *In theory: Classes, nations, literature* (pp. 159-219). London: Verso.

Akehurst, F. R. P. (1989). Troubadour, Trouvère. In *Dictionary of the Middle Ages* (*12*, pp. 214-19). New York: Scribner.

Albanese, Ralph, Jr. (1992). Aristocratic ethos and ideological codes. In P. Henry (Ed.), *An inimitable example: The case of the Princesse de Clèves*. Washington, DC: Catholic University of America Press.

Albert, Robert (1983). Family position and the attainment of eminence: A study of special family positions and special family experiences. In R. S. Albert (Ed.), *Genius and eminence: The social psychology of creativity and achievement* (pp. 141-54). Oxford: Pergamon.

Albert, Robert S., & Runco, Mark A. (1986). The achievement of eminence: A model based on a longitudinal study of exceptional gifted boys and their families. In R. J. Sternberg & J. E. Davidson (Eds.), *Conception of giftedness* (pp. 332-57). New York: Cambridge University Press.

Allbutt, Clifford (1921/1970). *Greek medicine in Rome*. New York: Benjamin Blom.

Allen, John L., Jr. (1999, April 16). The Vatican's enforcer. *National Catholic Reporter* (cover story; pp. 1-20 online).

Alvarez, A. (2001, August 9). Ice Capades. *New York Review of Books*, 14-17.

Ancelet-Hustace, Jeanne (1979). *Goethe*. Paris: Seuil.

Anderson, William (1980). *Dante the maker*. London: Routledge & Kegan Paul.

Anglo, Sydney (1987, July 2). Courtly lessons for today's careerist. *The Listener*, 15-16.

Annan, Noel (1998). Foreword. In I. Berlin's *The proper study of mankind* (pp. ix-xv). New York: Farrar, Straus and Giroux.

Ariès, Philippe (1962). *Centuries of childhood*. London: Jonathan Cape.

Arima Tatsuo (1969). *The failure of freedom: A portrait of modern Japanese intellectuals*. Cambridge, MA: Harvard University Press.

Arnold, Benjamin (1997). *Medieval Germany, 500-1300. A political interpretation*. Toronto: University of Toronto Press.

Arnold, David (2003, Jan. 3). Calcutta's secrets. *Times Literary Supplement*, 22.

Arnold, Thomas (1835/1970). Education in school. In T. W. Bamford (Ed.), *Thomas Arnold on education* (pp. 106-43). Cambridge University Press.

Ayduk, Ozlem, & Mischel, Walter (2002). When smart people behave stupidly: reconciling inconsistencies in social emotional intelligence. In R. J. Sternberg (Ed.), *Why smart people can be so stupid* (pp. 86-104). New Haven, CT: Yale University Press.

Backès, Jean-Louis (1978). *Pouchkine*. Paris: Seuil.

Bandello, Matteo (1554/1910). *Le novelle*. Bari: Laterza.

Barnouw, Dagmar (1979). *Elias Canetti*. Stuttgart: J. B. Metzler.

Baroja, Pío (1911/1974). *The tree of knowledge*. New York: Howard Fertig.

Bartolo da Sassoferrato (1340/1964). De Guelphis. In E. Emerton (Ed.), *Humanism and tyranny: Studies in the Italian Trecento*. Gloucester, MA: Peter Smith.

Barzini, Luigi (1965). *The Italians*. New York: Atheneum.

Barzini, Luigi (1983). *The Europeans*. New York: Simon and Schuster.

Basbanes, Nicholas (1999, August). Mount Athos: At this remote sanctuary, art and religion have intertwined for 1,000 years. *Smithsonian Magazine*, 90-99.

Batts, Michael S. (1971). *Gottfried von Strassburg*. New York: Twayne.

Becker, Marvin, B. (1981). *Medieval Italy*. Bloomington: Indiana University Press.

Belgrano, Manuel (ca. 1815/1965). The making of an insurgent. In R.A. Humphreys & J. Lynch (Eds.), *The Origins of the Latin American Revolutions 1808-1826* (pp. 275-80). New York: Knopf.

Benét, William R. (1987). *Reader's Encyclopedia*. New York: Harper & Row.

Benigni, U. (1912). Saint Zita. In *The Catholic Encyclopaedia* (*15*, p. 762). New York: Robert Appleton.

Bennassar Bartolomé (1979). *The Spanish character: Attitudes and mentalities form the sixteenth to the nineteenth century.* Berkeley: University of California Press.

Bennassar, Bartolomé (1981). *Inquisición Española: Poder político y control social.* Barcelona: Crítica.

Benoist-Méchin, Jacques G. (1964). *Alexandre le Grand, ou le rève dépassé*. Lausanne: Clairefontaine.

Bergin, Thomas G. (1981). *Boccaccio*. New York: Viking Press.

Bernard de Morval (fl. 1150/1929). *De contemptu mundi*. London: Bernard Quaritch.

Berlin, Isaiah (1953/1967). *The hedgehog and the fox*: An essay on Tolstoy's view of history. London: Weidenfeld and Nicolson.

Berlin, Isaiah (1960/1998). The concept of scientific history. In Berlin's *The proper study of mankind* (pp. 1-58). New York: Farrar, Straus and Giroux.

Berlin, Isaiah (1998, May 14). The first and the last. *New York Review of Books*, 52-60.

Berry, André F. (1970). Troubadours. In *Encylopaedia Britannica* (*22*, pp. 263-64). Chicago: Encyclopaedia Britannica.

Bevan, Edwyn (1927/1968). *The house of Ptolemy: A history of Egypt under the Ptolemaic dynasty*. Chicago: Argonaut.

Bhatia, S. (1985). *Arthur Miller: Social drama as tragedy*. New Delhi: Arnold-Heinemann.

Binyon, T. J. (2002). *Pushkin: A biography*. London: HarperCollins.

Blades, William (1971). *The biography and typography of William Caxton, England's first printer*. Totowa, NJ: Rowman and Littlefield.

Blake, N. F. (1969). *Caxton and his world*. London: Andre Deutsch.

Boase, Roger (1977). *The origin and meaning of courtly love: A critical study of European scholarship*. Manchester: Manchester University Press.

Boase, Roger (1983). Courtly Love. In *Dictionary of the Middle Ages* (*3*, pp. 667-68). New York: Scribner.

Boccaccio, Giovanni (ca. 1351/1998). *The Decameron*. Oxford: Oxford University Press.

Boorstin, Daniel J. (1983). *The discoverers*. New York: Random House.

Borg, Mark (1999). *Conferment of the Degree of Doctor (Honoris Causa) on Chiara Lubich by the University of Malta*. Malta: Transcript.

Borkenau, Franz (1947-57/1981). *End and beginning: On the generations of cultures and the origins of the West*. New York: Columbia University Press.

Bornstein, Diane (1983a). *The lady in the tower*. Hamden, CN: Archon.

Bornstein, Diane (1983b). Courtly Love. In *Dictionary of the Middle Ages* (*3*, pp. 668-74). New York: Scribner.

Bouwsma, William J. (1990). *A usable past: Essays in European cultural history*. Berkeley: University of California Press.

Bovini, Giuseppe (1991). *Ravenna: Art and history*. Ravenna: Longo.

Bowle, John (1971/2000). *The English experience: A survey of English history from early times to the end of Empire*. London: Phoenix Press.

Braudel, Ferdinand (1979). *The structures of everyday life*. New York: Harper & Row.

Braunstein, Philippe (1988). Toward intimacy: The fourteenth and fifteenth century. In G. Duby (Ed.), *A history of private life. II. Revelations of the Medieval world* (pp. 535-630). Cambridge, MA: Harvard University Press.

Brayer, Elizabeth (1996). *George Eastman: A biography*. Baltimore: Johns Hopkins University Press.

Brent, Peter (1981). *Charles Darwin: "A man of enlarged curiosity"*. New York: Harper & Row.

Brewer, Derek S. (1973). John Capgrave. In *Encyclopaedia Britannica* (*4*, pp. 829-30). Chicago: Encyclopaedia Britannica.

Brewer, Derek S. (1973). Geoffrey Chaucer. In *Encylopaedia Britannica* (*5*, pp. 354-59). Chicago: Encyclopaedia Britannica.

Brion, Marcel (1940). *Michelangelo*. New York: Bonanza.

Broadie, Alexander (1997). *The Scottish Enlightenment*. Edinburgh: Canongate.

Brooke, Christopher N. L. (1991). *The medieval idea of marriage*. Oxford: Oxford University Press.

Brooke, Zachary N. (1964). Lay investiture and its relation to the conflict of Empire and Papacy. In W. Schafer (Ed.), *The Gregorian epoch: Reformation, revolution, reaction?* Boston: Heath.

Bull, George (1976). Introduction. In Baldesar Castiglione's *The book of the courtier*. London: Penguin.

Bunkley, Allison Williams (1952). *The life of Sarmiento*. Princeton, NJ: Princeton University Press.

Burg, David, & Feifer, George (1972). *Solzhenitsyn*. London: Hodder and Stoughton.

Burgess, Glyn S. (1987). Marie de Champagne. In *Dictionary of the Middle Ages* (*8*, p. 135). New York: Scribner.

Burke, Peter (1996). *The fortune of the courtier: The European reception of Castiglione's cortegiano*. University Park: Pennsylvania State University Press.

Burnaby, John (1978). Saint Augustine of Hippo. In *Encyclopaedia Britannica* (*2*, pp. 364-68). Chicago: Encyclopaedia Britannica

Burton, Richard F. (transl.) (1885-88/1978). *Tales from Arabian Nights, selected from the book of the thousand nights and a night*. New York: Avenel.

Buruma, Ian (1991, Feb. 14). Sign of life. A review of Naipaul's "India: A million mutinies now". *New York Review of Books*, 3-5.

Buruma, Ian (2000, February 24). Divine killer [Mao]. *New York Review of Books*, 20-25.

Butcher, William (1999). Introduction. In J. Verne's *Around the world in eighty days*. Oxford: Oxford University Press.

Butscher, Edward (1976). *Sylvia Plath: Method and madness*. New York: Seabury.

Byron, Robert, & David Talbot Rice (1930/1968). *The birth of Western painting*. New York: Hacker Art Books.

Byron, William (1979). *Cervantes: A biography*. London: Cassell.

Cabaniss, Allen (1972). *Charlemagne*. New York: Twayne.

Calderón de la Barca, Pedro (1635/1931). *El medico de su honra*. Madrid: Prensa Moderna.

Calderón de la Barca, Pedro (1635/1960). *The surgeon of his honour*. Madison: University of Wisconsin Press.

Calvino, Italo (1991). *Perché leggere i classici*. Milan: Arnoldo Mondadori.

Canavaggio, Jean (1992). *Cervantes: En busca del perfil perdido*. Madrid: Espasa-Calpe.

Canetti, Elias (1935/1946). *Auto-da-fé*. New York: Stein and Day.

Canetti, Elias (1960/1973). *Crowds and power*. New York: Continuum.

Canetti, Elias (1977/1983). *The tongue set free*. New York: Continuum.

Canetti, Elias (1980/1982). *The torch in my ear*. New York: Farrar Straus Giroux.

Canetti, Elias (1985/1986). *The play of the eyes*. New York: Farrar Straus Giroux.

Canning, J. (1996). *A history of Medieval political thought 300-1450*, London: Routledge.

Cantor, Norman F. (1993). *The civilization of the Middle Ages*. New York: HarperCollins.

Cantor, Norman (Ed.) (1995). *The Medieval reader*. New York: HarperPerennial.

Capellanus, Andreas (ca. 1174/1959). *The art of courtly love*. New York: Frederick Ungar.

Capgrave, John (ca. 1445/1999). *The life of Saint Katherine*. Kalamazoo: Medieval Institute Publications/Western Michigan University.

Carter, William H. (1973). Spain. In *Encyclopaedia Britannica* (*20*, pp. 1107-13). Chicago: Encyclopaedia Britannica

Cartwright, Julia (1927). *The perfect courtier: Baldassare Castiglione—His life and letters, 1478-1529*. New York: E. P. Dutton.

Castiglione, Baldesar (1507-28/1976). *The book of the courtier*. London: Penguin.

Castro, Américo (1966). *Cervantes y los casticismos españoles*. Madrid: Alfaguara.

Castro, Américo (1972). *De la edad conflictiva: Crisis de la cultura española en el siglo XVII*. Madrid: Taurus.

Cave, Terence (1994, Feb. 25). A coiner of terms. *Times Literary Supplement*, p. 25.

Caxton, William (1483/1900). *The "Golden Legend" or lives of the saints, as englished by William Caxton*. London: J. M. Dent.

Caxton, William (1483/2002 (ex Robert Blackmon, 1900). *The Golden Legend (Aurea Legenda) compiled by Jacobus de Voragine, 1275; englished by William Caxton, 1483*. (in *Medieval Sourcebook*: http://www.fordham.edu/halsall/basis/goldenlegend/htm.

Cerezo Carrasco, Enrique (1952, August 10). Sindacalismo vertical. *Alcalá*, 11.

Chamberlin, Russell (1986). *Charlemagne: Emperor of the Western World.* London: Grafton Books.

Chaves, Julio Cesar (1964). *Unamuno y America*. Madrid: Ediciones Cultura Hispanica.

Cheetham, Nicolas (1992). *A history of the popes*. New York: Dorset Press.

Chesterton, G. K. (1923). *St. Francis of Assisi*. London: Hodder and Stoughton.

Chiellini, Monica (1988). *Cimabue*. Florence: Scala.

Childe, Edmund (1976). *William Caxton: A portrait in a background.* London: Northwood Publications.

Churchill, Winston S. (1932/1965). *Painting as a pastime.* New York: Cornerstone Library.

Clarence-Smith, Gervase (2002, June 28). Property in people. *Times Literary Supplement*, 7-8.

Clough, Cecil H. (1978). Francis I and the courtiers of Castiglione's courtier. *European Studies Review*, *8*, 23-70.

Clugnet, Leon (1908). Catherine of Alexandria. In *The Catholic Encyclopaedia, (5,* 445-46). New York: Robert Appleton.

Collins, Roger, (1998). *Charlemagne*. Toronto: Toronto University Press.

Commager, Henry Steele (1977). *The empire of reason.* Garden City, NY: Anchor/Doubleday.

Cook, Robert Francis (1987). *The sense of the Song of Roland.* Ithaca, NY: Cornell University Press.

Crary, David (1999, December 14). Painting depicts a modern Jesus. *Houston Chronicle*, 4D.

Cronin, Vincent (1969). *The flowering of the Renaissance*. London: The History Book Club.

Cronin, Vincent (1972). *Napoleon Bonaparte: An intimate biography.* New York: William Morrow & Co.

Cronin, Vincent (1978). Paolo Sarpi. In *Encyclopaedia Britannica* (*16*, pp. 251-52). Chicago: Encyclopaedia Britannica.

Crosby, Alfred W. (1997). *The measure of reality: quantification and Western society, 1250- 1600*. Cambridge: Cambridge University Press.

Crowley, Frances G. (1972). *Domingo Faustino Sarmiento*. New York: Twayne.

Csikszentmihalyi, Mihaly (1988). Society, culture, and person: a system view of creativity. In R.J. Sternberg (Ed.), *The nature of creativity* (pp. 325-39). New York: Cambridge University Press.

Csikszentmihalyi, Mihaly (1991, April). *The antecedents of intrinsic motivation and creativity in childhood.* Invited address to the biennial meeting of the Society for Research in Child Development, Seattle, WA. (Audio Production magnetic tape #H671).

Csikszentmihalyi, Mihaly (1999). Implications of a system perspective for the study of creativity. In R. J. Sternberg (Ed.), *Handbook of creativity* (pp. 313-35). Cambridge: Cambridge University Press.

Davis, David Brion (1966). *The problem of slavery in Western culture*. Ithaca, NY: Cornell University Press.

Davis, David Brion (1984). *Slavery and human progress*. New York: Oxford University Press.

Dawson, C. (1964). Monastic reform and Christian culture. In W. Schafer (Ed.), *The Gregorian epoch: Reformation, revolution, reaction?* (pp. 47-54). Boston: D. C. Heath.

DeAngelis, Tori (2003, February). Why we overestimate our competence. *Monitor on Psychology*, 60-62.

Dego, Giuliano (1967). *Moravia*. New York: Barnes & Noble.

Denomy, Alex J. (1944). An inquiry into the origins of courtly love. *Medieval Studies*, *6*, 175-260.

Denomy, Alex J. (1945). Fin' amour: The pure love of the troubadours, its amorality, and possible source. *Medieval Studies*, *7*, 139-207.

Denomy, Alex J. (1946). The *De Amore* of Andreas Capellanus and the condemnation of 1277. *Medieval Studies*, *8*, 107-49.

Denomy, Alex J. (1953). Concerning the accessibility of Arabic influences to the earliest Provençal Troubadours. *Medieval Studies*, *15*, 147-260.

Devisse, Jean (1970). Charlemagne. In *Encyclopaedia Britannica* (*5*, pp. 288-91). Chicago: Encyclopaedia Britannica.

Diaz-Guerrero, Rogelio (1990). ¿Existen rasgos básicos en la personalidad de los Mexicanos? *Revista Mexicana de Psicologia, 7*, 121-29.

Diaz-Playa, Fernando (1967). *The Spaniard and the seven deadly sins.* New York: Scribner's.

Donahue, William A. (2001). *The end of modernism: Elias Canetti's "Auto-da-fé"*. Chapel Hill: University of North Carolina Press.

Donald, D. H. (1996). *Lincoln.* New York: Simon & Schuster.

Douglas, David C. (1964). *William the Conqueror.* London: Eyre Methuen.

Drake, H. A. (2000). *Constantine and the bishops: The politics of intolerance.* Baltimore: Johns Hopkins University Press.

Duby, Georges (1978). *Medieval marriage.* Baltimore: Johns Hopkins University Press.

Dworetzky, John P. (1985). *Psychology.* St. Paul, MN: West Publishing.

Dwyer, Richard A. (1987). Ovid in the Middle Ages. In *Dictionary of the Middle Ages* (*9*, pp. 312-14). New York: Scribner.

Dyson, Freeman J. (2002, March 28). Science & religion: No ends in sight. *New York Review of Books* (4-6).

Earle, Peter G. (1971). *Prophet in the wilderness: The works of Ezequiel Martínez Estrada.* Austin: University of Texas Press.

Edelstein, Ludwig (1932/1967). The history of anatomy in antiquity. In L. Edelstein *Ancient Medicine* (pp. 247-301). Baltimore: Johns Hopkins Press.

Eiduson, Bernice T. (1962). *Scientists: Their psychological world.* New York: Basic Books.

Elias, Norbert (1939/1978). *The civilizing process: The history of manners.* New York: Urizen Books.

Elias, Norbert (1939/2000). *The civilizing process.* Oxford: Blackwell.

Eliot, T. S. (1951). *Selected essays.* London: Faber and Faber.

Elliott, J. H. (1964). *Imperial Spain.* New York: St. Martin's Press.

Elon, Amos (1999, Jan. 10). Exile's return. [A book review of "Out of place: A memoir" by Edward Said]. *New York Review of Books*, 12-15.

Encyclopedic Dictionary of Psychology (1983). (R. Harré & R. Lamb, Eds.). Cambridge, MA: MIT Press.

Endo, Shusaku (1978). *A life of Jesus.* New York: Paulist Press.

Endo, Shusaku (1979). Preface. In his *When I whistle* (pp. 7-10). London: Peter Owen.

Evans, R. I. (1969). *Psychology and Arthur Miller.* New York: E. P. Dutton.

Evergates, Theodore (1987). Nobility and nobles. In *Dictionary of the Middle Ages* (*9*, pp. 147-52). New York: Scribner.

Eyck, Frank (1998). *Religion and politics in German history - From the beginnings to the French revolution.* New York: St. Martin's Press.

Fairbank, John K. (1990, May 31). From the Ming to Deng Xiaoping. *New York Review of Books*, 16-18.

Fairbank, John King, & Goldman, Merle (1998). *China: A new history.* Cambridge, MA: Harvard University Press.

Falk, Thomas H. (1993). *Elias Canetti.* New York: Twayne.

Farmer, David (1997). *The Oxford Dictionary of Saints.* Oxford: Oxford University Press.

Faulkner, Harold U., & Kepner, Tyler (1934). *America: Its history and people.* New York: Harper & Brothers.

Federn, Karl (1902/1970). *Dante and his time.* New York: Haskell House.

Fermi, Laura (1954). *Atoms in the family: My life with Enrico Fermi.* Chicago: University of Chicago Press.

Ferns, H. S. (1969). *Argentina.* New York: Frederick A. Praeger.

Ferri, Elsa (1976). *Growing up in a one-parent family: A long-term study of child development.* Slough, UK: NFER Publishing.

Ferroni, Giulio (1991). *Storia della letteratura italiana (1). Dalle origini al Quattrocento.* Turin: Einaudi.

Figes, Orlando (2002). *Natasha's dance.* London: Penguin.

Fillitz, H. (1993). Die Gruppe der Magdeburger Elfenbeinplatten. In *Bernward von Hildesheim und das Zeitalter der Ottonen — Katalog der Ausstellung Hildesheim 1993* (*2*, pp. 40-48). Hildesheim: Bernard Verlag.

Flanagan, Owen (2000). *Dreaming souls: Sleep, dreams and the evolution of the conscious mind.* New York: Oxford University Press.

Folz, Robert (1964/1974). *The coronation of Charlemagne 25 December 800.* London: Routledge & Kegan Paul.

Forester, C. S. (1935/1980). *The African Queen.* London: Penguin Books.

Fournier, P. (1964). The Roman Canonical collections of the period of Gregory VII. In W. Schafer (Ed.), *The Gregorian epoch: Reformation, revolution, reaction?* (pp. 19-23). Boston: D. C. Heath.

Frame, Donald M. (1965). *Montaigne.* New York: Harcourt, Brace and World.

Frankfort, Henry (1948). *Kingship and the gods: A study of Ancient Near Eastern religion and the integration of society & nature.* Chicago: University of Chicago Press.

Freedman, Estelle B. (2002). *No turning back: The history of feminism and the future of women.* New York: Ballantine.

Freeman, Michelle A. (1983). Chrétien de Troyes. In *Dictionary of the Middle Ages* (*3*, pp. 308-11). New York: Scribner.

French, Roger (1999). *Dissection and vivisection in the European Renaissance.* Aldershot, England: Ashgate.

Frend, W. H. C. (1984). *The rise of Christianity.* Philadelphia: Fortress Press.

Freud, Sigmund (1926/1959). Address to the Society of B'nai B'rith, In *The standard edition of the complete psychological works of Sigmund Freud* (*20*, pp. 273-74). London: Hogarth.

Friedenthal, Richard (1965). *Goethe: His life and times.* New York: World Publishing.

Galbraith, John Kenneth (1986, August 14). Truly the last tycoons. *New York Review of Books*, 17.

Gallagher, Jim (1997). *A woman's work — Chiara Lubich: A biography of the Focolare Movement and its founder.* Hyde Park, NY: New City Press.

Gallagher, John J. (1938). *Church and state in Germany under Otto the Great (936-973).* Washington, DC: Catholic University of America.

Gálvez, Manuel (1913/1960). *El solar de la raza*. Buenos Aires: La Facultad.

Gálvez, Manuel (1938/1994). *Hombres en soledad*. Buenos Aires: El Ateneo.

Games, Gwilym (1997). Quakers. In J. P. Rodriguez (Ed.), *The historical encyclopedia of world slavery* (*2*, pp. 531-32). Santa Barbara, CA: ABC-Clio.

García Cárcel, Ricardo (1992). *La leyenda negra: Historia y opinión*. Madrid: Alianza Editorial.

García, Carlos (1617/1992). Quoted in R. García Cárcel's *La Leyenda negra: Historia y opinión*, pp. 56-60. Madrid: Alianz Editorial.

Gardner, Helen (1959). *Art through the ages*. New York: Harcourt, Brace.

Gardner, Howard (1993). *Creating minds: An anatomy of creativity*. New York: Basic Books.

Garmezy, Norman (1985). Stress-resistant children. The search for protective factors. In J. E. Stevenson (Ed.), Recent research in developmental psychopathology. *Journal of Child Psychology and Psychiatry Book Supplement (4*, pp. 213-33).

Gay, Peter (1969). *The Enlightenment: An interpretation. Vol. 2: The science of freedom*. New York: Alfred A. Knopf.

Gay, Peter (1988). *Freud: A life for our time*. New York: W.W. Norton

Geertz, Clifford (1975). On the nature of anthropological understanding. *American Scientist, 63,* 47-53.

Gianni, Angelo (1967). *Pulci uno e due*. Florence: "La Nuova Italia" Editrice.

Gilot, Françoise, & Lake, Carlton (1964). *Life with Picasso*. New York: MacGraw-Hill.

Glubb, John B. (1971). *Life and times of Muhammad*. New York: Stein and Day.

Goethe, Johann Wolfgang (1774/1983). *The sorrows of young Werther*. New York: Suhrkamp.

Goethe, Johann Wolfgang, von (1811-22/1969). *Goethe's autobiography: Poetry and truth*. New York: Public Affairs Press.

Goethe, Johann Wolfgang (1811-22/1897). *The autobiography of Goethe - Truth and poetry: From my own life*. London: George Bell.

Gordon, Benjamin Lee (1959). *Medieval and Renaissance medicine*. New York: Philosophical Library.

Gordon, Daniel (1994). *Citizens without sovereignty: Equality and sociability in French thought, 1670-1789*. Princeton, NJ: Princeton University Press.

Gordon, Lyndall (1978). *Eliot's early years*. Oxford: Oxford University Press.

Goustad, Edwin S., & Schmidt, Leigh E. (2002). *The religious history of America*. San Francisco: HarperSanFrancisco.

Grabar, André (1968). *Early Christian art: From the rise of Christianity to the death of Theodosius*. New York: Odyssey Press.

Grant, Michael (1975/2000). *The twelve Caesars*. New York: History Book Club.

Grant, Michael (2000). *Constantine the Great: The man and his times*. New York: History Book Club.

Green, R. H. (1981). Courtly love. In *New Catholic Encyclopedia* (*4*, pp. 393-99). Washington, DC: Catholic University of America.

Gregory VII (1073-1085/1932). *The correspondence of Pope Gregory VII*. Translated and introduced by E. Emerton. New York: Columbia University Press.

Gruber, Howard E., & Wallace, Doris B. (2001). Creative work: The case of Charles Darwin. *American Psychologist, 56*, 346-49.

Guenoun, Pierre (1968). Introduction. In Tirso de Molina, *El burlador de Sevilla y convidado de piedra (Don Juan)* (pp. 7-18) Paris: Aubier.

Guenoun, Pierre (1971). *Cervantès par lui-même*. Paris: Editions du Seuil.
Gurevich, Aron J. (1985). *Categories of Medieval culture*. London: Routledge & Kegan Paul.
Hackel, Sergei (1990). The Orthodox churches of Eastern Europe. In J. McManners (Ed.), *The Oxford Illustrated History of Christianity* (pp. 519-49). Oxford: Oxford University Press.
Hacket, Francis (1937). *Francis the First*. Garden City, NY: Doubleday, Doran.
Halford, Aubrey S., & Halford, Giovanna M. (1956). *The Kabuki handbook*. Tokyo: Charles S. Tuttle.
Halphen, Louis (1977). *Charlemagne and the Carolingian empire*. Amsterdam: North-Holland Publishing.
Hamer, Richard (1998). Introduction. In Jacobus de Voragine's *The golden legend: Selections*, selected and translated by C. Stave (pp. ix-xxxi). London: Penguin.
Hampe, Karl (1909/1973). *Germany under the Salian and Hohenstaufen emperors*. Totowa, NJ: Rowman & Littlefield.
Hanawalt, Barbara A. (1993). *Growing up in Medieval London: The experience of childhood in history.* New York: Oxford University Press.
Hanke, Lewis (1951). *Bartolomé de Las Casas*. The Hague: Martinus Nijhoff.
Hansen, Elaine Tuttle (1992). *Chaucer and the fictions of gender.* Berkeley: University of California Press.
HarperCollins Encyclopedia of Catholicism (1995). R. P. Mc Brien (Ed.). San Francisco: HarperCollins.
Harwood, Robin L., & Lucca Irizarry, Nydia (1992). Anglo and Puerto Rican mothers' perception of attachment: Preliminary findings from the island. *Revista Puertoriqueña de Psicologia*, *8*, 209-26.
Harwood, Robin L., & Miller, Joan G. (1991). Perception of attachment: A comparison of Anglo and Puerto Rican mothers. *Merrill-Palmer Quarterly, 37*, 583-99.
Hausheer, Roger (1983). Isaiah Berlin. In A. Bullock & R. B. Woodings (Eds.), *Twentieth-century culture: A biographical companion* (pp. 66-67) New York: Harper & Row.
Hawthorne, Julian (1884/1968). *Nathaniel Hawthorne and his wife*. Hamden, CT: Archon Books.
Hawthorne, Nathaniel (1836/1937). The Minister's Black Veil. In *The complete novels* (pp. 872-82). New York: Modern Library.
Hawthorne, Nathaniel (1843/1950). The Birthmark. In *Selected tales and sketches* (pp. 203-21). New York: Rinehart.
Hawthorne, Nathaniel (1843/1950). Egotism; Or the Bosom Serpent. In *Selected tales and sketches* (pp. 188-202). New York: Rinehart.
Hawthorne, Nathaniel (1850/1935). *The Scarlet Letter*. New York: The Heritage Press.
Hayman, Ronald (1990). *Proust: A biography*. New York: Carol and Graf.
Haymes, Edward R. (1986). *The Nibelungenlied*. Urbana: University of Illinois Press.
Haywood, John A. (1973). Thousand and one nights. In *Encyclopaedia Britannica* (*21*, pp. 1083-84). Chicago: Encyclopaedia Britannica.
Hebert, Hugh (1979, April 14). At the shrine of the fox. *The Guardian*, 10.
Heine, Steven J.; Kitayama, Shinobu; Lehman, Darrin R.; Markus, Hazel Rose (1999). Is there a universal need for positive self-regard? *Psychological Review*, *106*, 766-94.
Heine, Steven J.; Kitayama, Shinobu; Lehman, Darrin R.; Takata, Toshitake; Leung, Eugene Ide; Leung, Cecilia; Matsumoto, Hisaya (2001). Divergent consequences of success and failure in Japan and North America: An investigation of self-improving

motivation and malleable selves. *Journal of Personality and Social Psychology, 81*, 599-615.

Heinen, Heinz (1978). Ptolemy II Philadelphus. In *Encyclopaedia Britannica (15*, p. 182). Chicago: Encyclopaedia Britannica.

Hellie, Richard (1998). Byzantium. In *Macmillan Encyclopedia of World Slavery* (*1*, pp. 751-55). New York: Macmillan.

Herodotus (ca. 460 BC/1958). *The histories of Herodotus of Halicarnassus*. Haarlem: The Limited Editions Club.

Herrin, Judith (1987). *The formation of Christendom*. Princeton, NJ: Princeton University Press.

Hesse, Everett W. (1960). Introduction. In Calderón de la Barca's *The surgeon of his honour* (pp. vii-xxx). Madison: University of Wisconsin Press.

Hibbard, Howard (1974). *Michelangelo*. New York: Harper & Row.

Hill, Christopher (1994). *The English Bible and the Seventeenth-Century Revolution*. London: Penguin.

Hilton, Tim (2002, July 5). Was any other artist as good? *Times Literary Supplement*, 18-19.

Hitti, Philip K. (1973). *Capital cities of Arab Islam*. Minneapolis: University of Minnesota

Ho, David, Y. F. (1986). Chinese patterns of socialization: A critical review. In M. H. Bond (Ed.), *The Psychology of the Chinese People*. Oxford: Oxford University Press.

Hodges, Donald C. (1991). *Argentina's "Dirty War"*. Austin: University of Texas Press.

Holland Smith, John (1972). *Francis of Assisi*. New York: Scribner.

Holton, Gerald (2002, January 11). The miracle of the two tables: Enrico Fermi, piece of paraffin and the way toward nuclear fission. *Time Literary Supplement*, 12-13.

Humboldt, Alexander von (1811/1965). Creoles or Americans? In R.A. Humphreys & Lynch, J. (Eds.), *The origins of the Latin American revolutions 1808-1826* (pp. 269-74). New York: Knopf.

Hume Brown, P. (1920/1971). *Life of Goethe*. New York: Haskell House.

Humphreys, R. A., & Lynch, John (1965). *The origins of the Latin American revolutions 1808-1826*. New York: Knopf.

Hurtado, Larry W. (2000). *At the origins of Christian worship: The context and character of earliest Christian devotion*. Grand Rapids, MI: William B. Eerdmans.

Hyde, Janet S. (1985). *Half the human experience*. Lexington, MA: D.C. Heath.

Ignatieff, Michael (1998). *Isaiah Berlin: A life*. New York: Henry Holt.

Isaacson, José (1983). *La Argentina como pensamiento*. Buenos Aires: Ediciones Marymar.

Ishiguro, Kazuo (1989). *The remains of the day*. London: Faber and Faber.

Jackson, W. T. H. (1968). Faith unfaithful: The German reaction to courtly love. In, F. X. Newman (Ed.), *The meaning of courtly love* (pp. 55-76). Albany: State University of New York Press.

Jackson, W. T. H. (1971). *The anatomy of love: The Tristan of Gottfried von Strassburg*. New York: Columbia University Press

Jackson, W. T. H. (1983). Andreas Capellanus. In *Dictionary of the Middle Ages* (*3*, pp. 88-89). New York: Scribner.

Jaeger, Stephen C. (1977). *Medieval Humanism in Gottfried von Strassburg's Tristan und Isolde*. Heidelberg: Carl Winter — Universitätsverlag.

Jaeger, Stephen C. (1985). *The origins of courtliness: Civilizing trends and the formation of courtly ideals 939-1210.* Philadelphia: University of Pennsylvania Press.

James, Clive (2002, September 27). Master of paradise: The libertine Pushkin, mythologized by liberals, fans and lovers. *Times Literary Supplement*, 3-4.

Janson, H. W. (1986). *History of art.* New York: Harry N. Abrams.

Jardim, Anne (1970). *The first Henry Ford: A study in personality and business leadership.* Cambridge, MA: MIT Press.

Jeremy, Mary (1946). Caxton's "golden legend" and Varagine's "legenda aurea". *Speculum, 21*, pp. 212-21.

Ji, Li-Jun; Nisbett, Richard E.; Su, Yanjie (2001). Culture, change and prediction. *Psychological Science, 12,* 450-56.

Johnson, E. (1969). Charles Dickens: The dark pilgrimage. In E. W. F. Tomlin (Ed.), *Charles Dickens - 1812-1870: A centenary volume* (pp. 41-63). New York: Simon & Schuster.

Johnson, Paul (1976). *A History of Christianity.* New York: Atheneum.

Jones, Arnold H. M. (1973). Augustus. In *Encyclopeadia Britannica (2*, pp. 758-61). Chicago: Encyclopeadia Britannica.

Jones, Ernest (1953). *The life and work of Sigmund Freud.* New York: Basic Books.

Jones, George F. (1968). *Walther von der Vogelweide.* New York: Twayne.

Jones, George F. (1989). Walther von der Vogelweide. In *Dictionary of the Middle Ages* (*12*, pp. 536-41). New York: Scribner.

Jourcin, Albert (1978). The Medici family. In *Encyclopaedia Britannica (11*, pp. 815-22). Chicago: Encyclopaedia Britannica.

Juan Arbó, Sebastián (1963). *Pío Baroja y su tiempo.* Barcelona: Planeta.

Judge, E. A. (1986). The quest for mercy in late antiquity. In P.T. O'Brien & D.G. Peterson (Eds.), *God who is rich in mercy: Essays presented to D.B. Knox* (pp. 107-21). Sidney: Macquarie University Press.

Kagawa, Hiroshi (1997). *The inscrutable Japanese.* Tokyo: Kodansha International.

Kantorowicz, E. H. (1957). *The king's two bodies: A study in medieval political theology.* Princeton, NJ: Princeton University Press.

Katz, Bernard (1994). Planning and following the unexpected in scientific research. *Creativity Research Journal, 7*, 225-38.

Kazhadan, Alexander (1991). Eunuchs. In *Oxford Dictionary of Byzantium* (pp. 746-47). Oxford: Oxford University Press.

Keller, Hiltgart L. ((1968). *Reclams Lexikon der Heiligen und der biblischen Gestalten.* Stuttgart: Philipp Reclam, Jun.

Kertzer, David I. (1997). *The kidnapping of Edgardo Mortara.* New York: Knopf.

Kessler, H. L. (1977). *The illustrated bibles from Tours.* Princeton, NJ: Princeton University Press.

Kipling, Rudyard (1989). *Complete verse.* New York: Doubleday.

Kitayama Shinobu; Duffy, Sean; Kawamura, Tadashi; Larsen, Jeff T. (2003). Perceiving an object and its context in different cultures. *Psychological Science, 14*, 201-06.

Kitzinger, Ernst (1956/1976). *The art of Byzantium and the Medieval West: Selected studies by Ernst Kitzinger*, edited by W. Eugene Kleinbauer. Bloomington: Indiana University Press.

Kleinman, Arthur, & Kleinman, Joan (1991). Suffering and its professional transformation: Toward an ethnography of interpersonal experience. *Culture, medicine, and psychiatry, 15,* 275-301.

Knowles, David (1930). *The Benedictines*. New York: Macmillan.

Koestler, Arthur (1959/1968). *The sleepwalkers*. New York: Macmillan.

Koestler, Arthur (1960). *The lotus and the robot*. London: Hutchinson of London.

Koestler, Arthur (1964). *The act of creation*. New York: Macmillan.

Koestler, Arthur (1967). *The ghost in the machine*. London: Hutchinson of London.

Kohn, Melvin L. (1969). *Class and conformity: A study of values*. Homewood, IL: Dorset.

Krutch, Joseph W. (1934). Introduction. In M. Proust's *Remembrance of things past (1*, pp. vii-xv). New York: Random House.

Kuhn, Thomas S. (1970). *The structure of scientific revolutions*. Chicago: University of Chicago Press.

Kurosawa, Akira (1982). *Something like an autobiography*. New York: Alfred A. Knopf.

La Bruyère, Jean de (ca. 1690/1951). De la cour. In *Oeuvres complètes* (pp. 215-47). Paris: Gallimard.

Lacey, Robert (1986). *Ford: The men and the machine*. Boston: Little Brown.

La Fayette, Marie-Madeleine (Madame de) (1678/1951). *The Princess of Cleves*. New York: New Directions.

Landes, David S. (1998). *The wealth and poverty of nations: Why some are rich and some are poor*. New York: W.W. Norton.

Lane, Frederic C. (1973). *Venice: A maritime republic*. Baltimore: John Hopkins University Press.

Lasko, P. (1972). *Ars Sacra 800-1200*. London: Penguin.

Lastarria, José Victorino (1836-49/2000). *Literary memoirs*. New York: Oxford University Press.

Latourette, Kenneth S. (1953). *A history of Christianity*. New York: Harper & Brothers.

Lea, Henry Charles (1908). *The Inquisition in the Spanish dependencies*. New York: Macmillan.

Lee, Brian (1976). *Theory and personality: The significance of T. S. Eliot's criticism*. London: Athlone.

Le Goff, Jacques (1988). *Medieval civilization 400-1500*. Oxford: Basil Blackwell.

Leiris, Michel (1954). *Picasso and the human comedy*. New York: Random House.

Leonard, John (1982, Sept. 17). Books of the Times: [a book review of] Canetti's "The torch in my ear". *New York Times*, C21.

Lepie, Herta, & Minkenberg, Georg (1995). *Die Schatzkammer des Aachener Domes*. Aachen: Domkapitel Aachen.

Leslie, G. R., Larson, R. F., & Gorman, B. L. (1973). *Order and change: Introductory sociology*. New York: Oxford University Press.

Lewin, Boleslao (1967). *La Inquisición en Hispano America*. Buenos Aires: Paidos.

Lewis, Bernard (1999, April 23). From Babel to dragoman: The tortuous history of the interpreter in the Middle East. *Times Literary Supplement*, 12-14.

Lewis, C. S. (1937). *The allegory of love: A study on Medieval tradition*. Oxford: Oxford University Press.

Lewis, David L. (1976). *The public image of Henry Ford*. Detroit: Waynes State University.

Lewis, Katherine J. (2000). *The cult of St.Katherine of Alexandria in late Medieval England*. Woodbridge, Suffolk, UK: Boydell Press.

Lieberthal, Kenneth G. (1993). The People's Republic of China. In *Encyclopaedia Britannica (16*, pp. 36-230). Chicago: Encyclopaedia Britannica.

Lichtblau, Myron I. (1972). *Manuel Gálvez*. New York: Twayne.

Link, Perry (2001, October 1). China: The anaconda in the chandelier. *New York Review of Books*, 67-70.

Lipp, Frances R. (1989). William IX of Aquitaine. In *Dictionary of the Middle Ages* (*12*, pp. 630-33).

Little, Charles T. (1977). *The Magdeburg ivory group: A tenth century New Testament narrative cycle*; New York University: Ph.D. dissertation. University Microfilms International, 78-3115.

Loewe, Michael (1974). *Crisis and conflict in Han China 104 BC to AD 9*. London: George Allen & Unwin.

Longrigg, James (1993). *Greek rational medicine*. London: Routledge.

López Ibor, Juan José (1969). *El Español y su complejo de inferioridad* [ante la ciencia]. Madrid: Rialp.

Lowe, E. John (1999). Relativism. In S. Bullock & S. Trombley (Eds.), *The Norton Dictionary of Modern Thought* (p. 743). New York: W.W. Norton.

Lowenthal, Marvin (1935/1999). "Montaigne and his world", introduction to *The autobiography of Michel de Montaigne* (pp. xiii-xliv). Jaffrey, NH: David R. Godine.

Lowes, John Livingston (1927/1964). *The road to Xanadu*. Boston: Houghton Mifflin.

Lubich, Chiara (1999). *Acceptance speech on the conferment of the degree of Doctor (Honoris Causa) on Chiara Lubich by the University of Malta.* Malta Transcript.

Lukes, Steven (1998, March 27). Berlin's dilemma: The distinction between relativism and pluralism. *Times Literary Supplement,* 8-10.

Lumiansky, R. M. (1978). Geoffrey Chaucer. In *Encylopaedia Britannica* (*4*, pp. 62-65). Chicago: Encyclopaedia Britannica.

Lurie, Alison (2003, July 3). God's houses. *New York Review of Books*, 30-32.

Lynch, John (1993). From independence to national organization. In L. Bethell (Ed.), *Argentina since independence* (pp. 1-46). Cambridge: Cambridge University Press.

Lynch, Lawrence (1992). *Jules Verne*. New York: Twayne.

Maraini, Fosco (1960). *Meeting with Japan*. New York: Viking.

MacFarquhar, Roderick (1987). *The origins of the Cultural Revolution 3: The coming of the cataclysm 1961-1966*. Oxford: Oxford University Press.

Mack Smith, Denis (1988). *A history of Sicily: Medieval Sicily 800-1713*. New York: Dorset.

MacKay, Angus (1977). *Spain in the Middle Ages*. New York: Macmillan.

Madariaga, Salvador de (1928/1969). *Englishmen Frenchmen Spaniards*. New York: Hill and Wang.

Madariaga, Salvador de (1949). *The rise of the Spanish American Empire*. New York: Macmillan.

Madariaga, Salvador de (1962). *Latin America between the eagle and the bear*. New York: Praeger.

Maeztu, Ramiro de (1934/1952). *Defensa de la Hispanidad — Edición autorizada por la Señora Viuda del Autor para América y Filipinas*. Buenos Aires: Poblet.

Magarshack, David (1967). *Pushkin*. New York: Grove Press.

Magee, Bryan (1997). *Confessions of a philosopher: A journey through Western philosophy*. New York: Random House.

Maguire, Henry (1997). The Heavenly Court. In H. Maguire (Ed.), *Byzantine court culture from 829 to 1204* (pp. 247-58). Washington, D.C.: Dumbarton Oaks Research Library.

Male, Roy R. (1964). *Hawthorne's tragic vision*. New York: W.W. Norton.

Malevsky-Maletvitch, Peter (1937). A biographical sketch. In *Pushkin: The man and the artist* (pp. 30-57). New York: Paisley Press.

Mann, Thomas (1915-17/1983). *Reflections of a nonpolitcal man*. New York: Frederick Ungar.

Mann, Thomas (1933/1968). *Past masters*. Freeport, NY: Books for Libraries Press.

Mantel, Hilary (2002, Oct. 24). Naipaul's Book of the World [Review of Naipaul's "The writer and the world"]. *New York Review of Books*, 10-12.

Maraini, Dacia (1986). *Moravia: Il bambino Alberto*. Milan: Bompiani.

Maraini, Fosco (1960). *Meeting with Japan*. New York: Viking.

Maravall, José Antonio (1972/1990). *Teatro y literatura en la sociedad barroca*. Barcelona: Editorial Critica.

Marías, Julián (1941/1967). *History of philosophy*. New York: Dover.

Marías, Julián (1985). *España inteligible: Razón historica de las Españas*. Madrid: Alianza Editorial.

Marías, Julián (1987). *Ser español: Ideas y creencias en el mundo hispánico*. Barcelona: Planeta.

Marinis, Tammaro de (1934). Lucca: Storia. In *Enciclopedia Italiana* (*21*, pp. 560-63). Rome: Enciclopedia Italiana.

Markham, Clements R. (1968). *A history of Peru*. New York: Greenwood Press.

Marquis, Samuel S. (1923). *Henry Ford: An interpretation*. Boston: Little, Brown.

Martin, Terence (1983). *Nathaniel Hawthorne*. Boston: Twayne.

Massie, Robert K. (1981). *Peter the Great: His life and world*. New York: Ballantine Books.

Maugham, Robin (1949). *The servant*. New York: Harcourt, Brace.

Maxwell, Scott E. & Delaney, Harold D. (1990). *Designing experiments and analyzing data: A model comparison perspective*. Pacific Grove, CA: Brooks/Cole.

Mayr-Harting, Henry (1991). *Ottonian book illumination: An historical study*. London: Harvey Miller.

Mazlish, Bruce (Ed.) (1972). *Makers of modern thoughts*. New York: American Heritage Publishing.

McClelland, David C. (1967). *The achieving society*. New York: Free Press.

McCully, Marilyn (2002). Picasso. In *Encyclopaedia Britannica* (*25*, pp. 860-64). Chicago: Encyclopaedia Britannica.

McLellan, David (1973). *Karl Marx: His life and thought*. New York: Harper & Row.

McNeill, William H. (1963). *The rise of the West*. Chicago: University of Chicago Press.

McNeill, William H. (1974). *Venice: the hinge of Europe, 1081-1797*. Chicago: University of Chicago Press.

Medina, J. T. (ca. 1890/1945). *El tribunal del Santo Oficio de la Inquisicion en las Provincias del Plata*. Buenos Aires: Editorial Huarpes.

Menéndez y Pelayo, Marcelino (1944). *Estudios de critica histórica y literaria*. Buenos Aires: Espasa-Calpe Argentina.

Menéndez Pidal, Ramón (1946/1966). *The Spaniards in their history*. New York: W.W. Norton.

Merson, John (1990). *The genius that was China*. Woodstock, NY: Overlook Press.

Meyerhof, Max (1931). [Islamic] Science and Medicine. In T. Arnold & A. Guillaume (Eds.), *The legacy of Islam* (pp. 311-55). Oxford: Oxford University Press.

Miller, Arthur (1949/1976). *Death of a salesman*. New York: Penguin.

Miller, Edwin Haviland (1991). *Salem is my dwelling place*. Iowa City: University of Iowa Press.
Miller, W. L. (2002). *Lincoln's virtues: An ethical biography*. New York: Alfred A. Knopf.
Mirsky, Jonathan (1988, February 5). The mark of Cain. *New York Review of Books*, 31-33.
Mishima, Yukio (1977). *Yukio Mishima on Hagakure: The Samurai ethic and modern Japan*. London: Souvenir Press.
Mokyr, Joel (1990). *The lever of riches: Technological creativity and economic progress*. New York: Oxford University Press.
Moller, Herbert (1958). The social causation of the courtly love complex. *Comparative studies in society and history, 1*, 137-63.
Moloney, James C. (1954/1962). Child training and Japanese conformity. In B. S. Silberman (Ed.), *Japanese character and culture*. Tucson: University of Arizona Press.
Montanari, Luigi (1973). Bologna. In *Encyclopeadia Britannica (3*, pp. 889-890). Chicago: Encyclopeadia Britannica.
Montanelli, Indro, & Gervaso, Roberto (1967). *Italy in the golden centuries*. Chicago: Henry Regnery.
Monte, Christopher F. (1977). *Beneath the mask: An introduction to theories of personality*. New York: Praeger.
Moreau, Pierre (1966). *Montaigne: Connaissance des lettres*. Paris: Hatier.
Morris, Colin (1972). *The discovery of the individual 1050-1200*. New York: Harper & Row.
Motto, Anna Lydia (1973). *Seneca*. New York: Twayne.
Munz, Peter (1969). *Frederick Barbarossa*. Ithaca, NY: Cornell University Press.
Murray, E. (1967). *Arthur Miller, dramatist*. New York: Frederick Ungar.
Naipaul, Shiva (1982, June 10). 'I cannot disown Trinidad and it cannot disown me'. *The Listener,* 13-14.
Naipaul, V. S. (1999, Feb. 18). Reading & writing. *New York Review of Books*, 13-18.
Navarro Gerassi, Marysa (1968). *Los Nacionalistas*. Buenos Aires: Editoria Jorge Alvarez.
Nevins, Allan (1954). *Ford: The times, the man, the company*. New York: Scribner's.
Newton-Smith, W. H. (1981). *The rationality of science*. London: Routledge & Kegan Paul.
Nitobe, Inazo (1905/2001). *Bushido: The soul of Japan*. New York: ICG Muse.
Nykl, A. R. (1946). *Hispano-Arabic poetry: And its relations with the old Provençal troubadours*. Baltimore: J. H. Furst.
Oakes, Catherine (1996). Crucifix — Introduction. In *The dictionary of art* (*8*, pp. 210-11). London: Macmillan.
O'Callaghan, Joseph F. (1975). *A history of medieval Spain*. Ithaca, NY: Cornell University Press.
O'Donnell, T. J. (1967). Castration. In *New Catholic Encyclopaedia* (*3*, p. 194). New York: McGraw Hill.
Ortega y Gasset, José (1921/1937). *Invertebrate Spain*. New York: W.W. Norton.
Ortega y Gasset, José (1967). Prologo. In Ibn Hazm's *El collar de la paloma: Tratado sobre el amor y los amantes* (The Ring of the Dove), (pp. 13-92). Madrid: Sociedad de Estudios y Publicaciones.

Osborne, Harold (Ed.). (1970). *The Oxford companion to art*. Oxford: Oxford University Press.

Ott, Hugo (1988/1993). *Martin Heidegger: A political life*. New York: Basic Books.

Owen, D. D. R. (1975). *Noble lovers*. New York: New York University Press.

Oxford Dictionary of Saints (1997). D. Farmer Ed. Oxford: Oxford University Press.

Painter, George D. (1977). *William Caxton: A biography*. New York: Putnam's Sons.

Park, Robert E. (1928). Human migration and the marginal man. *American Journal of Sociology*, *33*, 881-93.

Parker, Harold T. (1983). *Three Napoleonic battles*. Durham, NC: Duke University Press.

Parmisano, S. F. (1967). Pride. In *New Catholic Encyclopedia* (*11*, pp. 765-66). Washington, DC: The Catholic University of America.

Patai, Raphael (1983). *The Arab mind*. New York: Scribner.

Pax Facta (1183/1976). *Pax facta per Dominum Fredericum Imperatorem in civitate Constancie cum Lombardis*. In G. D'Ilario, E. Gianazza, & A. Marinoni (Eds.), *Legnano e la battaglia* (pp. 22-35). Legnano: Banca di Legnano — Edizioni d'arte.

Paz, Octavio (1950/1985). *The labyrinth of solitude*. New York: Grove Press.

Pellat, Ch. (1990). Khasi (castrated men). In *Encyclopaedia of Islam* (pp. 1087-92). Leiden: E. J. Brill.

Pestelli, Leo (1961). *Dizionario delle parole antiche*. Milan: Longanesi.

Peters, Edward (1989). *Inquisition*. Berkeley: University of California Press.

Peterson, Thomas E. (1996). *Alberto Moravia*. New York: Twayne.

Petit-Dutaillis, Ch. (1964). *The feudal monarchy in France and England, from the tenth to the thirteenth century*. New York: Barnes and Noble.

Phillips, Kevin (1999). *The Cousins' War: Religion, politics, & the triumph of AngloAmerica*. New York: Basic Books.

Piaget, Jean (1952). Autobiography. In E.G. Baring et. al., *History of psychology in autobiography* (*4*, pp. 237-56). Worcester, MA: Clark University Press.

Pirenne, Henri (1925/1969). *Medieval cities: Their origins and the revival of trade*. Princeton, NJ: Princeton University Press.

Popper, Karl M. (1945/1971). *The open society and its enemies*. Princeton, NJ: Princeton University Press.

Popper, Karl M. (1957/1961). *The poverty of historicism*. London: Routledge & Kegan Paul.

Popper, Karl R. (1963/1972). *Conjectures and refutations: The growth of scientific knowledge*. London: Routledge & Kegan Paul.

Popper, Karl R. (1975). *Objective knowledge: An evolutionary approach*. Oxford: Oxford University Press.

Porras, W. M. (1967). Patronato Real. In *New Catholic Encyclopaedia* (*10*, pp. 1113-1116). New York: McGraw-Hill.

Porter, Roy (1997). *The greatest benefit to mankind*. New York: W. W. Norton.

Porter, Roy (2001). *Enlightenment: Britain and the creation of the modern world*. London: Penguin.

Potter, Paul (1976). Herophilus of Chalcedon: An assessment of his place in the history of anatomy. *Bulletin of the History of Medicine, 50*, 45-60.

Powell, Brian (1983, August 4). We may prefer to sigh with Ishiguro or groan with Endo. *The Listener*, 20-21.

Preston, Paul (2001, June 29). An awareness of guilt. *Times Literary Supplement*, 3-4.

Procacci, Giuliano (1971). *History of the Italian people*. New York: Harper & Row.
Prodi, Paolo (1973). The structure and organization of the church in Renaissance Venice. In J. R. Hale (Ed.), *Renaissance Venice*. Totowa, NJ: Rowman and Littlefield.
Pryce-Jones, David (1989). *The closed circle: An interpretation of the Arabs*. New York: Harper & Row.
Putnam, Robert D. (2000). *Bowling alone*. New York: Simon & Schuster.
Pye-Smith, Philip H., & Singer, Charles (1973). William Harvey. In *Encyclopaedia Britannica, (11*, pp. 142-44). Chicago: Encyclopaedia Britannica.
Quint, David (1996). Introduction. In Ludovico Ariosto's *Cinque Canti - five cantos* (pp. 1-44). Berkeley: University of California Press.
Raffel, Burton (1982/1991). *T. S. Eliot*. New York: Continuum.
Rebay, Luciano (1970). *Alberto Moravia*. New York : Columbia University Press.
Riasanovsky, Nicholas V. (1952). *Russia and the West in the teachings of the Slavophiles*. Cambridge, MA: Harvard University Press.
Rice, David Talbot (1963). *Art of the Byzantine era*. New York: Praeger.
Richardson, John (1984, July 19). The catch in the late Picasso. *New York Review of Books*, 21-28.
Richardson, John (1991). *A life of Picasso, I, 1881-1906*. New York: Random House.
Riché, Pierre (1993). *The Carolingians: A family who forged Europe*. Philadelphia: University of Pennsylvania Press.
Richey, M. F. (1943/1969). *Essays on medieval German poetry*. New York: Barnes & Noble.
Richie, Donald (1996). *The films of Akira Kurosawa*. Berkeley: University of California Press.
Richie, Donald (1996). Second preface. In Lafcadio Hearn's *Japan: An anthology of his writings on the country and its people*. Tokyo: Charles E. Tuttle.
Ricketts, Harry (1999). *Rudyard Kipling: A life*. New York: Carrol & Graf.
Ridolfi, Roberto (1978). Girolamo Savonarola. In *Encyclopaedia Britannica* (*16*, pp. 289-91). Chicago: Encyclopaedia Britannica.
Robert, Monty (1997). *The man who listens to horses*. New York: Random House.
Robertson, J. G. (1970). *A history of German literature*. Elmsford, NY: London House & Maxwell.
Rock, David (1985). *Argentina 1516-1982*. Berkeley: University of California Press.
Rodriguez, Junius P. (1997). *Chronology of world slavery*. Santa Barbara, CA: ABC-Clio.
Roe, Anne (1952). *The making of a scientist*. New York: Dodd, Med.
Roediger, Henry L. III; Rushton, J. Philippe; Capaldi, Elizabeth D.; & Paris, Scott G. (1984). *Psychology*. Boston: Little, Brown.
Rogers, Jack D. (1978). Henry Ford. In *Encyclopaedia Britannica* (*7*, p. 520-22). Chicago: Encyclopaedia Britannica.
Rosenstock-Huessy, Eugen (1938/1950). *The driving power of Western Civilization: The Christian revolution of the Middle Ages*. Boston: Beacon Press.
Rosenthal, Erwin, I. J. (1978). Averroës. *Encyclopeadia Britannica (2*, pp. 538-540). Chicago: Encyclopeadia Britannica.
Ross, Joan, & Freed, Donald (1972). *The existentialism of Alberto Moravia*. Carbondale: Southern Illinois University Press.
Rougemont, Denis de (1973). Love. In *Dictionary of the history of ideas* (*3*, pp. 94-108). New York: Scribner.
Rowdon, Maurice (1970). *The silver age of Venice*. New York: Praeger.

Runciman, Steven (1975). *Byzantine style and civilization.* London: Penguin.

Rupp, Ernest G. (1978). Martin Luther. In *Encyclopaedia Britannica* (*11*, pp. 188-96). Chicago: Encyclopaedia Britannica.

Russell, Bertrand (1945). *A history of western philosophy.* New York: Simon and Schuster.

Russell, James, C. (1994). *The germanization of Early Medieval Christianity: A sociohistorical approach to religious transformation* [376-754]. New York: Oxford University Press.

Ruthven, Malise (1999, Oct. 8). A dream of Jerusalem [Review of Edward Said's "Out of Place: A memoir"]. *Times Literary Supplement*, 29.

Ryan, Alan (1998, December 17). Wise man [Review of M. Ignatieff's "Isaiah Berlin: A life" and I. Berlin's "The proper study of mankind: An anthology of essays"]. *New York Review of Books*, 29-37.

Ryan, Granger, & Ripperger, Helmut (1941/1969). Foreword. In J. De Voragine's *The golden legend* (pp. v-xvi). New York: Longman, Green.

Sacchetti, Franco (1378/1908). *Tales from Sacchetti.* London: J. M. Dent.

Safranski, Rüdiger (1998). *Martin Heidegger: Between good and evil.* Cambridge, MA: Harvard University Press.

Said, Edward W. (1978). *Orientalism.* New York: Pantheon Books.

Said, Edward W. (1999). *Out of place: A memoir.* New York: Alfred Knopf.

Salter, Elizabeth (1970). Courtly love. In *Encyclopaedia Britannica* (*6*, pp. 664-66). Chicago: Encyclopaedia Britannica.

Sánchez-Albornoz, Claudio (1959/1991). *España, un enigma histórico.* Barcelona: Edhasa.

Santillana, Giorgio de (1955). *The crime of Galileo.* Chicago: University of Chicago Press.

Sarmiento, Domingo F. (1845/1961). *Life in the Argentine Republic in the Days of the Tyrants* (or, *Civilization and Barbarism*). New York: Collier.

Sarmiento, Domingo, F. (1883/1953). *Conflicto y Armonias de las razas en America.* Buenos Aires: Editorial Luz del Dia.

Sarton, George (1947). *Introduction to the history of science.* Washington, D.C.: Carnegie Institution of Washington.

Saunders, J. J. (1965). *A history of Medieval Islam.* London: Routledge.

Schaeffer, Pamela, & Allen, John l. Jr. (1999, December 24). Jesus 2000 winners chosen. *National Catholic Reporter*, 3-4.

Schmemann, Alexander (1973). Aleksey Stepanovich Khomyakov. In *Encyclopaedia Britannica* (*13*, pp. 332-33). Chicago: Encyclopaedia Britannica.

Schneider, Ben R., Jr. (1995). *King Lear in its own time: The difference that death makes.* Internet: http://www.humanities.ualberta.ca/elms/01-1/schnlear:html.

Scholz, Piotr O. (2001). *Eunuchs and castrati: A cultural history.* Princeton, NJ: Markus Wiener.

Schuchert, Richard A. (1978). Translator preface. In S. Endo's *A Life of Jesus* (pp. 3-5). New York: Paulist Press.

Scott Wilson, William (1983). Introduction. In T. Yamamoto (ca. 1700), *Hagakure: The book of the samurai* (pp. 9-16). Tokyo: Kodansha International.

Seligman, Martin E. P., & Csikszentmihalyi, Mihaly (2000). Positive Psychology: An introduction [to 15 articles]. *American Psychologist*, *55*, 5-14.

Serjeant, Robert B., & Ghul, Mahmud Ali (1978). History of Arabia. In *Encyclopaedia Britannica* (*1*, 1043-51). Chicago: Encyclopaedia Britannica.

Seton-Watson, Hugh (1967). *The Russian Empire 1801-1917*. Oxford: Oxford University Press.

Shaw, Bernard (1903/1930). *Man and superman*. New York: Wm. H. Wise.

Sherrard, Philip (1966). *Byzantium*. New York: Time-Life Books.

Shiba, Ryotaro (1967/1998). *The last shogun: The life of Tokugawa Yoshinobu*. New York: Kodansha.

Shioya, Sakae (1956). *Chushingura: An exposition*. Tokyo: Hokuseido.

Shipley, Graham (2000). *The Greek world after Alexander*. London: Routledge.

Short, Philip (2000). *Mao: A life*. New York: Henry Holt.

Shumway, Nicolas (1991). *The invention of Argentina*. Berkeley: University of California Press.

Siciliano, Enzo (1982). *Alberto Moravia: Vita, parole e idee di un romanziere*. Milan: Bompiani.

Sighinolfi, Lino (1930). Introduzione. In Mondino de' Liucci's *Anatomia* (pp. 1-22). Bologna: L. Cappelli.

Simmons, Ernst J. (1937). A biographical study of Pushkin. In S. H. Cross & E. J. Simmons (Eds.), *Centennial essays for Pushkin* (pp. 3-44). Cambridge, MA: Harvard University Press.

Simonton, Dean Keith (1984). *Genius, creativity, and leadership*. Cambridge, MA: Harvard University Press.

Simonton, Dean Keith (1994). *Greatness: Who makes history and why*. New York: Guilford.

Simpson, Joyce G. (1962). *Le Tasse et la littérature et l'art baroques en France*. Paris: Librairie A. G. Nizet.

Singer, Charles (1925/1957). *A short history of anatomy from the Greeks to Harvey*. New York: Dover.

Singleton, Charles (1958). *An Essay on the Vita Nuova*. Cambridge, MA: Harvard University Press.

Skidmore, Thomas E., & Smith, Peter H. (1997). *Modern Latin America*. New York: Oxford University Press.

Sneyd, Charlotte A. (translator) (1497/1847). Notes to "A relation, or rather a true account of the island of England about the year 1500". *Camden Society*, *37*. London: Camden Society

Soderlund, Jean R. (1985). *Quakers & slavery*. Princeton, NJ: Princeton University Press.

Soderlund, Jean R. (1998). Quakers. In *Macmillan Encyclopedia of World Slavery* (2, pp. 751-55). New York: Macmillan.

Stark, Rodney (1997). *The rise of Christianity*. San Francisco: HarperSanFrancisco.

Stein, Stanley J., & Stein, Barbara (1970). *The colonial heritage of Latin America*. New York: Oxford University Press.

Stewart, Bob (1988). *Charlemagne: Founder of the Holy Roman Empire*. Poole, UK: Firebird.

Stewart, Randall (1948). *Nathaniel Hawthorne: A biography*. New Haven, CT: Yale University Press.

Stewart, Randall (1973). Nathaniel Hawthorne. In *Encyclopaedia Britannica* (*11*, pp. 188-89). Chicago: Encyclopaedia Britannica.

Stokes, William S. (1965). Violence as a power factor in Latin American politics. In J. D. Martz (Ed.), *The dynamics of change in Latin American politics* (pp. 140-49). Englewood Cliffs, N.J.: Prentice-Hall.

Storr, Anthony (1999, July 23). Maladjust your mindset. *Times Literary Supplement*, 31.

Straus, Bernard (1980). *Maladies of Marcel Proust: Doctors and disease in his life and work*. New York: Holmes & Meier.

Stuckenberg, J. H. W. (1882). *The life of Immanuel Kant*. London: Macmillan.

Tales from the Arabian Nights (R. F. Burton transl. 1885-88/1978). New York: Avenel.

Tarn, W. W. (1927/1961). *Hellenistic civilization*. New York: World Publishing.

Tasso, Torquato (1586/1973). *Il Malpiglio: A dialogue on the court*. Hanover, NH: Darthmouth College.

Taylor, Ronald J. (1973). Minnesinger. In *Encyclopaedia Britannica* (*15*, pp. 533-34). Chicago: Encyclopaedia Britannica.

Tedlow, Richard S. (2001). *Giants of enterprise: Seven business innovators and the empires they built*. New York: HarperBusiness.

Tellenbach, Gerd (1936/1991). *Church, state and Christian society at the time of the Investiture Controversy.* Toronto: University of Toronto Press.

Temperley, Howard (1981). The ideology of antislavery. In D. Eltis & J. Walvin (Eds.), *The abolition of the Atlantic slave trade: Origins and effects in Europe, Africa, and the Americas* (pp. 21-35). Madison: University of Wisconsin Press.

Texte, Joseph (1899). *Jean-Jacques Rousseau*. London: Duckworth.

Therivel, William A. (1993). The challenged personality as a precondition for sustained creativity. *Creativity Research Journal*, *6*, 413-24.

Therivel, William A. (1998). Creative genius and the GAM theory of personality: Why Mozart and not Salieri? *Journal of Social Behavior and Personality*, *13*, 201-34.

Therivel, William A. (1999a). Why are eccentrics not eminently creative? *Creativity Research Journal*, *12*, 47-55.

Therivel, William A. (1999b). Why Mozart and not Salieri. *Creativity Research Journal*, *12*, 67-76.

Therivel, William A. (2001, 2002, 2003). *The GAM/DP Theory of Personality and Creativity; vols. 1, 2, 3*. Minneapolis, MN: Kirk House.

Tierney, B. (1964). *The crisis of church & state 1050-1300, with selected documents*. Englewood Cliffs, NJ: Prentice-Hall.

Timerman, Jacobo (1981). *Prisoner without a name — cell without a number*. New York: Knopf.

Tirso de Molina (c. 1620/1968). *El burlador de Sevilla: Don Juan*. Paris: Aubier-Flammarion.

Tougher, Shaun F. (1997). Byzantine eunuchs. In James, Liz (Ed.), *Women, men and eunuchs* (pp. 168-84). London: Routledge.

Tovar, Antonio (1953, March 25). Quoted in "Sumar y no restar". *Alcalá*, 3.

Toynbee, Arnold J. (1972). *A study of history (a new edition revised and abridged by the author and Jane Caplan)*. New York: Weathervane.

Trafton, Dain A. (1973). Introduction in T. Tasso's *Il Malpiglio: A dialogue on the court* (pp. 1-13). Hanover, NH: Darthmouth College.

Trevor-Roper, H. R. (1967). *Religion, the Reformation and social changes*. London: Macmillan.

Troyat, Henri (1974). *Pushkin*. Garden City, NY: Doubleday.

Tsai, Shih-shan Henry (1996). *The eunuchs in the Ming dynasty*. Albany: State University of New York Press.

Turnbull, S. R. (1977). *The samurai: A military history*. New York: Macmillan.
Tursi, Renee (August 9, 2002). The sailor's son who laboured under a veil. *Times Literary Supplement*, 8.
Uitti, Karl (1988). Provençal Literature to 1200. In *Dictionary of the Middle Ages* (*10*, pp. 159-73). New York: Scribner.
Unamuno, Miguel de (1906/1974). La Venda. In *Ver con los ojos y otros relatos novelescos* (pp. 47-51). Madrid: Espasa-Calpe.
Unamuno, Miguel de (1906/1951). Sobre la europeización. In *Ensayos* (*1*, pp. 901-20). Madrid: Aguilar.
Unamuno, Miguel de (1913/1954). *Tragic sense of life*. New York: Dover.
Urbina, I. Ortiz de (1963). *Nicée et Constantinople*. Paris: L'Orante.
Valdés, Mario J. (1966). *Death in the literature of Unamuno*. London: University of Illinois Press.
Valignano, Alessandro (1580/1962). *Sumario* (selected pages). In B. S. Silberman (Ed.), *Japanese character and culture: A book of selected readings* (pp. 286-88). Tucson: University of Arizona Press.
Varazze, Jacopo da (ca. 1266/1998). *Legenda Aurea*. Florence: Simsel, Edizioni del Galluzzo.
Veblen, Thorstein (1919). The intellectual pre-eminence of Jews in Modern Europe. *Political Science Quarterly*, *34*, 33-42.
Véliz, Claudio (1980). *The centralist tradition of Latin America*. Princeton, NJ: Princeton University Press.
Véliz, Claudio (1994). *The new world of the gothic fox: Culture and economy in English and Spanish America.* Berkeley: University of California Press.
Verne, Jules (1872/1999). *Around the world in eighty days*. Oxford: Oxford University Press.
Veyne, Paul (1987). The Roman Empire. In P. Veyne (Ed.), *A history of private life: From pagan Rome to Byzantium* (pp. 5-233). Cambridge, MA: Harvard University Press.
Vicinelli, Augusto (1943). *Scrittori nostri: Dalle origini alla fine del secolo XIV*. Milan: Arnoldo Mondadori.
Vickery, Walter N. (1970). *Alexander Pushkin*. New York: Twayne.
Voltaire (1764/1927). *Philosophical dictionary*. Vol. 3 in Lord Morley (Ed.), *The works of Voltaire - A contemporary version*. New York: Dingwall-Rock.
Voltaire (1733/1999). *Letters concerning the English Nation [Lettres anglaises*]. Oxford: Oxford University Press.
Voragine, Jacobus de (ca. 1266/1941/1969). *The golden legend* (trans. by Granger Ryan & Helmut Ripperger). New York: Longman, Green.
Voragine, Jacobus de (ca. 1266/1993). *The golden legend: Reading on the saints* (trans. by William Granger Ryan). Princeton, NJ: Princeton University Press.
Voragine, Jacobus de (ca. 1266/1998). *The golden legend* (select. & trans. C. Stace). London: Penguin.
Waerden, Bartel L., van, & Heath, Thomas L. (1978). Archimedes. In *Encyclopaedia Britannica* (*1*, pp. 1087-88). Chicago: Encyclopaedia Britannica.
Waldstein, Michael (1995). Mysticism. In R. P. McBrien (G.Ed.), *Encyclopedia of Catholicism* (pp. 901-02). San Francisco: HarperSanFrancisco.
Waley, Daniel (1969/1988). *The Italian city-republics*. London: Longman.
Wallace, Robert (1967). *The world of Leonardo, 1452-1519*. New York: Time-Life Books.

Wang Liu, Hui-Cheng (1959). An analysis of Chinese clan rules. In D. S. Nivison & A. F. Wright (Eds.), *Confucianism in action* (pp. 63-96). Stanford, CA: Stanford University Press.

Ware, Kallistos (1993). Eastern Christendom. In J. McManners (Ed.), *The Oxford History of Christianity* (pp. 131-66). Oxford: Oxford University Press.

Watson, Burton (1958). *Ssu-Ma Ch'ien grand historian of China.* New York: Columbia University Press.

Weeks, David, & James, Jamie (1995). *Eccentrics.* London: Weidenfell & Nicolson.

Weinstein, Donald, & Bell, Rudolph M. (1982). *Saints & society: The two worlds of Western Christendom, 1000-1700.* Chicago: University of Chicago Press.

Weisstein, Eric (2003). Mondino de Luzzi (1275-1326). *World of scientific biography.* http://scienceworld.wolfram.com/biography.MondinodedL.html.

Weiten, Wayne (1989). *Psychology: Themes and variations.* Pacific Grove, CA : Brooks/Cole.

Werblowsky, Zwi R. J. (1973). Prophet. In *Encyclopaedia Britannica* (*18*, pp. 633-38) Chicago: Encyclopaedia Britannica.

Werner, Emmy E. (1989). *Against odds: A longitudinal study of high risk children from birth to adulthood.* Final Report Project Grant #86-1110—86. William T. Grant Foundation.

Werner, Emmy E., Smith, Ruth S. (1982). *Vulnerable but invincible: A longitudinal study of resilient children and youth.* New York: McGraw-Hill.

Werner, Robert (1978). Ptolemy I Soter. In *Encyclopeadia Britannica (15*, pp. 180-82). Chicago: Encyclopeadia Britannica.

Wickersheimer, Ernest (1926). *Anatomies de Mondino dè Luzzi et de Guido de Vigevano.* Paris: Éditions E. Droz.

Wiggins, Peter DeSa (2000). *Donne, Catiglione and the poetry of courtliness.* Bloomington: Indiana University Press.

Wills, Garry (1998, February 19). The Vatican monarchy. *New York Review of Books*, 20-25.

Wills, Garry (2001). *Venice: Lion city — The religion of empire.* New York: Simon & Schuster.

Wilson, A. N. (1988). *Tolstoy.* New York: W.W. Norton.

Winstead, Karen A. (1999). Introduction. In J. Capgrave's (ca. 1445/1999) *The life of Saint Katherine* (pp. 1-9). Kalamazoo: Medieval Institute Publications/Western Michigan University.

Winston, Richard (1956/1969). *Charlemagne: From the hammer to the cross.* London: Constable.

Wittberg, Patricia (1994). *The rise and fall of Catholic religious orders: A social movement perspective.* New York: The State University of New York Press.

Wojda, Paul J. (1995). Charity. In *The HarperCollins Encyclopedia of Catholicism* (pp. 300-01). San Francisco: HarperSanFrancisco.

Woodcock, Matthew (2002, March 8). World of interiors [a review of P. DeSa Wiggins's "Donne, Castiglione, and the poetry of courtliness"]. *Times Literary Supplement*, 24.

Woodhouse, J. R. (1978). *Castiglione: A reassessment of "The Courtier".* Edinburgh: Edinburgh University Press.

Yamamoto, Tsunetomo (ca. 1700/1983). *Hagakure: The book of the samurai.* Tokyo: Kodansha International.

Zum Felde, Alberto (1943). *El problema de la cultura americana.* Buenos Aires: Losada.

Author Index (*)

(*) If an author is not listed in this section, please look in the Subject section

Y

Z

Subject Index